National Defense and the Environment

National Defense and the Environment

Stephen Dycus

University Press of New England
Hanover and London

University Press of New England, Hanover, NH 03755

Printed in the United States of America
5 4 3 2 1
CIP data appear at the end of the book

To Elizabeth

Contents

Preface

In his 1961 inaugural address, President John F. Kennedy made this commitment: "Let every nation know, whether it wishes us well or ill, that we shall pay any price, bear any burden, meet any hardship, support any friend, oppose any foe, in order to assure the survival and the success of liberty."[1] Measured in American lives and financial resources, that price has been staggering. But it has been well recognized, widely debated, and willingly paid. What has hardly been acknowledged until recently is the related price of injuries to public health and the environment. This price remains largely unpaid. With the bill comes a reminder that in the future we need to consider carefully the environmental implications of decisions about how to defend the nation.

Throughout almost half a century of Cold War we polluted the water and air, made noise, defaced the landscape, and generated millions of tons of hazardous and radioactive wastes, all in the name of national security. Early on, we acted at least partly out of ignorance of the environmental risks. More recently, we simply disregarded those risks, assuming that it would be impossible to maintain a strong defense if we had to worry about protecting the environment. Sometimes our actions were reckless, even unconscionable, as when we deliberately exposed unsuspecting civilians to releases of radioactivity and biological agents.[2] When we were forced into actual combat, we never reckoned systematically the environmental costs of war itself.

Now that the Cold War is over, we have begun to realize that the values supported by a healthy environment—life, liberty, and freedom from fear and want—are the same ones for which we stand ready to fight and die. There is a growing consensus in this country that the environment itself is worth defending at home and abroad—that environmental protection is an aspect of national security.

Yet even with the end of the Cold War, it will not be easy to maintain a strong defense that is also environmentally sound. As the United States redefines its military role in a politically unstable and often hostile world, it must be prepared not only to turn aside but also to deter any threat. The challenges are unpredictable and ambiguous. Meeting them may require some environmental compromises.

Current domestic politics may produce other environmental compromises. The federal environmental laws are under sharp attack in Congress, where control has recently shifted to the Republican Party. Proposed amendments would relax standards and procedures designed to protect public health and safety. While the changes are aimed primarily at saving money for regulated industries and avoiding restrictions on the use of private lands, they would also affect the operation of the federal government and its defense agencies. Perhaps even more troubling, both President Clinton and the congressional leadership have vowed to lower the federal budget deficit and cut taxes for the middle class by reducing defense expenditures for environmental compliance and cleanup. This would mean reduced emphasis on protection of health and safety in current operations. It would also mean that a vast Cold War legacy of hazardous wastes and radioactive contamination at defense facilities from coast to coast would be left to imperil and impoverish future generations of Americans.

This book presents a framework for determining when environmental sacrifices are necessary to protect us from sovereign aggression or terrorism, as well as for assessing the implications of proposed changes in the environmental laws. The current federal environmental statutes provide a starting point for our analysis. In the first part of the book we examine each of these laws in turn, emphasizing ways in which they apply differently to national defense activities than they do to, say, textile mills or municipal sewage plants. We look carefully at current efforts by the Pentagon and the Department of Energy to clean up and dispose of toxic and radioactive wastes generated during the Cold War. In a separate chapter, we focus on the ongoing controversy surrounding the closure of military bases across the country. Another chapter addresses legal constraints on wartime environmental destruction. We then examine the role of courts in enforcing these laws, and the parties—citizens, states, and the federal government—who can appear in court to insist on their enforcement. Finally, we consider the impact of national security activities on individual citizens, and prospects for compensation when they are injured. Our study includes a number of case histories showing how the environmental laws have actually been applied to national defense efforts.

This book is intended to provide a thoroughgoing introduction to the relationship between defense and environmental issues. It is meant to inform and to provoke further inquiry. The text is written in language accessible to the lay reader. Every effort has been made to avoid the acronyms and technical jargon that so afflict the literature in this field. Yet what follows hardly qualifies as light reading. The technology of military preparedness and war-making is complicated. So is cleaning up afterward. More important, the process of deciding how to defend ourselves without destroying the environment that sustains us can be infuriatingly complex. That is because, as with other important political questions, we have trouble agreeing on common goals, and even more difficulty agreeing on strategies for achieving those goals. Nevertheless, the story of our efforts to date is extremely interesting, and it concerns some of the most important issues confronting our society. Extensive notes point to additional readings, making this a unique research tool for military and civilian lawyers, government officials, politicians, scientists, and teachers.

Although our focus here is on the United States experience, much of what is said will be of interest to readers elsewhere. Many nations have adopted environmental laws modeled on ours. All share a common interest in avoiding war while protecting the earth for future generations. All can benefit from our successes and failures.

This work is dedicated to the thousands of men and women in government and the defense industries who have devoted their professional lives to keeping this nation strong and free, who believe in the importance of the rule of law, and who are determined to protect the natural environment for future generations of Americans. It is written, too, for citizens who want to join in the debate about some of the gravest matters of state, when the government makes decisions about how to protect us from foreign aggression, and when those decisions may profoundly affect the environment in which we all live, work, and play.

This book would not have been started and could not have been completed without the encouragement and collaboration of Dan W. Reicher, formerly Senior Attorney at the Natural Resources Defense Council (NRDC) in Washington, D.C., now Deputy Chief of Staff and Counselor to Secretary Hazel R. O'Leary at the United States Department of Energy. While at NRDC, he contributed to early drafts of chapters 3 and 4, and he reviewed other parts of the work during its preparation, although he bears no responsibility for any errors in the finished product. Dan is a fine and devoted public servant, as well as a friend, and I am deeply grateful for his help.

I also wish to acknowledge the generous financial support of the John Merck Fund and the Samuel Rubin Foundation, and the special advice of George Perkovich. In particular, I want to recognize the contributions of the Natural Resources Defense Council, where I spent part of a year as Visiting Scholar, and its Washington, D.C., staff, especially Thomas B. Cochran, Richard W. Fieldhouse, David O'Very, Christopher Paine, S. Jacob Scherr, Jean Reynolds, and James Werner, all of whom helped make this work possible. I am also grateful for the support of Colonel Dennis R. Hunt, Head of the Law Department at the United States Military Academy, where I taught as a Visiting Professor, and Maximillian W. Kempner, Dean of Vermont Law School. Several of my colleagues on the Vermont Law School faculty, including Richard O. Brooks, Celia Campbell-Mohn, Patrick A. Parenteau, and Karin P. Sheldon, reviewed portions of the manuscript and provided helpful suggestions. Finally, I am indebted to a succession of bright, enthusiastic student research assistants at Vermont Law School for their help in completing this work: Molly M. Brown, Charles S. Conerly, Carol Conragon, James C. Crowley, Dana Drukker, Gina M. Godfrey, A. Morgan Goodson, J. Patrick Kennedy, Steven F. Lachman, James N. Langell, Trevor R. Lewis, Matthew J. Matule, Elaine C. Schwartz, and Karen M. Willis.

April 1995
South Royalton, Vermont

S.D.

Note on Citation of Legal Authority

A brief word of introduction to legal citations may prove helpful to some readers. References in the notes to legal sources are set out in substantially the form used by almost all lawyers and judges. Aside from books and periodicals, those sources generally fall into one of three categories: judicial decisions, statutes and regulations, and other government materials.

Judicial Decisions

All cases cited in this book were decided or are pending in the federal court system. In general, the lowest level or trial court is the District Court. An example of a reported federal District Court case is *Barcelo v. Brown,* 478 F. Supp. 646 (D.P.R. 1979). The plaintiff in this case was named Barcelo, the defendant was Brown. The court's written decision appears in volume 478 of the Federal Supplement Reporter at page 646. The case was decided in 1979 in the District of Puerto Rico.

The intermediate level federal court is the Court of Appeals. The same case on appeal is reported as *Romero-Barcelo v. Brown,* 643 F.2d 835 (1st Cir. 1981). Romero-Barcelo (called Barcelo below) is the appellant here, having lost in the lower court. To find the opinion one would look in volume 643 of the Federal Reporter (Second Series) at page 835. The decision of the First Circuit Court of Appeals was published in 1981.

When the same case was appealed to the United States Supreme Court, the Court's decision was reported as *Weinberger v. Romero-Barcelo,* 456 U.S. 305 (1982). (Note that Secretary of Defense Weinberger replaced Secretary Brown as a party.) The decision, handed down in 1982, appears in volume 456 of the United States Reports at page 305.

For decisions not published in one of the reporters just mentioned, citations are given to Envtl. L. Rep. (Environmental Law Reporter), WL (Westlaw, a computer data base), or the docket number of the court in which the case was decided.

Statutes and Regulations

A citation to 33 U.S.C. §1251 (1988) refers to volume 33 of the United States Code, section 1251 (part of the federal Clean Water Act). Statutes not appearing in U.S.C. (typically because they were enacted too recently to have been codified, or because they have been replaced) are cited by their public law number and location in the United States Statutes at Large. Thus, the 1990 Base Closure and Realignment Act is cited as Pub. L. No. 101-510, 104 Stat. 1808 (1990). Regulations promulgated by executive agencies appear in the Code of Federal Regulations, as for example in 40 C.F.R. §1421(b).

Other Government Materials

Executive orders, issued by the President, are another important source of authority in this field. They are cited by number, along with the volume and page of the Federal Register where they were published—for example, Exec. Order No. 12,856, 58 Fed. Reg. 41,981 (1993). Document numbers are provided for agency publications wherever possible.

. . .

For detailed rules of usage see *The Bluebook: A Uniform System of Citation* (15th ed.) (Cambridge, Mass.: Harvard Law Review Association, 1991). Most of the legal materials cited here, other than agency publications, can be found in any large law library. Most federal government documents are held by Government Depository Libraries, which are open to the public, or may be obtained by contacting the agencies listed in Appendix B.

National Defense and the Environment

Chapter 1

National Defense vs. Environmental Protection: Can We Have It Both Ways?

If the recent war in the Persian Gulf was one of the shortest in history, it was also one of the costliest. It exacted a dreadful toll in human life and suffering, it wasted precious resources, and it ruined irreplaceable cultural and archeological treasures. It also showed how destructive a "conventional" war can be to the natural environment. Iraqi leader Saddam Hussein, ignoring the cost to his own people and to future generations throughout the Middle East, ordered unprecedented acts of environmental vandalism. United States and coalition forces, unlike Saddam, took elaborate precautions to avoid "collateral" damages, yet they too caused extensive environmental harm in Iraq, Kuwait, and Saudi Arabia. Whether and how that harm might have been avoided are questions that urgently need answers.

The war and its aftermath come at a time of heightened concern about the environmental costs of peacetime preparations for war. With increased use of the Freedom of Information Act to gain access to government records, and with the enactment of environmental laws that require monitoring and disclosure of federal facility activities, the public is better informed and less tolerant of many defense-related environmental injuries that earlier would have been dismissed, without question, as unavoidable. Even before the end of the Cold War, our focus had begun to shift from the external threat of an "Evil Empire" to the internal threat of a defense establishment sometimes heedless of its impacts on nature and human health. In particular, revelations about nearly a half-century of environmental neglect at America's military bases and nuclear weapons plants have brought an angry reaction from a public exposed, without its knowledge or consent, to serious health risks, and saddled with the financial burden of cleaning it all up.

Defense officials have now pledged to repair the damage and to be diligent stewards of the environment in the future. They have done so not only because the environmental laws require it, but also because they recognize that it is in the national interest. As the Pentagon's top environmental lawyer once put it:

> Environmental law is an area where compliance is a must. Not only is it the law, pure and simple, but compliance is smart, too. . . . Compliance protects people, whether military or civilian. It also protects the Air Force's ability to perform its mission.[1]

His statement mirrors the commitment of national security professionals throughout the government to learn from past mistakes, and to treat environmental protection as a top priority. Yet their commitment may not be enough to protect public health and the environment from needless injuries in the name of national security. A new Republican Party leadership in Congress has pledged to enact weakening amendments to federal environmental laws that regulate government and private activities alike. Regulations to implement existing laws, many of them intended to guide operations of federal defense facilities, would be put on hold. And both President Clinton and a majority in Congress have said they would like to reduce the federal deficit and provide funds for a middle-class tax cut by slashing Pentagon and Department of Energy budgets for environmental compliance and cleanup.

These proposed changes have nothing whatever to do with the nation's defense. They have everything to do with the immediate struggle for domestic political supremacy. It is important to bear in mind that the real issues in the current debate are money, the environment, and the health and safety of the American people. The discussion that follows largely avoids this debate. It assumes instead that we are determined to prevent defense-related injuries to the environment unless the security of the nation really requires it. It assumes further that we are willing to clean up lands and waters contaminated by a half-century of preparation for war, and that we will not pass that burden on to future generations.

Environmental Protection: A National Defense Priority

Is it possible to maintain a strong national defense, even to win a war, while at the same time protecting the environment for future generations? Former Defense Secretary Dick Cheney answered this way: "Defense and the environment is not an either/or proposition. To choose between them is impossible in this real world of serious defense threats and genuine environmental concerns."[2] "I want every command," he wrote, "to be an environmental standard by which Federal agencies are judged."[3] President Bush even linked "environmental security" with economic growth as a basic national security objective.[4]

Despite such resolve, it will not be easy to abandon bureaucratic habits developed over generations, or to change the way we have long thought about what is needed to defend the nation. We first have to recognize that environmental security is itself an important national interest, one worth fighting to protect. One writer's definition of a threat to the national security includes "whatever threatens to significantly degrade the quality of life of the people."[5] Threats to the environment satisfy this definition, for the environment includes all those natural resources that sustain life and make it meaningful, such as air, water, land, plants, and animals. According to the Pentagon, "[d]efending our national security and protecting our environment are closely linked. Both share the ultimate goals of ensuring our well-being and preserving our rich national heritage."[6] Rhetoric must be coupled with action, however. Decision-makers at every level of the defense establishment have to be rewarded for their environmental sensitivity, just as they are for performance of their military missions. And they must be given the resources needed to carry out their missions in an environmentally responsible way.[7]

There will almost surely be occasions when environmental sacrifices will have to be made to preserve the nation. The problem comes in deciding when such sacrifices are really necessary. Because the consequences of a wrong choice could be catastrophic, our tendency has been to resolve any doubts, without extensive analysis or public discussion, in favor of security. More often we have simply ignored the environmental risks altogether. Many examples will be found in the following chapters.

We need a settled procedure for determining *when* we must choose between environmental protection and national defense. This procedure should include a clear articulation of the issues and evaluation of the stakes. It should describe *who* will be entrusted with the fateful decision and *how* he or she will go about making it. Because the choices are apt to present themselves in times of great urgency and stress, we should plan ahead, laying the groundwork for decisions in a system of rules that will yield wise and, insofar as possible, predictable results. Because these decisions affect the fate of the nation and every one of its citizens, the public should be informed and invited to participate in both formulating and enforcing the rules.

Many of the rules and procedures needed to reconcile national defense with environmental protection will be found in the United States environmental laws. These laws have not yet been applied directly to armed conflict, and their application to other national defense activities has been controversial. Nevertheless, these laws can help protect the lives and health of Americans by providing government officials and citizens alike with information about environmental risks, when trade-offs are proposed in the name of national defense. The same laws may help protect the nation's economic base by preventing the needless destruction of natural resources that make us productive. In war and in peace the consistent, judicious enforcement of these laws may offer the best way to keep America strong, free, healthy, and beautiful, and to make it an environmentally responsible member of the community of nations.

The Defense Establishment at Home and Abroad

With the collapse of the Soviet empire and severe budget constraints at home, the United States military establishment is experiencing drastic cutbacks for the first time since the end of the Vietnam War. A drawdown in personnel will reduce the number of men and women in uniform by at least 20 percent before the end of 1995. Future peacekeeping roles will be performed by smaller, more mobile forces equipped with modern, high-tech weapons. The Defense Department has canceled high-priced weapons systems that have been on the drawing board for years, and has withdrawn requests for the development of new nuclear weapons.[8] Existing stocks of nuclear weapons will probably be cut by at least half in the next few years, in part to encourage the dismantling of former Soviet nuclear arsenals. The Department of Energy is busy cleaning up its weapons manufacturing complex. The Pentagon is also closing or scaling back operations at hundreds of military bases at home and abroad.[9]

Do these developments mean that United States national defense activities no longer represent a significant hazard to the environment? Absolutely not. Even after the reductions just described, the American military will remain the costliest and most powerful on earth. To fulfill its role as chief enforcer of the new world order, the United States military will continue to be deployed more widely than any other, with air and ground forces on six continents and a navy sailing around the globe. Approximately 1.5 million men and women in uniform and about the same number of civilian employees will continue preparing for war. When employees of defense contractors are added to these numbers, we can see that the impact on national and local economies, the domestic political process, and the environment will still be enormous.

Military operations affect the environment in a variety of ways in peacetime and in war. In addition to nonmilitary civil works projects of the Army Corps of Engineers, United States armed forces drain swamps and build roads, dams, and other structures. Like many large industrial companies, they discharge wastes into lakes and streams, groundwaters, the ocean, and the air. They make noise, destroy archeological sites, threaten endangered species, and deplete the ozone layer. Over the years, they have contaminated vast areas with toxic wastes. Military training exercises, utilizing tracked vehicles, high explosives, and massed troops, can be nearly as environmentally destructive as war itself. Military forces are sometimes used for domestic law enforcement as well, especially in the war on drugs, with significant environmental consequences.[10]

The Pentagon directly controls some 25 million acres of land in this country alone, an area nearly the size of Virginia. While these lands are exposed to injury from defense-related activities, they are simultaneously spared from commercial development, an irony that has not escaped the notice of neighbors to many closing military bases. Moreover, about three-fourths of Defense Department lands represent important fish and wildlife habitat managed under cooperative agreements with the Department of the Interior and state governments to provide public access for hunting, fishing, outdoor recreation, and natural and historical research. These lands also provide important reservoirs of biological diversity. In addition to their activities on the ground, all three of DOD's service branches (Army, Air Force, and Navy, including the Marines) carry out pilot training in the skies above millions of additional acres of public and private lands.[11]

The most dangerous national security activity since World War II, other than war itself, has been the manufacture of nuclear weapons. For more than forty years, at facilities spread across 2.4 million acres in thirty-four states, the Department of Energy (DOE) and its contractors were locked in a dreadful contest with their Soviet counterparts, polluting the air, land, and waters around them with toxic and radioactive wastes. The Department of Energy has admitted that until recently it had a policy of resisting efforts by the Environmental Protection Agency (EPA) and state governments to make it comply with applicable federal environmental statutes.[12] Some DOE sites are so contaminated that they may be off-limits to human use forever. The rest will take at least thirty years to clean up, at a cost of between $200 and $300 billion. In the end, we will be left with an immense quantity of highly radioactive waste. So far, we have no clear idea of how to dispose of it safely. In the wake of recent arms control agreements, the disposal of nuclear weapons components and weapons-grade materials presents additional technological and political challenges.[13]

Until recently, most of these military activities were shrouded in secrecy, even though in many cases public disclosure of their environmental impacts would in no way have prejudiced the national security. For example, homeowners downwind from some nuclear weapons plants were not told when radioactive materials were released into the air. And the Defense Department has not yet revealed how or even whether United States strategic planners have considered the risk of a nuclear winter.[14]

Finally, American combat forces have torn up and polluted a great deal of land overseas, from Korea to Indochina to the Persian Gulf. Whether so much destruction was necessary to accomplish the political objectives of the wars we fought in those places is a matter for continuing debate. With improvements in the accuracy of some weapons, and international constraints on the use of others, there is reason to hope that future wars, if they must be fought, will be less environmentally destructive. We also hope that military planners will carefully and systematically take into account the environmental costs of peacetime operations abroad.

A Role for the Environmental Laws?

In recent years, the federal environmental laws have profoundly changed the way American private industry does business. Congress and the courts have made it clear that the environmental laws generally apply in the same way to activities of the federal government, including those concerned with national defense. Yet despite the assurances of former Defense Secretary Cheney and ex-President Bush, some in the defense establishment still seem determined to carry out their military missions without regard for the environmental costs. Others may not clearly understand their obligations under the law. One top military environmental lawyer explains that the Army is like a very large industrial company that is lagging five to ten years behind its corporate counterparts in compliance with the environmental laws. Its failures, he insists, have mostly been "stupid mistakes." The military services are learning from experience to improve internal communications and to centralize responsibility for environmental matters. Local base commanders are becoming more sensitive to the environmental impacts of their maintenance and training activities. One training manual contains this instruction:

> As a commander, you are entrusted with the stewardship of the land, air, water and natural and cultural resources associated with performing your military mission. These resources must be carefully managed to serve both the Army's, and the nation's, short and long-term needs. Today, environmental considerations must be part of your decisions.[15]

There is also evidence of a generational change within the ranks, as younger officers more attuned to environmental issues move into positions of leadership. What is true for the Defense Department appears to be true as well for the Department of Energy.

In some instances, Congress has not been perfectly clear in explaining when its environmental statutes are applicable. For example, we cannot be entirely certain, from an examination of either the statutory text or the legislative history, that Congress "intended" for the National Environmental Policy Act (NEPA)[16] to apply to military activities abroad. It seems more likely that no one in Congress gave the matter much thought when the statute was enacted. Given this ambiguity, the President has predictably interpreted the statute as applying only to domestic military actions. The courts, charged with enforcement of the statute and eager to make principled decisions, have failed to rule consistently on this issue. Despite two decades of intense controversy, Congress has yet to clarify its intent.

At other times, Congress's meaning is unmistakable. Even so, agency officials responsible for the nation's defense sometimes claim that the environmental laws do not apply to them, or that they apply differently because of their special mission. They take this position not in a spirit of lawlessness, but because they think the national security requires it. Yet under our Constitution, only Congress is authorized to make laws. The duty of the executive branch (the President and federal agency officials) is to execute the laws that Congress makes, not to ignore the plain requirements of a congressional mandate.

Sometimes agencies concerned with the national defense *say* they are in compliance with the law, but *act* as if the law does not apply to them in the same way it does to other government agencies. Agency officials know that few potential environmental plaintiffs have the resources, expertise, and sheer determination needed to successfully challenge their behavior. The courts have often supported the agencies in these claims by deferring to the agencies' own interpretations of their statutory duties, or by refusing to order a halt to statutory violations. The ball is thrown back to Congress, where it frequently rests unplayed.

Does all this mean that the environmental laws do not affect the formulation and execution of national security policy? Far from it. In the day-to-day routines of most agencies charged with the nation's defense, the environmental laws seem to apply in much the same way they do in other agencies. But there are some spectacular exceptions, as we see in the following chapters. These exceptions provide special insight into the relevant legal and political processes. They also point the way to needed improvements in the law.

We are concerned here with two broad types of federal environmental laws. One type requires all federal agencies to carefully analyze and document the environmental impacts of their proposed actions, and to take those impacts into account when deciding whether to go ahead with their proposals. The National Environmental Policy Act (NEPA) and the Endangered Species Act[17] are perhaps the best-known examples. The environmental review they prescribe must be open and public. Indeed, members of the public are invited to participate, unless the disclosure of relevant information would threaten the national security.

There has never been any serious doubt that these planning statutes apply generally to domestic national security activities. However, agencies have tried to limit the kinds of environmental impact that must be considered before a project goes forward. To mention just one example, the Air Force has insisted that in deploying a nuclear weapon system the environmental effects of a nuclear war need not be considered, since the primary purpose of the weapon system is to deter war, not to fight one.[18] In other instances, agencies have struggled to keep the environmental implications of a proposed action secret, as when the Navy planned to berth ships carrying nuclear weapons in New York Harbor.[19] The need for a public environmental analysis of military actions abroad, either in peacetime or in war, has always been controversial. The application of these environmental planning laws to national security activities is examined in some detail in chapter 2.

The other environmental laws are designed to protect specific resources, such as air and drinking water; to limit exposure to particular environmental threats, such as pesticides; or to clean up environmental contamination. The Clean Water Act,[20] for example, requires anyone (including a government agency) who discharges anything into the nation's waters to obtain a permit. A permit will not be issued unless the discharge conforms to standards set out in the statute or in EPA regulations. Discharge of some substances is forbidden altogether. Public involvement is an important part of the permitting and enforcement process. The Comprehensive Environmental Response, Compensation and Liability Act (CERCLA),[21] by contrast, is concerned with the cleanup of sites already befouled by pollution.

Because strict compliance with these laws has sometimes been perceived as hampering accomplishment of an agency's national defense mission, enforcement efforts have met with resistance from the military and the Department of Energy (DOE). For example, the Navy was reluctant to apply for a Clean Water Act permit when it discharged ordnance into the water at a target practice range in Puerto Rico.[22] Resistance to compliance may stiffen when the laws are administered by state governments with EPA-approved programs, sometimes applying state standards more stringent than the federal ones. So when the State of Colorado wanted to apply its own stricter standards to the cleanup of forty years' accumulation of chemical weapons wastes at a plant near Denver, the Army argued that more lenient federal standards should apply.[23] And when the State of Ohio sought to collect civil penalties from the Department of Energy for polluting the land and waters around its Fernald Feed Materials Plant with uranium, DOE was able to persuade the Supreme Court that the applicable statutes did not allow such state sanctions for federal agency misbehavior.[24]

Many of these regulatory environmental laws provide for a waiver of their strictures when the "paramount interest of the United States" requires it. However, the waiver provisions are almost never invoked. As with the planning statutes, the applicability of these laws to military activities abroad remains a subject of great confusion and controversy. Chapter 3 explores the application of the regulatory environmental laws to control national security actions that might injure the environment. Chapter 4 addresses environmental statutes and programs used to clean up after forty-five years of Cold War.

With the end of the Cold War and the establishment of new priorities for the United States military, dozens of military bases across the country and overseas are being closed or reorganized, affecting the jobs, health, and environmental quality of millions of Americans. Chapter 5 describes the environmental implications of these changes, and shows how the environmental laws may influence the closure process. For example, in one case currently in litigation, the Air Force transferred lands at a closed base to a civilian agency for use as an international airport and industrial park.[25] Nearby homeowners and the local town government have complained that the transfer was not preceded by a proper environmental analysis under NEPA, and that air pollution from the proposed new use would violate provisions of the Clean Air Act. They also claim that the Air Force violated a congressional ban on transfer of lands contaminated with hazardous wastes, even though the law was subsequently amended to permit just such a transfer.

It has been widely assumed that the environmental laws do not apply to armed conflict or to other wartime activities of the military. One military lawyer has insisted that environmental laws "do not apply in wartime, period." While there is no statutory basis for this assumption, it seems to grow out of a belief that no legal constraints are tolerable when the fate of the nation hangs in the balance. This view ignores the fact that the natural environment is itself a national interest we would fight to protect. It also overlooks two centuries of congressional limitations on the President's power during foreign wars.[26] Even during wartime, however, consistent, flexible compliance with the environmental laws can actually enhance national defense efforts. Besides, the nation is already limited in its use of armed force by a host of international laws, some of them specifically aimed at environmental protection, others having that effect. Chapter 6 looks at these issues.

Chapter 7 examines the critical role of courts in interpreting and enforcing the environmental laws. Of special interest here is the deference that courts have often shown to the military when the national security is implicated. For example, when the Navy violated the Clean Water Act by dropping ordnance into the ocean without a permit at its target practice range in Puerto Rico, the Supreme Court allowed the violation to continue while the Navy applied for a permit, because it accepted the Navy's argument that an injunction "would cause grievous, and perhaps irreparable harm . . . to the general welfare of this Nation."[27] Citizens and states play an important part in securing judicial enforcement of the environmental laws, especially because of a Justice Department policy holding that EPA cannot bring suit against other federal agencies for violations.

Chapter 8 considers the aftermath of national defense activities that injure the environment and the people who live in it. The issue is whether and when the government and defense contractors will have to compensate those harmed. Cases involving victims of above-ground nuclear weapons testing in Nevada, and complaints by soldiers that they were injured by the use of Agent Orange in Vietnam, are examined. So are claims by citizens unwittingly exposed to radioactivity in experiments by the Department of Energy and its predecessors. The rules for financial liability are sometimes linked to the environmental protection laws. In one recent case, a court found that the Department of Energy's contractor at the Fernald Feed Materials Plant could not escape liability for damages caused by its pollution, because it operated in plain violation of pertinent environmental statutes.[28] The courts have played a critical role in these cases, sometimes turning back claims based on sensitive evidence because of the courts' own perception of threats to the national security.

Human nature being what it is, we cannot expect that nations will soon abandon their warring ways in resolving their political disputes. Nor can we ignore the likelihood that the United States and other military powers will be increasingly called on to intervene in the internal struggles of nations to protect the rights of noncombatants, or even the environment. Because we live in a dangerous and contentious world, we must remain strong and vigilant. But in preparing for a fight, we must not destroy the very thing we would fight to protect. And if war is unavoidable, we have to be mindful of the political values for which we go to war.

Chapter 2

Environmental Planning for National Defense

No one likes surprises where the national security is concerned. That is why for many years we have kept two million Americans in uniform and spent about a quarter of each year's federal budget preparing for war. We have spared almost no expense in developing new and more efficient weapons, training relentlessly, and imagining worst case scenarios. We want to anticipate any significant threat to the nation and be ready to meet it.

Neither do we like to be surprised by threats to the environment, even when those threats come from the government's efforts to protect the nation. If environmental trade-offs must be made in the interest of national security, or if the environment is to be placed at risk, we want to be informed in advance. We want to be able to say that the environmental harm was unavoidable, that there were no viable alternatives, and that we planned it that way.

This chapter is about rules for predicting environmental harms while planning for the nation's defense. Of central importance here is the resolution of real and apparent conflicts between national security and environmental protection during the planning process.

The National Environmental Policy Act (NEPA)[1] requires the federal government to plan ahead for the environmental consequences of its actions, and it sets out a procedural framework for doing so. In principle, NEPA applies equally to all agencies and all activities of the federal government. In practice, it has often been applied less rigorously to America's defense establishment, especially when fear of foreign aggression has displaced worries about environmental consequences. Still, NEPA has had an enormous influence on national security decision-making, not only by increasing the environmental sensitivity of government planners, but also by providing members of the public with a window into the planning process.

A variety of other federal statutes require planning for the protection of particular environmental amenities, including endangered plants and animals, marine mammals, coastal zones, and historic sites. These statutes, too, play an important role in providing for the nation's defense.

This chapter begins with a brief description of NEPA, emphasizing those parts of the statute that have special relevance to national security matters. We then look at how the executive branch and the courts apply the statute to national defense activities during peacetime. We also examine the government's reluctance to analyze the environmental consequences of using new weapons systems in combat, and its efforts to protect sensitive information. The first section of this chapter concludes with an analysis of the controversy surrounding NEPA's application to national defense activities abroad. At the end of the chapter we review several other statutes that address planning for the environmental consequences of military preparedness.

The National Environmental Policy Act (NEPA)[2]

The National Environmental Policy Act of 1969 is the briefest of the major environmental statutes but also the broadest in its application. It is "our basic national charter for protection of the environment."[3] By requiring all federal agencies to carefully consider the environmental consequences of their actions, NEPA has had a pervasive effect throughout the executive branch of the government.

Briefly, NEPA does three things:

1. It establishes a national policy for all federal agencies, requiring them to use "all practicable means . . . to create and maintain conditions under which man and nature can exist in productive harmony." It also sets out a number of broad environmental protection goals, acknowledging the "responsibilities of each generation as trustee of the environment for succeeding generations."
2. It requires all federal agencies (including those charged with national security) to "utilize a systematic, interdisciplinary approach . . . in planning and decisionmaking which may have an impact on man's environment." Specifically, it calls on agencies to prepare a written study, an environmental impact statement (EIS), before proceeding with any major federal action that might significantly affect the human environment.
3. It creates the Council on Environmental Quality (CEQ), whose job it is to oversee the application of NEPA.[4]

NEPA declares that it is "the continuing responsibility of the Federal Government to use all practicable means, consistent with other considerations of national policy, to improve and coordinate Federal plans, functions, programs, and resources" to protect and enhance a variety of environmental values.[5] According to the Supreme Court,

> NEPA promotes its sweeping commitment to "prevent or eliminate damage to the environment and biosphere" by focusing Government and public attention on the environmental effects of proposed agency action. By so focusing agency attention, NEPA ensures that the agency will not act on incomplete information, only to regret its decision after it is too late to correct.[6]

While each agency is bound to measure the environmental costs of its actions,

> NEPA itself does not mandate particular results, but simply prescribes the necessary process. If the adverse environmental effects of the proposed action are adequately identified and evaluated, the agency is not constrained by NEPA from deciding that other values outweigh the environmental costs.[7]

In other words, NEPA's requirements are "essentially procedural."[8] An agency is free in principle to make a foolish decision but not an uninformed one. Moreover, a court will not "substitute its judgment for that of the agency as to the environmental consequences of [the agency's] actions. The only role for the court is to insure that the agency has taken a 'hard look' at environmental consequences."[9]

Detailed requirements are spelled out in the regulations of the Council on Environmental Quality[10] and in hundreds of judicial decisions. Most federal agencies, including the Defense Department and its three service branches (Army, Air Force, and Navy, including the Marines), have adopted their own NEPA regulations generally conforming to the CEQ regulations, but tailored to their own specific missions.[11]

NEPA directs each federal agency to prepare environmental impact statements for all "proposals for legislation and other major Federal actions significantly affecting the quality of the human environment."[12] An EIS must describe adverse environmental effects which cannot be avoided if the proposal is implemented. According to CEQ regulations, it must also set forth the relationship between short-term uses of the environment and long-term productivity, as well as any irreversible or irretrievable commitments of resources that would result from the proposed action. Both direct and indirect effects must be spelled out, including energy requirements, possible conflicts with land use plans and controls, and impacts on urban quality, and historic and cultural resources.[13]

The EIS also must discuss all reasonable alternatives to the proposed action, along with the environmental consequences of each alternative, so both the decision-maker and the public can make an educated choice among the options. The list of alternatives must include different actions that might achieve the same objective, different ways of carrying out the same action that might be less harmful to the environment, and the consequences of doing nothing.[14] In addition, the EIS must address possible measures to mitigate environmental harms.[15]

CEQ regulations provide that some agency actions may be categorically excluded from formal NEPA review if they ordinarily produce no significant environmental impact. For example, NEPA regulations for the Army categorically exclude recruiting, recordkeeping, and the routine procurement of goods and services.[16] If an action is not categorically excluded, an agency may prepare a brief environmental assessment (EA), to determine whether the impact will be "significant" enough to require preparation of a more extensive EIS. If, on the basis of the EA, the agency decides that no further analysis is required, it will issue a finding of no significant impact (FONSI), briefly setting forth its reasons, sometimes following a period for public comment.[17] Between 10,000 and 20,000 environmental assessments are prepared each year by various agencies, compared with about 450 environmental impact statements.[18]

When an EIS is required, the agency announces in the Federal Register that it will conduct a scoping process, typically including one or more public hearings, to determine the range (or scope) of issues that will be analyzed in the impact statement. The agency then produces a draft environmental impact statement (DEIS) and circulates it for comment to other federal agencies that have special expertise or jurisdiction over the subject matter of the proposal, including the Environmental Protection Agency (EPA), as well as to state and local governments and the public.[19] CEQ regulations call for extensive public involvement at each step in the review process, to improve the quality of decisions and to make public officials more accountable. For example, the agency must solicit and carefully consider comments from the public on the draft EIS, then include responses to those comments in the final EIS (FEIS).[20] The agency announces its decision to abandon or go forward with its proposal in a record of decision (ROD). This explains the environmental considerations and other factors, including any "essential considerations of national policy," and any measures the agency has adopted to mitigate environmental harms.[21]

There are three other kinds of impact statements. If an agency develops a proposal for several related actions, or for a broad program or policy, it may have to prepare a programmatic environmental impact statement (PEIS), even though separate environmental impact statements are required for individual components of the larger program.[22] According to the regulations, whether actions should be combined for NEPA analysis usually depends upon whether each action could stand alone.[23] In one case involving related projects, however, the Navy was allowed to prepare two separate environmental impact statements, one for a new home port facility for the battleship *Iowa,* and another for associated housing. A court reasoned that the port could be put into service without the housing, even though, paradoxically, the housing would not have been constructed without the port.[24]

If substantial changes are made in a proposed activity after the FEIS is completed, or if new information comes to light that bears on the project's environmental effects, the agency must prepare a supplemental environmental impact statement (SEIS) that analyzes the changes.[25] An agency may also have to prepare a legislative environmental impact statement (LEIS) if the agency is involved in the proposal of a bill in Congress (other than an appropriation).[26]

Most NEPA litigation raises questions about either the need for an EIS or the adequacy of one that has already been prepared. Of special interest here are cases testing the significance of a particular environmental impact, that is, whether the impact is substantial enough to require preparation of an EIS. According to CEQ regulations, significance depends upon such concerns as effects on public health and safety, the uniqueness of affected resources, threats to endangered species, and the degree to which the effects might be highly controversial.[27] The Supreme Court has long held that an agency must make a "searching and careful" inquiry into whether an EIS should be prepared. Still, in 1989, the Court indicated that it would not disturb an agency's fact-based finding that a particular impact was not significant unless that finding was arbitrary and capricious.[28] One critical contrary view holds that

> [j]udicial solicitude for agency discretion is proper when, based on whatever record the law requires, the agency exercises *informed* discretion. An agency's decision not to prepare an EIS, however, is a decision not to inform its discretion and therefore invites more exacting judicial scrutiny. An agency should not be enabled to bypass the entire EIS requirement with a cursory assessment to which a court gives an equally cursory review.[29]

The Administrator of the Environmental Protection Agency is directed to review and comment publicly on each EIS to determine whether the proposed action or legislation (or agency regulation) would be "unsatisfactory from the standpoint of public health or welfare or environmental quality."[30] Specifically, she must consider the severity, duration, and geographical scope of environmental effects, their importance as precedents, the availability of preferable alternatives, and possible violations of other environmental laws, such as the Clean Water Act or the Resource Conservation and Recovery Act.[31] (These other laws are described in chapters 3 and 4.) If the EPA Administrator has objections and is unable to resolve them with the agency that prepared the EIS, she can refer the matter to the Council on Environmental Quality (CEQ). If CEQ cannot bring about an agreement between the agencies, it will refer the matter to the President for resolution.[32]

NEPA Applies to National Security Activities

It is well settled in principle that NEPA applies to national security activities of the federal government, even though the statute does not say so explicitly.[33] It is the continuing policy and responsibility of the "Federal Government," not just non-defense segments of the government, to protect and enhance the quality of the human environment. The action-forcing language of the statute is likewise directed to "all agencies of the Federal Government."[34] NEPA does not explicitly exempt any agency or any kind of activity from compliance. Since other statutes *do* expressly exempt certain national defense activities from strict compliance with NEPA,[35] we can assume that Congress intended no blanket exception.

Qualifying language in two NEPA sections might be read as permitting national defense activities to be treated differently. One section states that "[i]n order to carry out the policy set forth in this chapter, it is the continuing responsibility of the Federal Government to use *all practicable means, consistent with other essential considerations of national policy*, to improve and coordinate Federal plans, functions, programs and resources." Another includes the qualification that agency review of environmental consequences shall be carried out "*to the fullest extent possible*."[36] National security might be regarded as such an essential consideration of national policy, or so urgent, that the environmental analysis prescribed by NEPA could be avoided. However, no court has so held,[37] and nothing in NEPA's legislative history supports such a conclusion.[38]

CEQ regulations for NEPA make no specific mention of national security activities; they are uniformly "applicable to and binding on all Federal agencies."[39] A 1977 executive order by President Carter likewise directs every federal agency to comply with the CEQ regulations.[40] However, some agencies with national defense responsibilities have promulgated NEPA regulations that depart significantly from the CEQ prescription. A Defense Department regulation states that in "no event shall DOD Components delay an emergency action necessary to the national security, or for preservation of human life, for the purpose of complying with . . . the CEQ regulations."[41] Others indicate that NEPA simply does not apply to actions taken "in the course of armed conflict."[42] Whether these regulations conform to NEPA or impermissibly deviate from it, they leave no room for doubt that the statute does generally apply to national defense activities.

The courts have repeatedly confirmed the obligation of military agencies to evaluate the environmental impacts of their activities.[43] For example, when the Navy began construction of a home port facility for the Trident submarine in Puget Sound, it argued that "'NEPA cannot possibly apply' to strategic military decisions made by the Department of Defense-Navy."[44] That contention was firmly rejected by the court:

> We view this as a flagrant attempt to exempt from the mandates of NEPA all such military actions under the overused rubric of "national defense." This effort to carve out a defense exemption from NEPA flies in the face of the clear language of the statute, Department of Defense and Navy regulations, Council on Environmental Quality ("CEQ") Guidelines, and case law. . . .
>
> There is no support in either the statute or the cases for implying a "national defense" exemption from NEPA. The Navy, just like any federal agency, must carry out its NEPA mandate "to the fullest extent possible" and this mandate includes weighing the environmental costs of the Trident Program even though the project has serious national security implications.[45]

In cases involving national security activities that range from the storage of nuclear weapons near Honolulu airport to mock amphibious landings on a state park beach, a number of other courts have reached the same conclusion.[46] More often, in dozens of cases testing military compliance with NEPA, the statute's applicability has simply been taken for granted by courts and litigants alike.[47]

Perhaps more important, NEPA has influenced military planning, although the extent of that influence is uncertain.[48] In one study, the Navy reported that while "[n]o projects have been terminated because of NEPA considerations, . . . almost all projects have been changed to some degree because of factors discovered during the course of the NEPA process."[49] On the other hand, the Navy, unlike its sister branches, has argued that NEPA's environmental analysis is not required at all important decision points in major weapons acquisition programs, and that environmental agencies and the public need not be involved in those decisions, a position rejected by the DOD Inspector General.[50]

One persistent complaint from military planners is that NEPA's environmental analysis unnecessarily delays their work. "The very quickest the Air Force has managed to do an EIS," one official remarked, "is about 270 days, and most decisions simply cannot wait that long." By contrast, Navy officials report that "NEPA planning is so built into the planning process and the administrative process that no delays have been caused by NEPA procedures."[51] Another worry is that NEPA review is expensive. The Air Force says preparation of a full-blown EIS costs between $100,000 and $70 million (the latter figure for the MX missile).

Some military planners respond to these concerns by first deciding whether to go ahead with a proposed action, then preparing an EIS as a kind of "disclosure document."[52] Others conduct the required environmental analysis simultaneous to, but separate from, the principal decision-making process. For example, in planning for the development of its new F-22 aircraft, the Air Force decided not to begin preparation of an EIS until 1997, fully three years after a so-called critical design review, which was nearly the last opportunity to make adjustments based on environmental considerations.[53] Such responses make a mockery of the statute, since the NEPA review can only fulfill its purpose of informing decision-makers if it is completed before decisions are made.

Yet even when the final judgment on a proposal is made by a single individual who acts in good faith and who is publicly accountable for his or her actions, NEPA's goal of a decision fully informed by environmental considerations may not be achieved. The larger and more complex the proposed action, the more likely it is that critical decisions leading up to a final resolution will have been made by various people who did not communicate with one another and who did not have the benefit of the collected wisdom reflected in the final EIS.[54]

None of these concerns is unique to decision-making about national defense matters. But they present a special challenge when both national security and environmental security hang in the balance. The following case study illustrates these difficulties.

Case Study: Finding a Home for the Trident Submarine[55]

In the late 1960s, U.S. military planners began work on a new strategic submarine system, the Trident, that would need a single base close to deep water. After reviewing eighty-nine potential sites on the East and West coasts, they chose the rural community of Bangor, Washington, on Puget Sound, as the leading candidate.

The Navy prepared a total of twenty-seven environmental assessments or impact statements for the project, including individual evaluations of the submarine, its new missile, and a variety of supporting facilities at Bangor. Risks to air, water, flora and fauna, historical and archeological sites, and visual quality, as well as social and economic impacts, were extensively analyzed. These studies reportedly resulted in a number of design changes intended to mitigate harmful impacts. Following public hearings and the publication of a draft EIS for the base, the Navy prepared a final EIS, on the basis of which it decided that "the environmental disturbance . . . caused by Trident [would] not outweigh the military benefits."

Local residents, joined by several environmental organizations, went to court to challenge the adequacy of the final EIS. They claimed that the Navy had failed to discuss the environmental consequences of feasible alternatives to the single dedicated submarine base. The court agreed, pointing out that such information was required to make a reasoned choice among alternatives. The environmental plaintiffs also contended that the Navy's projections of environmental effects did not extend past the time the base would first become operational. The court called such an arbitrary cutoff date unreasonable, and it ordered the Navy to extend its forecast of impacts into the foreseeable future.

However, despite clear violations of NEPA, the court refused to halt construction of the base while the Navy developed the information missing from its EIS. The court was convinced that the Navy had taken "no arbitrary or capricious action," and declared, paradoxically, that in spite of the inadequacy of the EIS "the Navy gave proper weight to environmental considerations in deciding to proceed with this strategically important project." While it must have been clear to the court that the additional data would have little or no influence on the construction project, national security concerns seemed to trump strict compliance with the statute.

In a serious emergency, it may be important for the government to be able to act quickly to protect the nation without waiting to prepare a formal environmental impact analysis. While NEPA contains no special provision for exigent circumstances, CEQ regulations do provide for a limited waiver of the EIS requirement. When such waivers are necessary, undesirable environmental consequences will be minimized if the acquisition of weapons systems, training programs, policies, and planning have been subjected to the NEPA review process. NEPA waivers are examined in some detail at the end of chapter 6.

Even in the absence of an emergency, there may be some question about how to apply NEPA when the national security is implicated. This uncertainty arises precisely because an agency like DOD may find it difficult to protect the nation, its citizens, and its allies from aggression while at the same time protecting the environment. The courts have been extremely sensitive to this apparent dilemma, often deferring broadly to the judgment of military officials about just what kinds of environmental risks need to be addressed in an EIS, or shielding the analysis itself from public view, as we see in the following sections.

Thinking About the Unthinkable: "Worst Case" Analysis

The business of defending the country is fraught with uncertainty. Predicting the onset of any particular armed conflict, the course of a battle, or the final political outcome is almost impossible. However, we can forecast at least some of the environmental consequences of using particular weapons or strategies. When military planners and weapons designers created a new fuel-air bomb at the beginning of the Persian Gulf War, for example, they had a very specific environmental impact in mind. Can we afford to build weapons and adopt strategies for their use without also carefully anticipating the unintended results?

One of NEPA's goals is to avoid "undesirable and unintended consequences."[56] For any proposed major federal action, the EIS must describe not only environmental consequences that are certain to occur, but also risks of significant harm.[57] If the likelihood of a particular impact is extremely small, it can be ignored.[58] But what if an agency lacks the information it needs to describe either the magnitude of an impact or the probability of its occurrence? One court answered this way:

> The agency need not foresee the unforeseeable, but by the same token neither can it avoid drafting an impact statement simply because describing the environmental effects of and alternatives to particular agency action involves some degree of forecasting. And one of the functions of a NEPA statement is to indicate the extent to which environmental effects are essentially unknown. . . . Reasonable forecasting and speculation is thus implicit in NEPA.[59]

CEQ regulations at first required agencies to describe the "worst case" outcome from each proposed action. Where information about the worst case was incomplete or unobtainable, an agency was to indicate the "probability or improbability" of a worst case occurrence, then "weigh the need for the action against the risk and severity of possible environmental impacts were the action to proceed in the face of uncertainty."[60] But to stop what it called "endless hypothesis and speculation," the CEQ amended its regulation, confining the scope of impact statements to "reasonably foreseeable significant adverse impacts." The impact analysis must be supported by credible scientific evidence and be within the rule of reason; it must not be based on "pure conjecture."[61] The new regulation was upheld by the Supreme Court three years later.[62] Nevertheless, an EIS still must address "impacts which have catastrophic consequences, even if [the] probability of occurrence is low."[63]

The few courts that have addressed the question have found that NEPA does not require analysis of the environmental risks of using U.S. weapons of mass destruction in combat. Yet as much as we hope that such weapons will never have to be used, it is in just such cases that the potential conflict between national defense and environmental security is brought most sharply into focus, as the following case studies demonstrate.

Case Study: Command and Control for a Nuclear War

In 1988, suit was filed against the Air Force to stop construction of the Ground Wave Emergency Network (GWEN), an array of 300-foot radio towers, stretching from coast to coast, designed to withstand a nuclear attack.[64] The towers can only be used to transmit a President's orders to launch atomic weapons. The plaintiffs argued that the Air Force's EIS for the network should have described the environmental effects of a nuclear war, since the use of the towers as designed would surely result in the onset or prolongation of such a war. They also claimed that each GWEN tower would be a primary target for an attacking enemy.[65] The Air Force countered that the purpose of the towers was not to make war but to deter it by assuring a reliable U.S. nuclear response to a Soviet nuclear first strike. (Following the Air Force's logic, it might be argued that NEPA would not require analysis of the destructive potential of any weapon system or military policy, because the primary purpose of each is to deter enemy aggression.)

Citing the new CEQ rule that requires only analysis of "reasonably foreseeable impacts," the court characterized the plaintiffs' arguments as "merely speculative," even though the plaintiffs offered expert testimony to show that the Air Force policy of deterrence—based on psychological assessments of Soviet strategic planners—was equally speculative.[66] Despite the fact that the deterrent value of the GWEN system depends entirely upon United States resolve to use it if attacked, the court called the nexus between construction of GWEN and nuclear war "too attenuated to require discussion."[67]

The court also declared that while "both experts and laymen disagree on the precise environmental impact of a nuclear exchange, everyone recognizes that these effects would be catastrophic. Detailing these results would serve no useful purpose."[68] Thus, the Air Force was allowed to proceed without making a thorough environmental analysis of a system that is designed to be activated under circumstances permitting no further environmental review.

On several occasions Congress has decided that a particular action is of such vital importance to the nation that its environmental consequences can be ignored. For example, in the 1988 Base Realignment and Closure Act,[69] it explicitly waived NEPA's environmental review requirements in the selection of individual military installations for closure. Congress' instructions are not always so clear, however, and the courts must then decide what Congress intended. As the next case study suggests, in interpreting the sometimes ambiguous statutory language, a court may be influenced by its own judgment of what is in the nation's best interests.

Case Study: Preserving the "Peacekeeper"

In 1983, Congress appropriated funds to build twenty-one new ten-warhead MX intercontinental ballistic missiles (dubbed "Peacekeeper" by President Reagan) and deploy them in existing Minuteman silos.[70] The State of Colorado and several environmental organizations filed suit challenging the adequacy of the EIS for the proposed deployment.[71] Among the complaints was a charge that the Air Force should have discussed the environmental impact of the *wartime* use of the missiles.

The court noted that in the authorizing legislation Congress had called for preparation of an EIS "on the proposed deployment and *peacetime* operations of MX missiles in the Minuteman silos."[72] Invoking the principle that explicit prescription of one thing amounts to an implicit waiver of another not mentioned, the court ruled that wartime impacts need not be addressed. The court then decided that the scope of NEPA review was similarly restricted for another thirty-nine MX missiles authorized in later years, even though the quoted language was not repeated in subsequent legislation.

Circuit Judge Arnold, concurring, had a different reason for limiting the scope of review:

> It is really expecting too much of the National Environmental Policy Act (NEPA) and the EIS process to ask them to grapple with issues of that scope and magnitude, issues that would inevitably involve large amounts of classified information, and as to which environmental effects as the term is normally used would be unlikely to be of decisional significance. Nuclear war would probably destroy anything worthy of the name of environment. No one needs an EIS to make that clear. Defendants cannot reasonably be expected to reinvent the world or re-examine strategic doctrine from the ground up every time they compile an EIS on a weapons-related decision.[73]

But is it too much to expect defense planners to grapple with the environmental implications of their plans, no matter what the scope or magnitude? Are we prepared to say that any federal action will be exempt from NEPA's requirements if it is sufficiently large and complex, or sufficiently damaging to the environment? Judge Arnold's statement flies in the face of NEPA's command to "use all practicable means, consistent with other essential considerations of national policy," to protect the environment for present and future generations of Americans.[74]

In one remarkable episode toward the end of the Cold War, without relying on NEPA, Congress passed special legislation directing the Pentagon to prepare a public environmental analysis of one of the most frightening prospects of a nuclear war. As the following case study indicates, Congress' mandate was repeatedly ignored.

Case Study: Buttoning Up for a Nuclear Winter

Throughout most of the Cold War, it was believed that even a "total" nuclear war would spare a significant fraction of the U.S. population, and that life in some form, albeit drastically altered, would go on. In 1981, for example, the Federal Emergency Management Agency (FEMA) asserted that a civil defense program of moderate expense could protect 80 percent of American citizens in a heavy nuclear attack.[75]

More recent research suggests that a large-scale nuclear war (5,000 to 10,000 megatons of explosive force) could result in 750 million immediate deaths worldwide from the blast alone. Another 1.1 billion persons might eventually die from the combined effects of blast, fire, and radiation, while 1.1 billion more might suffer injuries requiring medical attention (which would be largely unavailable).[76]

Even more troubling, survivors could face the prospect of a "nuclear winter" caused by shading of the earth's surface by dust and soot lofted into the upper atmosphere by nuclear explosions. According to one study, detonation of just half of the superpowers' nuclear arsenals (at 1986 levels) could reduce the temperature over northern continental land masses by 40 to 70 degrees Fahrenheit, with sunlight dropping to only a fraction of its normal intensity, and near-cessation of normal rainfall, depending on the season and targeting strategies of the adversaries.[77] But even a much smaller dip in average temperature, say 5 to 10 degrees Fahrenheit, would virtually eliminate agricultural production in the Northern Hemisphere and would devastate rice production elsewhere. There is spirited disagreement over the magnitude of the risk of a nuclear winter, but not about its existence. In one study the authors conclude optimistically that "there does not seem to be a real potential for human extinction."[78]

Alarmed by these studies, Congress decided not to wait for a NEPA environmental impact statement to describe the new threat. In 1984 it ordered the Pentagon to prepare an unclassified report containing "a detailed review and assessment of the current scientific studies and findings on the atmospheric, climatic, environmental, and biological consequences of nuclear explosions and nuclear exchanges."[79] The report was supposed to show how those findings affect U.S. plans for fighting a nuclear war, as well as for arms control and civil

> defense. In response, the Defense Department produced a slender document just seventeen pages in length.[80] When Congress repeated its order the following year, the result was a five page report.[81] The third time, the Pentagon told Congress it would have to refer to the two earlier reports.[82] Congress has not taken steps to require a more meaningful review.
>
> Meanwhile, no EIS for any nuclear weapons system or policy has included a public discussion of the risk of a nuclear winter. Although the direst predictions are controversial, they are based not on "pure conjecture" but on the work of dozens of Nobel laureates and other researchers. In the language of the CEQ regulation described earlier, a nuclear winter is a "reasonably foreseeable" consequence of using atomic weapons, the impact of which could be "catastrophic."[83] While the recently ratified Strategic Arms Reduction Treaty (START) and follow-on discussions for further reductions in nuclear weapons seem to have reduced the likelihood of a nuclear conflagration, many thousands of weapons remain in the hands of the former superpowers. Still, the public has no way of knowing whether or how the prospect of a nuclear winter has influenced United States strategic planning.

Without addressing the issue of wartime environmental impacts of weapons systems, a 1993 DOD Inspector General's audit report found that DOD "had not effectively integrated environmental management" or properly assessed trade-offs in its weapons acquisition programs.[84] Such concerns have been relegated to a "distant second" to program performance, schedule, and cost, according to the report. For example, an environmental assessment for the M1A2 Abrams tank considered the environmental impacts of testing and peacetime deployment of the tank, but ignored the tank's development, manufacturing, and disposal. This failure, the report concluded, could result in choices that were not the most effective either in terms of cost or operationally, and could result in unanticipated and unfunded costs. The Inspector General also criticized DOD components for neglecting to involve the public in the NEPA review process for weapons acquisitions.

Secret Environmental Impact Statements

Congress was well aware, when it enacted NEPA, that some environmental impact statements would contain classified information, and that public dissemination of that information might endanger the national security. It addressed that concern by directing that each EIS (along with comments received from various federal, state, and local agencies) be made available to the public "as provided by section 552 of Title 5."[85] The statutory cross-reference is to the Freedom of Information Act (FOIA), which makes most federal government agency records available to members of the public for the asking, but which exempts properly classified data from disclosure.[86] Thus, information that an agency could refuse to turn over to a FOIA requester can also be kept secret during the environmental analysis prescribed by NEPA.

One way to keep sensitive information secret is to separate it from the body of an EIS and attach it as a classified appendix. This way it will be available only to members of Congress, agency officials, and congressional staff who have appropriate security clearances. CEQ and agency regulations call for just such treatment.[87] As a result, the public EIS may reveal relatively little about the proposed project or its environmental consequences. And with the loss of open discussion and expert public criticism, the quality of the agency's final decision may be seriously compromised. But that is the trade-off Congress made in providing for both national security and environmental security. As Justice Rehnquist put it in one case,

> Congress has . . . effected a balance between the needs of the public for access to documents prepared by a federal agency and the necessity of nondisclosure or secrecy. . . . Congress intended that the public's interest in ensuring that federal agencies comply with NEPA must give way to the Government's need to preserve military secrets.[88]

Nevertheless, the Court went on to point out, the military "must consider environmental consequences in its decisionmaking process, even if it is unable to meet NEPA's public disclosure goals."[89] However, the question remains: how can members of the public ensure agency compliance with NEPA when national security secrets are involved?

Case Study: Nuclear Weapons in Paradise?[90]

In the mid-1970s, the Navy decided to construct forty-eight earth-covered ammunition magazines on the Hawaiian island of Oahu that were admittedly *capable* of storing nuclear weapons. Following a long-standing policy, however, the Navy said it could neither confirm nor deny plans to *actually* store nuclear arms there. It prepared a public environmental assessment and issued a finding of no significant impact (FONSI) for the construction without describing the environmental implications of storing nuclear weapons. A local citizens' group then filed suit calling for an EIS that would analyze: (1) the risk and consequences of a nuclear accident, (2) the effect of a plane from nearby Honolulu International Airport crashing into one of the magazines, and (3) the hazard to local residents from low-level radiation.

When the case eventually reached the Supreme Court, it was dismissed. The Court noted that virtually everything having to do with nuclear weapons is secret. Under FOIA, it said, the Navy was entitled to keep from the public any plans it had to store nuclear weapons at the new facility. The plaintiffs, therefore, were unable to show that the Navy "proposed" the storage of nuclear weapons, and could only show that it "contemplated" such storage.[91] Since an agency proposal is NEPA's trigger mechanism for preparation of an EIS, the question of NEPA compliance in this instance was placed "beyond judicial scrutiny."[92] The Court's holding did not mean that the Navy could avoid preparing an EIS if it actually proposed to store nuclear weapons in the new magazines, but merely that under such conditions of secrecy no judicial enforcement of the Navy's NEPA obligation would be available.[93]

Like the court in the *GWEN* case, the Supreme Court showed that it was willing to consider the construction of the facility separate from its designed use. Recall that in the *Trident* case the court said the Navy violated its NEPA obligation when its EIS failed to discuss the future operation of the Trident submarine base as designed.[94] In any event, neither the Court nor the Navy could explain why the Navy would go to all the trouble and expense of building the magazines to such rigorous specifications if it did not propose to store nuclear weapons there. Nor could they say how the national security would be compromised by a prospective public analysis of the environmental risks, since all environmental impact statements are, by their very nature, prospective. The Court simply decided that since Congress had not explicitly prescribed a "hypothetical" EIS, it would not read NEPA as requiring one.

Within months, the Supreme Court precedent in the Hawaii case was followed in another case involving apparent plans to store both nuclear and conventional arms at the Seal Beach Naval Weapons Station in California.[95] It was followed again in 1989, when the Navy sought to establish a homeport for the battleship *Iowa* on Staten Island. Because of the Navy's policy to "neither confirm nor deny" the presence of nuclear weapons on its ships, the court ruled that the public EIS for that project did not have to address the environmental risks of bringing such weapons into New York Harbor, even though ships in the *Iowa* group normally carried them.[96]

Applying NEPA to Military Activities Abroad

Does NEPA apply to federal agency actions affecting the natural environment beyond the nation's borders?[97] The question is an important one, since so much of our national security effort is expended overseas, in peacetime as well as in war.

Congress imposed no geographical limits on the applicability of NEPA when it called for efforts to "prevent or eliminate damage to the environment or biosphere and stimulate the health and welfare of man,"[98] and it made no distinction between domestic and foreign impacts in calling for preparation of an EIS. While one of NEPA's announced goals is to "assure for all *Americans* safe, healthful, productive, and esthetically and culturally pleasing surroundings," Congress also directed agencies to

> recognize the worldwide and long-range character of environmental problems and, where consistent with the foreign policy of the United States, lend appropriate support to initiatives, resolutions, and programs designed to maximize international cooperation in anticipating and preventing a decline in the quality of mankind's world environment.[99]

NEPA's legislative history also strongly hints at the statute's applicability abroad.[100]

Early on, several courts either ruled that NEPA applies outside the nation's borders or assumed that it does.[101] In one case, the court ordered the military to prepare an EIS for a simulated nuclear explosion on a Pacific atoll. But it based its decision in part on a finding that the blast site was located in a U.S. Trust Territory where there was little concern for conflict with "foreign policy or the balance of world power."[102] In another case, this one involving the export of nuclear power plant components to the Philippines, the court worried about possible interference with U.S. foreign policy or with Philippine sovereignty.[103] Unable to find a clear expression of congressional intent, the court labored to harmonize NEPA with the Atomic Energy Act[104] and the Nuclear Non-Proliferation Act,[105] concluding "only that NEPA does not apply to NRC nuclear export licensing decisions—and not necessarily that the EIS requirement is inapplicable to some other kind of major federal action abroad."[106]

In 1979, President Carter issued Executive Order No. 12,114,[107] calling for a very limited preliminary review of foreign environmental impacts. The order claims to be based entirely on the President's "independent authority," rather than on NEPA's statutory mandate. By its terms, the order would exempt many—perhaps most—U.S. national security activities abroad from any meaningful environmental review. While it calls for a NEPA-style EIS of impacts on the "global commons," such as the oceans or Antarctica, a much less rigorous "environmental study" or "review" is prescribed for impacts in countries not "participating with the United States [or] otherwise involved" in a proposed agency action. Most environmental impacts inside a "participating" nation escape review entirely.[108] Thus, a joint military exercise with NATO forces in Germany would require no consideration of the environmental effects in Germany, no matter how severe, while a "study" or "review" would be conducted of impacts in a nonparticipating nation like Switzerland. According to the order, no environmental analysis is required for:

a. actions taken by the President;
b. actions involving the national security directed by the President or a cabinet member;
c. actions taken in the course of an armed conflict;
d. intelligence activities and arms transfers; or
e. export licenses or permits or approvals.[109]

Even the limited review called for in the order can be modified or waived in emergencies or in the interest of "national security," or when there is a need for "governmental confidentiality." Draft reviews need not be publicized, and public hearings are not required. The order purports to foreclose citizen suits for its enforcement by claiming that it creates no "cause of action."[110]

Case Study: Planning for the Disposal of Chemical Weapons

For more than two decades, in anticipation of an attack by Warsaw Pact forces, the U.S. Army stored 100,000 8-inch and 105-millimeter artillery shells containing nerve gas agents GB and VX in Germany. In 1986, President Reagan entered into an agreement with Chancellor Helmut Kohl to remove these weapons from German soil. Congress also ordered the destruction of all such chemical weapons by 1997.[111] The Army decided to transport the shells in special trucks to a German seaport, then send them by container ship to Johnston Atoll in the Pacific for incineration.

When environmental plaintiffs filed suit in 1990 to require preparation of a "comprehensive" EIS covering the entire operation,[112] the court expressed concern about the potential for disruption of American foreign policy, since the President had negotiated agreements with the German government for removal of the weapons, and about possible interference with the sovereignty of another state, namely Germany. Citing both NEPA's mandate in §102(2)(F) to act consistently with "the foreign policy of the United States," and the statute which ordered destruction of the weapons, the court balanced the "environmental goals of NEPA against the particular foreign policy concerns" and concluded that NEPA was inapplicable to the movement of munitions inside Germany. It speculated that it was "perhaps precisely this potential conflict which caused Congress to leave open the question of whether NEPA applies to the environmental impacts of federal action abroad."[113]

> The court also found that the Army's preparation of an environmental assessment under Executive Order No. 12,114 for the transoceanic shipment of the weapons, rather than a full-blown EIS, did not violate NEPA. Nevertheless, it specifically rejected the government's claim that Executive Order No. 12,114 "preempts application of NEPA to all federal agency actions taken outside the United States. Such an application of an Executive Order," it decided, "would be inappropriate and not supported by law." The court was careful to limit its holding to the "specific and unique facts" presented. "In other circumstances," it said, "NEPA may require a federal agency to prepare an EIS for action abroad."[114]

More recently, in 1993, the D.C. Circuit Court of Appeals decided that the National Science Foundation must prepare an EIS before deciding to operate an incinerator at a research station in Antarctica.[115] The government conduct regulated by NEPA, said the court—decision-making that precedes the proposed action—is "uniquely domestic," since it occurs almost exclusively inside the United States. Because NEPA imposes no substantive requirements that would govern conduct abroad, the usual presumption against extraterritorial application of a statute does not apply. Moreover, according to the court, the effect of the decision would be felt in Antarctica, a place without a sovereign; therefore, there would be no clashes between United States laws and those of other nations.

Like the decision to build an incinerator in Antarctica, most major military decisions, even those that are to be implemented overseas, are made in Washington. Where a proposed action would be carried out within the territory of a sovereign ally, it would naturally have to conform to local laws, or to any applicable status of forces agreement. However, an environmental impact analysis would have no effect whatever on those obligations. Nor would compliance with NEPA impinge on the sovereignty of a host nation, unless perhaps by producing embarrassing disclosures of environmentally destructive actions by the host government. That chiefly diplomatic concern was apparently behind a federal district court's decision in another recent case not to require the preparation of environmental impact statements for certain U.S. military operations in Japan.[116] The outcome in that case might have been justified by the need to avoid interference with U.S. treaty obligations, since compliance may be excused if it would result in "irreconcilable and fundamental conflict" with other agency duties.[117] But the court surely exceeded its judicial authority when it baldly declared that "no EISs would be required because U.S. foreign policy interests outweigh the benefits from preparing an EIS."[118] Such policy judgments have been entrusted by Congress neither to the courts nor to the executive.

Recent proposed amendments to NEPA would make it expressly apply to environmental impacts abroad, but would exempt extraterritorial actions "taken to protect the national security of the United States, votes in international conferences and organizations, actions taken in the course of armed conflict, strategic intelligence actions, [or] armament transfers.[119] These exemptions would mark a significant retreat from the United States longstanding commitment to act as an environmentally responsible world citizen and to set an example for other nations to follow in the environmental field.

Without amending NEPA, the Clinton administration has proposed revisions to Executive Order No. 12,114 that would create opportunities for public notice and comment, order environmental analysis of "international agreements," and require review of impacts in "participating" nations.[120]

Other Planning Statutes

A variety of federal statutes require federal officials to engage in planning or to coordinate with other federal, state, and local agencies before carrying out actions that might affect particular environmental amenities, such as endangered species or historic sites. Like NEPA, these statutes apply to national defense activities. And as with NEPA, they provide opportunities for pubic involvement in decisions about defense policy. Under the National Wildlife Refuge System Administration Act, for example, the Air Force cannot conduct pilot training over a National Wildlife Refuge without the approval of the Secretary of the Interior.[121] The Archeological Resources Protection Act requires permits for any disturbance of archeological sites on federally owned or managed lands.[122] Four other statutes in this category are briefly described below.

Endangered Species Act

The purpose of the Endangered Species Act of 1973[123] is to protect not only endangered and threatened animals and plants but also the habitats upon which they depend for their survival. An "endangered species" is one that is in danger of extinction throughout all or a significant portion of its range. A "threatened species" is one that is likely to become endangered within the foreseeable future.[124] The act calls on the Secretary of the Interior, or the Secretary of Commerce for marine species, to identify species that are endangered or threatened because of the condition of their habitats, overutilization by humans, disease or predation, or other factors, and to designate critical habitat for each listed species.[125]

Once a species is listed as endangered or threatened, the Secretary of the Interior must develop and implement a recovery plan for its conservation and survival.[126] Members of the public are given an opportunity to review and comment on each plan before it is completed. Because the Pentagon and the Department of Energy control or use lands harboring many listed species, they often play a key role in developing and carrying out these plans.

It is unlawful for anyone to hunt, capture, kill, wound, harass, collect, or trade in any endangered species of plant or animal, or to violate any agency regulation adopted for its protection. Violators are subject to civil and perhaps criminal penalties. The act also provides for citizen suits to enjoin violations or to compel the Secretary of the Interior to perform any nondiscretionary function.[127]

The Endangered Species Act requires federal agencies to ensure that their actions are not "likely to jeopardize the continued existence of any endangered or threatened species or result in the destruction or adverse modification" of their critical habitat.[128] To satisfy this mandate, an agency must take the following steps:

1. An agency proposing to take an action must inquire of the Fish and Wildlife Service (F&WS) whether any threatened or endangered species "may be present" in the area of the proposed action.
2. If the answer is affirmative, the agency must prepare a "biological assessment" to determine whether such species "is likely to be affected" by the action. The biological assessment may be part of an environmental impact statement or environmental assessment.
3. If the assessment determines that a threatened or endangered species "is likely to be affected," the agency must formally consult with the F&WS. The formal consultation results in a "biological opinion" issued by the F&WS. If the biological opinion concludes that the proposed action would jeopardize the species, or destroy or adversely modify critical habitat, then the action may not go forward unless the F&WS can suggest an alternative that avoids such jeopardy, destruction, or adverse modification. If the opinion concludes that the action will not violate the act, the F&WS may still require measures to minimize its impact.[129]

If the Fish and Wildlife Service issues a "jeopardy opinion," the agency proposing the action may ask for an exemption from a committee of high-ranking government officials commonly referred to as the "God Squad." An exemption may be granted if the committee finds that there are no reasonable and prudent alternatives to the proposed agency action, that the action is regionally or nationally significant and in the public interest, and that the benefits of the action clearly outweigh the benefits of alternatives that might avoid extirpation of the species or destruction of its habitat. The Secretary of Defense also may exempt an agency from compliance "for reasons of national security,"[130] although he has not yet been called upon to do so.

The following case study shows how Endangered Species Act procedures were followed in planning one defense-related agency action.

Case Study: Saving the Cui-ui

Fallon Naval Air Station is an important training base in the middle of the Nevada desert. In order to reduce the risk of damage to aircraft from sand storms and debris lofted by desert winds, the Navy leases buffer zones around its runways to local farmers who use them for irrigated crops or pasture. Water for the irrigation comes from the Truckee River, which is the principal source of water for Pyramid Lake.

The cui-ui is an endangered species of fish that lives in Pyramid Lake and spawns in the Truckee River. Before entering into each of a series of one-year leases for the buffer zones and associated water rights, the Navy consulted with the Fish and Wildlife Service. Each year Fish and Wildlife issued a biological opinion stating that the leases would not jeopardize the cui-ui, and the Navy relied on these opinions in executing the leases.

Over time, diminished flows in the Truckee River created a "precarious condition" for the cui-ui, which needs proper water temperatures and volume to spawn in the river, sufficient clearance to cross the Truckee River delta into Pyramid Lake, and enough water to maintain appropriate levels and water quality in the lake. A local Indian tribe, the Pyramid Lake Paiute, brought suit in federal court alleging that the Navy's leases violated the Endangered Species Act by authorizing the withdrawal of too much water from the river.[131]

The court upheld the leases, finding that the Navy's reliance on Fish and Wildlife's biological opinions was not arbitrary or capricious, which is the standard for review in such cases. Each federal agency must act affirmatively in the interest of listed species, the court pointed out, even if it results in frustration of the agency's accomplishment of its primary mission. In this case, however, the court found that the tribe had offered no data to undermine either the Fish and Wildlife Service's opinion or the Navy's reliance on it. According the usual deference to the Navy's interpretation of its statutory duties, the court decided that the Navy was entitled to reject the tribe's proposed alternatives, which it said would have only an "insignificant" effect on the availability of water in the river.[132]

In an earlier case, a different federal court of appeals found that the Navy had violated the procedural requirements of the Endangered Species Act when it failed to obtain a biological opinion from the Fish and Wildlife Service concerning five endangered or threatened species at its Vieques Island, Puerto Rico training area.[133] In two other reported cases, complaints of violations were dismissed on grounds that the citizen plaintiffs lacked standing to sue. One sought to enjoin the Navy from shooting feral goats on its property. The other involved Navy and Department of the Interior plans to drain a marsh harboring a colony of endangered salt-marsh harvest mice.[134]

How has compliance with the Endangered Species Act affected the day-to-day operations of defense agencies? The Army says it has tried to balance mission requirements with those of endangered species. For example, new guidelines for management of habitat for the endangered red-cockaded woodpecker have forced the closure or restricted use of some training areas at Fort Bragg, North Carolina. Yet while realistic training has become more difficult and costly, there have reportedly been no reductions in unit readiness.[135]

Marine Mammal Protection Act

The Marine Mammal Protection Act[136] is aimed primarily at the prevention of commercial whaling, as well as fishing that incidentally kills or injures ocean mammals. However, the act makes any hunting, capture, killing, harassment, or trade of a marine mammal unlawful without a permit from the Secretary of Commerce. Thus, the act applies to national defense activities that might threaten such creatures.

In one reported case, an environmental group sued to enjoin the Navy's "deployment" of Atlantic bottlenose dolphins at its Trident submarine base in Bangor, Washington without complying with NEPA's environmental review requirements.[137] From 1973 to 1987, the National Marine Fisheries Service, the Commerce Secretary's delegate in these matters, had issued three permits under the Marine Mammal Protection Act allowing the Navy to capture dolphins in the wild for military uses. In 1986, Congress enacted special legislation allowing the Secretary of Defense to authorize the taking of up to twenty-five marine mammals each year for "national defense purposes," without a permit but with the "concurrence" of the Secretary of Commerce.[138] Two years later, the Secretary of Defense exercised this authority and obtained the Commerce Secretary's concurrence to take up to twenty-five marine mammals in each of five succeeding years. Without deciding whether the Navy had complied with the Marine Mammal Protection Act, the court ruled that the Navy had to prepare an EIS under NEPA for its proposed use of the dolphins.

Another suit invoking the Marine Mammal Protection Act sought to stop the transfer of a dolphin from the New England Aquarium to a Navy research facility without a permit.[139] That case was dismissed on standing grounds when members of the plaintiff organization were unable to distinguish between Kama, the conscripted dolphin, and other dolphins at the aquarium.

Coastal Zone Management Act

The Coastal Zone Management Act[140] requires planning for activities that affect the nation's coastal waters and adjacent shorelands. Each coastal state is encouraged through federal financial assistance to develop a management program for its coastal zone and get its program approved by the Secretary of Commerce. Any federal agency activity affecting the coastal zone of a state with an approved program must be "consistent with" that program "to the maximum extent practicable."[141] Because so many naval and other military facilities are arrayed along the nation's coasts, this provision could have an important influence on defense efforts. However, the ambiguity of the requirement, as well as the fact that federal lands are specifically excluded from the act's definition of "coastal zone," have made its application to national defense activities controversial.[142] The controversy has been fueled by a different provision that allows a variance from the consistency requirement for a federal permit or financial assistance, over a state's objection, when "necessary in the interest of national security."[143]

One case applying the Coastal Zone Management Act involved the Navy's plans to construct a homeport for the aircraft carrier *Nimitz* and its attendant support vessels in Everett, Washington.[144] The project called for the dredging of 3.4 million cubic yards of sediment, and for disposal of the dredged spoil in water up to 430 feet deep. The State of Washington issued a water quality certification required by the Clean Water Act,[145] but conditioned its approval upon the Navy's obtaining a permit under the state's approved coastal zone management program. A federal court of appeals rejected the Navy's claim of sovereign immunity from regulation under the act. It pointed out that the act's federal lands exclusion had no application where the state program was aimed, as here, at environmental protection rather than land use planning. Thus, the court enjoined the Navy from proceeding with construction until a state permit was obtained.[146]

National Historic Preservation Act

In the National Historic Preservation Act,[147] Congress declared that "the historical and cultural foundations of the Nation should be preserved as a living part of our community life and development." The act provides for the creation of a National Register of Historic Places, listing sites and objects "significant in American history, architecture, archeology, engineering, and culture." Each federal agency is responsible for the preservation of historic properties under its ownership or control. To this end, it must establish a program, in consultation with state, local, and tribal governments, for identifying and nominating such properties to the National Register. Before taking any action, or issuing a permit, or providing federal financing for a project, an agency must first consider the effect on properties listed or eligible for listing on the National Register.[148] Federal agency responsibilities are spelled out in a presidential executive order.[149] Compliance may be waived in the event of an "imminent threat to the national security," or under other circumstances by an independent Advisory Council on Historic Preservation.[150]

The National Historic Preservation Act was invoked by the Commonwealth of Puerto Rico in an effort to stop the Navy's use of Vieques Island for training exercises.[151] Only after suit was filed did the Navy conduct a survey of the island, which disclosed a number of sites possibly eligible for listing on the National Register of Historic Sites. Even then, the court concluded, the Navy's survey was not sufficiently thorough, and no effort had been made to follow up on leads discovered in the survey.

An earlier case involved the Navy's conduct of bombing practice on a small island in Hawaii.[152] When the case reached the court, the Navy, working with state officials, had completed a survey of only 90 percent of the target area and 34 percent of the entire island, but it had discovered a number of candidate sites. The court admonished the Navy to carefully protect the candidate sites while it finished its survey, but refused to stop the use of live ordnance for practice, because of the Navy's insistence that the military readiness of the Third Fleet might be seriously reduced by an injunction.[153]

Chapter 3

Environmental Regulation of the Defense Establishment

The National Environmental Policy Act calls on government planners to carefully assess the environmental consequences of proposed actions, but it does not require any particular outcome. Most of the other federal environmental statutes operate differently. One type works prospectively, to reduce the number and amount of pollutants that might otherwise be released into the environment. The Resource Conservation and Recovery Act (RCRA)[1] does this by encouraging manufacturing processes that generate fewer hazardous wastes, and by promoting the recycling of wastes. Other prospective statutes regulate the release of pollutants that threaten important environmental values, such as clean air and safe drinking water. For example, the Clean Water Act[2] provides for the development of uniform national criteria for water quality, and prohibits the discharge of pollutants into the nation's waters without a permit. Still other statutes are aimed at particular environmental threats. The Toxic Substances Control Act[3] is an example. It regulates the manufacture and distribution of the most dangerous chemicals. These forward-looking preventive and regulatory environmental statutes are the subject of this chapter.

Another group of environmental statutes operates retrospectively, to clean up pollutants already released into the environment. The best known is the Comprehensive Environmental Response, Compensation and Liability Act (CERCLA), or Superfund law.[4] These statutes, along with defense programs developed to administer them, are described in chapter 4.

This chapter begins with a description of laws intended to eliminate pollution at its source. It then addresses common elements among the various statutes that regulate releases of pollutants into the environment. Next, individual regulatory statutes are described, with special emphasis on their application to national security activities. Issues surrounding the enforcement of these statutes overseas are then briefly examined. The chapter concludes with a description of defense programs meant to ensure compliance with these statutory mandates.

Pollution Prevention: Getting at the Source of the Problem

Several of the environmental regulatory statutes contain provisions aimed at reducing environmental contamination at its source. For example, the Clean Water Act calls on industries to adopt "best management practices" to reduce the likelihood that runoff, spills, or leaks will pollute the nation's waters.[5] The Resource Conservation and Recovery Act (RCRA) requires each generator of hazardous wastes to certify that it has a program in place to reduce the "quantity and toxicity of such waste to the degree determined by the generator to be economically practicable."[6] The same statute provides that goods procured by federal agencies must contain the highest practicable percentage of recycled materials.[7] Potential liability for cleanup costs at contaminated sites has furnished a financial incentive for industries to minimize the amount of waste they generate. In addition, the Toxic Substances Control Act[8] and the Federal Insecticide, Fungicide and Rodenticide Act[9] authorize EPA to prohibit the manufacture or distribution of unreasonably dangerous chemicals and pesticides.

When Congress enacted the Pollution Prevention Act[10] in 1990, however, it acknowledged that neither the source reduction requirements just described, nor the broader regulatory programs, had been fully effective in protecting human health and the environment. It ordered that increased emphasis be placed on the reduction of pollution at its source:

> Source reduction is fundamentally different and more desirable than waste management and pollution control. . . .
>
> The Congress hereby declares it to be the policy of the United States that pollution should be prevented or reduced at the source whenever feasible; pollution that cannot be prevented should be recycled in an environmentally safe manner, whenever feasible; pollution that cannot be prevented or recycled should be treated in an environmentally safe manner whenever feasible; and disposal or other release into the environment should be employed only as a last resort and should be conducted in an environmentally safe manner.[11]

Source reduction can be effected through modifications of equipment or technology, changes in processes and procedures, reformulation or redesign of products, substitution of raw materials, and improvements in housekeeping, maintenance, training, and inventory controls.

In the 1990 legislation, Congress ordered the Environmental Protection Agency to establish a new office to promote a "multi-media" approach to source reduction, rather than relying entirely on separate programs to deal with air, water, hazardous waste, and so on. It directed EPA to develop and implement a new strategy for promoting source reduction, including standards to measure reductions, new technologies, federal procurement practices to encourage reductions, and improved public access to information.[12] Every facility in the United States that is required to report to EPA on its releases of toxic chemicals into the environment must also report its efforts at reducing or recycling such chemicals.[13] These reports are available to members of the public from EPA. Still, the Pollution Prevention Act is largely aspirational, since it contains no specific performance standards or goals, and no provisions for the enforcement of its policy.

By contrast, a 1993 executive order provides clear goals and standards for application of the act to federal agencies.[14] Each agency must ensure that its facility management and procurement activities utilize source reduction and recycling to minimize the quantity of toxic chemicals entering any waste stream, and to help achieve compliance with the various regulatory environmental statutes. Each federal agency and every agency facility must develop a pollution prevention program, both for operations and for procurement of goods and services, that will reduce toxic chemical releases by half before the end of the century. Any wastes that are generated must be stored, treated, or disposed of in a manner that protects human health and the environment. All hazardous wastes released into the environment or shipped off-site must be cataloged and reported to EPA each year, beginning in mid-1995, for inclusion in EPA's Toxic Release Inventory, in accordance with the Emergency Planning and Community Right-to-Know Act.[15] All reports and program details are available to the public from the agency that generated them or from EPA. Each federal agency is directed to develop and test innovative pollution prevention technologies, as well. The executive order is "intended only to improve the internal management of the executive branch and is not enforceable against the government or any of its agencies or officials." Nevertheless, the President may exempt a particular federal facility from compliance with the executive order "in the interest of national security."[16]

Another 1993 executive order implements and adds to the RCRA requirements for use of recycled products.[17] That order establishes a new office within EPA to develop guidance for waste prevention, energy and water efficiency, and recycling for each federal agency. Contractor operators of government facilities must follow the same guidelines. When an agency procures goods and services, it must "consider" the use of recovered rather than virgin materials, life cycle costs, waste prevention, and the eventual disposal of whatever it buys.

The Department of Energy, manager of the United States nuclear weapons complex, has announced its strong commitment to source reductions through process modifications, material substitutions, recycling and reuse.[18] For example, it has begun to develop safe substitutes for toxic halogenated solvents used in metal cleaning and paint stripping, and new methods for recycling spent acids from metal plating and milling operations. Information on these new technologies is published for other industrial users. In 1994, DOE spent $17 million for waste reduction, and it expects to cut the total amount of waste it generates in half by 1999.

The Pentagon has also begun incorporating source reduction into its acquisition programs.[19] Environmental implications must be considered at each critical decision point, and life-cycle environmental costs projected, from initial research to disposal. In particular, the use of hazardous materials must be reduced as much as possible. When their use cannot be avoided, these materials must be tracked, handled, stored, and disposed of in an environmentally responsible manner. It is not enough, in other words, to buy a new battle tank that is reasonably fuel-efficient and quiet; some thought also must be given to the eventual safe disposal of the depleted uranium armor that protects the tank's crew. Procurement contracts are being rewritten accordingly. Because DOD is such a huge consumer of products, it has been able to create whole new markets for some recycled goods such as paper, construction materials, and tires. It expects to further reduce the amount of waste it generates through training of its personnel and research into new technologies.

Despite these good intentions, a recent audit by DOD's Inspector General revealed that existing environmental policies and procedures for acquisitions were not being followed consistently.[20] DOD reportedly was not exercising effective environmental oversight of its suppliers, and it had failed to adequately estimate the government's share of potential liability for cleaning up after defense contractor operations. Another DOD Inspector General's report indicates that the Pentagon has made significant progress in reducing the volume but not the sources of hazardous waste.[21] In other words, we can do better, and we must.

The Regulatory Statutes: Common Elements

When the total elimination of wastes is not possible or practicable, the environmental regulatory statutes are supposed to limit releases of those wastes to amounts, times, and locations that will not threaten human health or the environment. While NEPA applies only to federal agencies, these statutes regulate the actions of federal and nonfederal entities alike. They apply to the Department of Energy (DOE) and Department of Defense (DOD), to their commercial contractors operating defense facilities, and to industrial plants of every description.

To avoid any confusion about their applicability to the federal government, most of the regulatory statutes have provisions making them expressly enforceable against federal facilities. The Clean Water Act is typical:

> Each department, agency, or instrumentality . . . of the Federal Government . . . shall be subject to, and comply with, all Federal, State, interstate, and local requirements . . . respecting the control and abatement of water pollution in the same manner, and to the same extent as any nongovernmental entity. . . .[22]

Federal agencies are also directed by executive order to comply with all "applicable pollution control standards" set out in the various environmental laws, including the "same substantive, procedural, and other requirements that would apply to a private person."[23] The same executive order calls on each agency to prepare a plan each year for the control of environmental pollution.

The Defense Department currently operates under thousands of air emission permits, hundreds of wastewater discharge permits, and storm water permits at every installation. It holds more than three hundred permits to treat, store, or dispose of hazardous wastes under the Resource Conservation and Recovery Act (RCRA). DOD facilities contain approximately thirty thousand regulated underground storage tanks. In 1995, the military will spend about $2.8 billion complying with environmental regulations, as well as on pollution prevention, conservation, and environmental research and development.[24] DOE's 1995 budget for what it calls "waste management" is about $3 billion.[25]

The Environmental Protection Agency plays the central oversight and enforcement role for all federal facilities through its Office of Federal Facilities Enforcement. It promulgates regulations for implementation, and it is supposed to work closely with other federal agencies and state governments to ensure compliance. The key guidance document is the 1988 "Yellow Book," so called for the color of its cover.[26] In the years since its publication, the Yellow Book has been augmented and supplanted in part by a variety of new EPA policy directives.

Because Congress has stressed local solutions to local environmental problems, many of the regulatory statutes are implemented through delegated state programs approved by the Environmental Protection Agency. States also have their own independent environmental protection programs. According to EPA, it is "imperative that all Federal agencies recognize their obligation to comply with State and local environmental requirements and incorporate such requirements into their environmental compliance and auditing programs."[27] The states therefore have a critical role to play. They may even apply state controls more stringent than those adopted at the national level. However, while there is no doubt that federal agencies must comply with substantive "requirements" of both state and federal environmental laws, the question of how those requirements will be enforced, and by whom, has been the subject of much litigation and scholarly output.

The regulatory statutes generally provide for direct administrative or judicial enforcement of their terms by the Environmental Protection Agency. States that implement EPA-approved programs must have comparable enforcement powers. In addition, many of these statutes contain "citizen suit" provisions enabling citizens and states to seek enforcement in the federal courts. States have used both kinds of provisions to obtain declaratory and injunctive relief for violations by federal agencies and their contractors. They have had less success, however, in imposing administrative orders to force compliance and collecting penalties for federal agency violations.

The main obstacle to state enforcement against federal agencies lies in the principle of "sovereign immunity." The federal government will generally be insulated from lawsuits as well as from financial responsibility unless Congress has unequivocally waived its immunity. Even when suits are permitted for injunctive relief to halt continuing or threatened statutory violations, economic sanctions may be unavailable. In 1992, the Supreme Court found that neither the Clean Water Act nor the Resource Conservation and Recovery Act allowed states to levy punitive penalties for past federal agency violations.[28] Although Congress has since acted to expressly authorize state administrative orders and penalties against federal agencies under RCRA, the availability of these enforcement tools under other environmental laws remains in doubt.[29]

The major regulatory environmental statutes say that federal agencies are subject to "any process or sanction, whether enforced in Federal, State, or local courts or in any other manner."[30] Since early in the Reagan administration, however, the Justice Department has insisted that what it calls the "unitary executive" principle prevents EPA from issuing administrative compliance orders or filing suit against other federal agencies for violations. EPA enforcement efforts against other federal agencies have accordingly been confined to persuasion and negotiation, formalized through compliance agreements and consent orders.[31] Executive orders direct the EPA Administrator to resolve conflicts between agencies whenever possible, otherwise to refer them for resolution to the Director of the Office of Management and Budget, or to the Attorney General if the dispute involves a question of law.[32] Although one order indicates that these "conflict resolution procedures are in addition to, not in lieu of, other procedures, including sanctions, for the enforcement of applicable pollution control standards," the Justice Department has taken the position that the President has a right to resolve such problems internally without court interference.[33]

In the 1992 Federal Facility Compliance Act, Congress challenged the unitary executive principle by explicitly authorizing EPA to initiate "administrative enforcement actions" against other federal agencies for violations of the Resource Conservation and Recovery Act. The 1992 act also added the United States to the list of "persons" referred to in the act.[34] These changes allow EPA to issue administrative compliance orders and impose civil penalties, or initiate judicial enforcement actions, against agencies such as the Defense Department and the Department of Energy.

So long as the unitary executive policy is followed, mandatory enforcement against federal facilities of statutes other than RCRA can be achieved only through state action under EPA-approved programs or suits for injunctive relief or penalties under the citizen suit provisions of these statutes. Of course, EPA is free to use its full enforcement powers for actions against contractor operators of government facilities,[35] just as it does against defense contractors generally.

Some of the statutes described in this chapter provide criminal sanctions for violations. The defense environmental community was galvanized in 1989 by news that three civilian employees at the U.S. Army's Aberdeen Proving Ground in Maryland were convicted for illegally storing and dumping hazardous wastes from research on chemical weapons, in violation of the Resource Conservation and Recovery Act.[36]

Unlike NEPA, many of these statutes provide for waiver of their application when it is in the "paramount interest" of the United States to do so,[37] although the waiver provisions are almost never invoked. In fact, they have been used just once, during the Mariel boatlift in 1980, when President Carter relocated thousands of Cuban and Haitian refugees from Florida to a military base in Puerto Rico.[38] Although the temptation to waive compliance must have been strong in other cases, it would be surprising if any President wanted to explain publicly why the health and safety of local citizens should be compromised for government expediency, even when the national security was involved. In 1982, when the Navy proposed that military actions such as weapons tests, effluent discharges from ships, and hazardous waste disposal be exempted from compliance with federal and state environmental laws, EPA Administrator Gorsuch replied that it was inappropriate for the Defense Department and the Navy to be exempt, warning that such a proposal "could raise unnecessary state and congressional concerns."[39]

Some of the statutes call on regulated entities to keep careful records of their activities and make regular reports to EPA. Under the Clean Water Act, for example, a federal agency holding a discharge permit must monitor the pollutants it discharges into the environment and send its records to EPA at least annually.[40] Federal facilities also must comply with "any recordkeeping or reporting requirement" of EPA-approved state programs. Several statutes call for prompt reporting of leaks or spills. Most require regular inspections of regulated facilities by EPA or by states with EPA-approved programs.[41] EPA encourages all such facilities, including those operated by the federal government, to conduct their own periodic environmental audits—independent, systematic reviews of compliance with environmental requirements.[42]

These reports and records are generally available for public inspection at EPA or other federal agencies, or they may be obtained through a Freedom of Information Act request. Likewise, permit applications, transporter manifests, compliance orders, and correspondence between regulators and regulated agencies are public documents. Information from these sources may form the basis of an enforcement action if they reveal a violation,[43] or they may provide baseline data for preparation of an environmental impact statement.

Another way agency violations are sometimes publicized is through reports by employees of the agencies or their contractors. Several of the statutes include explicit protection for "whistleblowers," forbidding retaliatory firings or reassignments that might discourage the voluntary disclosure of illegal activities. For example, the Safe Drinking Water Act provides that an employer may not discharge or discriminate against an employee who initiates, assists, or testifies in an enforcement proceeding.[44] Whistleblower provisions in the environmental statutes are augmented by the broader federal Whistleblower Protection Act of 1989.[45]

Finally, a 1974 directive from the White House Office of Management and Budget (OMB), Circular A-106, calls on each agency to develop an annual five-year plan for preventing, controlling, and abating environmental pollution, along with a proposal for funding the plan. The A-106 process is designed to help EPA and affected states monitor agency compliance with laws and regulations, and to assist OMB in reviewing agency budget submissions.[46]

In the following sections, we briefly examine each of the major environmental regulatory statutes in turn, focusing especially on provisions relevant to national security activities. Each section begins with an overview of a statute as it applies to both federal and nonfederal entities, including defense contractors, then addresses its application to agencies with national security missions. Several case studies show how these laws have been applied in practice. At the end of the chapter we consider the controversy surrounding application of environmental controls abroad and defense agency programs for compliance. Issues of enforcement by EPA, the states, and citizens are dealt with in greater depth in chapter 7.

Clean Water Act

The goal of the Clean Water Act[47] is to "restore and maintain the chemical, physical, and biological integrity of the Nation's waters."[48] The act employs a two-pronged strategy to meet this goal. It calls for the achievement of water quality standards from coast to coast, and it prescribes technology-based, end-of-the-pipe limitations on discharges of pollutants, utilizing a permit system. This section first describes the operation of the Clean Water Act generally, then analyzes the act's application to federal defense facilities.

Overview of the Clean Water Act

The Clean Water Act requires the establishment of water quality standards for every body of water in the nation.[49] These standards describe criteria for various designated uses of lakes, streams, and wetlands. For example, a particular river designated as cold water fish habitat might be required to have at least 7 milligrams per liter (mg/l) of dissolved oxygen at all times. A lake used for a public water supply might be permitted to have no more than 18 organisms per 100 milliliters (ml) of E. coliform bacteria, and allowed no changes in color, odor, or taste from background conditions. Uses of individual bodies of water are set by the states. States may adopt their own water quality criteria, so long as they are at least as stringent as those set by EPA. Each state must review its use and criteria designations at least every third year. If a state fails to adopt appropriate standards, EPA must do so instead.[50] States are supposed to engage in planning to avoid violations of the standards. Although the Clean Water Act calls for standards that will "protect the public health and welfare, and enhance the quality of water," EPA has long followed a policy of allowing some degradation of waters that are cleaner than the standards require, in order to "accommodate important economic and social development in the area."[51]

The act also prohibits the discharge of any pollutant from a point source into the nation's waters without a permit. The term "pollutant" is defined broadly to include every kind of industrial, municipal, and agricultural waste, as well as radioactive materials and munitions. Permits are issued under the National Pollution Discharge Elimination System (NPDES) by EPA or by states that have their own permit programs approved by EPA.[52] State programs must be at least as stringent as EPA's program, and states must faithfully enforce their own programs.

Before a discharge permit will be issued, an applicant must show that it will conform to nationwide, technology-based effluent limitations. These end-of-the-pipe standards describe permissible amounts or concentrations of pollutants in a discharger's waste stream, and are set by EPA on an industry-by-industry basis. For example, a plant manufacturing explosives may discharge no more than one-quarter pound of total suspended solids into waterways for every thousand pounds of explosives produced.[53] Dischargers in existence in 1972 must meet limitations based on "best practicable control technology," which takes into account the costs of applying that technology to particular industries.[54] Over time, dischargers are required to apply "best conventional technology" for some of the most common water pollutants, such as biological oxygen demand (BOD), suspended solids, pH, coliform bacteria, oil, and grease.[55] All other kinds of pollutants, including toxic materials,[56] must be controlled using more stringent "best available control technology."[57] New dischargers also must conform to these stricter limitations.[58] Analogous rules cover publicly owned sewage treatment works. Industrial concerns that discharge their wastes through a publicly owned treatment works, rather than directly into public waters, need not obtain a permit but must comply with special "pretreatment standards" developed by EPA.[59]

Direct discharges of pollutants are limited in another way. Even though a discharge would conform to the nationwide uniform effluent limitations, it will be further curtailed if it would cause a violation of water quality standards or prevent the use of the receiving waters for public water supply, fishing, or swimming.[60] On the other hand, the usual end-of-the-pipe standards can be relaxed for nontoxic, nonconventional pollutants if no violation of any water quality standard would result.[61]

Many of the nation's current water quality problems are traceable to "nonpoint" pollution sources, such as farming, mining, construction, and disposal of wastes in wells.[62] The Clean Water Act directs EPA to develop guidelines for identifying and evaluating such nonpoint sources and methods for controlling them. It also orders states to identify waters threatened by such sources and adopt programs for their management.[63]

Special Clean Water Act rules apply to thermal discharges, such as heated water from nuclear reactors, to spills of oil and hazardous substances, and to discharges into the ocean.[64] The Clean Water Act also regulates uses of "wetlands" by prohibiting dredging or discharge of fill material into navigable waters without a permit from the Army Corps of Engineers.[65] Before issuing a permit, the Army must consult with EPA and the U.S. Fish and Wildlife Service regarding possible adverse environmental impacts. EPA may prevent the issuance of a permit if it determines that the discharge would "have an unacceptable adverse impact on municipal water supplies, shellfish beds and fishery areas . . . wildlife, or recreational areas."[66]

The Clean Water Act and its regulations, as well as the terms of discharge permits, may be enforced by administrative compliance orders or penalties, or by civil actions in state or federal courts for fines or injunctive relief. Criminal sanctions may be imposed on knowing or negligent violators.[67] Citizens and states may also bring suit under the act's "citizen suit" provision against any "person," including a federal government agency, for violations, or against EPA for failing to perform any nondiscretionary duty under the act.[68]

Federal Facility Compliance

The Clean Water Act directs federal facilities to comply with all federal, state, and local requirements to the same extent as private parties.[69] Thus, the Navy was ordered to apply for a permit to discharge contaminated storm water runoff at its Ships Parts Control Center in Pennsylvania.[70] Nevertheless, a study by the General Accounting Office several years ago found that federal facilities, including nuclear weapons plants, laboratories, and military bases, violated Clean Water Act regulations twice as often as private companies, and that 40 percent of those violations continued for a year or longer.[71]

Like the other major regulatory environmental statutes, the Clean Water Act permits the President to exempt a federal facility from compliance if he or she decides that it is in "the paramount interest of the United States to do so." Congress has limited the President's discretion, however, by providing that no exemption is available from new source, toxic, or pretreatment standards. The President also may by regulation exempt any "weaponry, equipment, aircraft, vessels, vehicles, or other classes or categories of property" from compliance.[72] In one recent case involving DOE's planned restart of its K Reactor at the Savannah River nuclear weapons facility, the court observed that the President may waive compliance with the NPDES system without having to show that compliance otherwise would be impossible.[73]

The following case study illustrates the reluctance defense agencies have sometimes felt either to submit to the regulatory process or to seek a presidential exemption. It also shows what a difficult time some courts have had fashioning remedies for statutory violations when the national security was implicated.

Case Study: Shooting into the Ocean

For many years the Navy has used Vieques Island, a small island off the Puerto Rico coast, for weapons training. Navy aircraft practice air-to-ground combat there, dropping bombs and shooting cannons and rockets at land and water targets. Marines make amphibious landings. Surface-to-air missiles are fired, and electronic warfare is simulated. In the course of these exercises, some ordnance invariably falls into the water. In 1978, the Governor of Puerto Rico filed suit alleging that the Navy was violating the Clean Water Act by discharging pollutants into the nation's waters without an NPDES permit.

The federal district court wasted little time finding that the Navy had indeed violated the statute. Each airplane was a "point source" discharging "pollutants" into the water. Therefore, a discharge permit was required. However, the court decided that these were merely "technical violations" not causing any "appreciable harm" to the environment. There was undisputed evidence that waters around the bombing range were "some of the best fishing grounds in Vieques." The court ordered the Navy to apply for a permit, but refused to enjoin the target practice in the meantime. The court reasoned that "the granting of the injunctive relief sought would cause grievous, and perhaps irreparable harm, not only to Defendant Navy, but to the general welfare of this Nation. . . . [T]he continued use of Vieques by Defendant Navy for naval training exercises is essential to the defense of the Nation."[74]

The Supreme Court agreed. Ignoring the fact that the Clean Water Act makes no provision for excusing strict compliance based on the kind or degree of harm to the environment, the Court decided that neither the language nor the purpose of the act indicated Congress' determination to require an immediate injunction to ensure compliance. The Court also rejected the Governor's argument that Congress had already made the policy choice undertaken by the Court, namely, by authorizing the President to waive compliance if he found it in the paramount national interest to do so. Instead, said the Court, refusal to grant an injunction would come closer to serving Congress' purpose by promoting compliance (though not immediately), rather than noncompliance, which a waiver would allow. In a scathing dissent, Justice Stevens called the Court's decision "an open-ended license to federal judges to carve gaping holes in a reticulated statutory scheme designed by Congress to protect a precious natural resource from the consequences of ad hoc judgments about specific discharges of pollutants."[75]

Before a permit could be issued to the Navy, the Commonwealth of Puerto Rico (acting as a state for this purpose) had to certify that the Navy's target practice would not cause a violation of any water quality standards.[76] It took several more years of litigation and negotiation before a permit finally was issued to the Navy.[77]

The Clean Water Act expressly forbids the discharge of any "radiological, chemical, or biological warfare agent, [or] any high-level radioactive waste." It also defines "pollutant" as including "radioactive materials."[78] However, in 1976, in a case that arose at the Department of Energy's Rocky Flats nuclear weapons plant, the Supreme Court decided that discharges of "source, byproduct, and special nuclear materials" regulated under the Atomic Energy Act[79] do not require Clean Water Act permits. "Special nuclear materials" include plutonium and enriched uranium used to make nuclear bombs. According to the Court, Congress did not intend for the Clean Water Act to significantly alter the pervasive regulatory scheme embodied in the Atomic Energy Act.[80]

That decision left a large class of very dangerous pollutants free not only from NPDES permitting but also free from opportunities afforded by the Clean Water Act for state and citizen oversight and enforcement. Nor can states impose more stringent discharge standards than those set by EPA under the Atomic Energy Act. The EPA regulations did not prevent DOE from releasing millions of curies of tritium into surface waters around its Savannah River facility during the 1980s, at times resulting in tritium concentrations more than 750 times greater than EPA's public drinking water standard.[81] In a single incident in 1991, the K Reactor at the same facility discharged cooling water contaminated with several thousand curies of tritium into the Savannah River, forcing the temporary closure of drinking water plants, food processors, and oyster beds on the river.[82]

Several doctrinal obstacles have hampered efforts to enforce the Clean Water Act against federal facilities. As with the other regulatory environmental statutes, the Justice Department has long argued that its "unitary executive" doctrine limits EPA to negotiated compliance agreements and consent orders.[83] No such constraint limits state or citizen enforcement actions, however. In one recent case a private public interest organization invoked the Clean Water Act's citizen suit provision to obtain an injunction against the Air Force for violations of discharge permit conditions at McGuire Air Force Base in New Jersey.[84] Nevertheless, the doctrine of sovereign immunity has been invoked to prevent states from assessing civil penalties for federal facility violations.[85] Both the unitary executive and sovereign immunity doctrines were called into play in a case involving the Navy's discharge of improperly treated wastes into San Francisco Bay, in violation of its Clean Water Act permit. The court found that only the EPA Administrator, and not the state, was authorized to impose civil penalties for violations. Thus, the state could not and EPA would not utilize this important enforcement tool to compel compliance with the act.[86]

In other litigation, a court decided that the Clean Water Act's citizen suit provision did not create a cause of action to enforce state water quality standards against the Air Force for polluting waters around the former Suffolk County Air Force Base with JP4 jet fuel.[87] However, another court has ruled that a private defense contractor operating a government ammunition plant is not a federal "department, agency, or instrumentality" shielded by sovereign immunity from state agency penalties for violations of state water pollution statutes.[88]

Clean Air Act

The Clean Air Act[89] provides a legal structure "to protect and enhance the quality of the Nation's air resources so as to protect the public health and welfare."[90] Signed into law in 1970, and extensively revised in 1977 and 1990, the act calls for the establishment of national air quality and air pollutant emission standards.

United States preparations for war contribute significantly to local, regional, national, and even global air pollution. Like many large industrial companies, the Defense Department operates boilers and incinerators, and it owns huge fleets of motorized vehicles and mobile generators to serve a variety of electrical needs. The Department of Energy emits radioactive materials into the air when it manufactures and tests nuclear weapons. The Clean Air Act generally applies to these agencies the same way it does to nongovernment concerns that pollute the air.

Overview of the Clean Air Act

The Clean Air Act calls on EPA to develop regulatory programs that prescribe ambient air quality standards and controls on stationary and mobile air pollution sources. Special programs are aimed at hazardous air pollutants, acid rain, and stratospheric ozone depletion.[91]

EPA is directed to draw up a list of the most common "criteria" pollutants that "may reasonably be anticipated to endanger public health or welfare." For each of these pollutants, EPA must develop national primary ambient air quality standards to protect "public health . . . allowing an adequate margin of safety,"[92] but "without regard to the economic and technical feasibility of attainment." Secondary ambient air quality standards are supposed to protect the public welfare more broadly, addressing such concerns as visibility and damage to wildlife, crops, and property.[93] After more than two decades, however, primary and secondary standards have been developed for only six pollutants: lead, sulfur oxides, ozone, carbon monoxide, nitrogen oxide, and small particulates.[94]

EPA is also responsible for issuing national emission standards for "hazardous" air pollutants, those which present a threat of adverse effects on human health (including substances known or suspected to cause cancer, tumors, genetic mutations, or reproductive dysfunctions) or the environment (for example, through bioaccumulation).[95] The 1990 Clean Air Act amendments require EPA to promulgate standards for 189 hazardous pollutants specifically named in the legislation, including radionuclides (radioactive particles), plus any other pollutants that display hazardous characteristics.[96] The standards for hazardous pollutants are intended to achieve the maximum reductions in emissions achievable in light of cost and available technology. The act also provides for evaluation of public health risks remaining after implementation of the technology-based standards.[97]

The primary responsibility for air pollution control and prevention rests with the states. Each state is required to develop a State Implementation Plan (SIP) that will enable it to achieve national air quality standards.[98] The plan must contain standards for individual pollutants at least as stringent as those developed by EPA. Each state's plan also must include a strategy to prevent any significant deterioration of air quality within areas that already meet or exceed the national standards, as well as a program to clean up the air in areas that fall below such standards (so-called "nonattainment" areas).[99] The activities of each federal agency, including permitting and financial assistance, must conform to its state's SIP.[100] For example, in ongoing litigation concerning the closure of Pease Air Force Base in New Hampshire, described in a case study in chapter 5, plaintiffs assert that planned commercial redevelopment of the base, which is located in a nonattainment area for ozone, will make compliance with the state's SIP impossible. The Clean Air Act also provides for state implementation of federally established standards for hazardous air pollutants and for new air pollution sources.[101]

The 1990 amendments to the Clean Air Act added a permit program for existing stationary air pollution sources, augmenting one already in place for new or modified sources. As with the Clean Water Act, the EPA may approve state permit programs that meet or exceed federal standards. Members of the public, as well as neighboring states whose air quality could be affected, must be given notice and opportunity to comment on permit applications. Permit holders are required to monitor their emissions and report them to state or federal regulatory authorities.[102] Such reports are generally available for public inspection.

Either EPA or a state with an approved program may enforce Clean Air Act requirements, including the terms of individual permits, through compliance orders, administrative penalties, judicial actions for injunctions or penalties, and criminal sanctions. In addition, a citizen suit provision enables any person or a state to seek judicial remedies for violations.[103]

The President may exempt any stationary source from compliance with rules for hazardous air pollutants for up to two years if he finds that the technology needed to comply is not available and that it is "in the national security interests of the United States to do so."[104]

Federal Facility Compliance

Federal agencies are required to comply with the Clean Air Act "in the same manner, and to the same extent as any nongovernmental entity."[105] Specifically, the act calls on federal facilities to comply with all permitting and record-keeping requirements, and to pay fees imposed by state and local regulatory programs. Thus, in one case the Air Force was ordered to obtain a permit for boilers at two installations under Ohio's EPA-approved Clean Air

Act program, and to halt its violations of Ohio rules for particulate emissions.[106] In other litigation, states have collected regulatory fees and costs from DOE facilities and military installations in New York and California.[107]

Air pollution from federal sources must be considered in each state's plan to implement federal standards (SIP), and federal facilities must comply with state plans. Early on, for example, the Navy was ordered to comply with state and local standards under California's SIP in its operation of several jet engine test cells at facilities around that state.[108]

In 1989, EPA adopted rules for radionuclide emissions from DOE facilities providing that releases must not exceed amounts that would cause any member of the public to receive more than ten millirems per year of radiation, roughly the equivalent of a chest x-ray.[109] To ensure compliance, DOE must use computer modeling, air sampling, and continuous monitoring of emission sources such as vents and stacks. Each year DOE is required to calculate the exposure of any member of the public off-site where there is a residence, school, or business, and report the results to EPA. Its report must include a list of radioactive materials used at the facility, as well as a description of release points and emissions controls. To improve accountability, each report must be certified and signed by the official in charge of the facility. Like most government records, such reports are available to members of the public for the asking under the Freedom of Information Act, except for portions that are properly classified.[110]

The following case study illustrates the importance of Clean Air Act rules for radionuclide emissions, as well as the difficulty that states may experience in trying to enforce those rules.

Case Study: Radioactive Air at Los Alamos[111]

The Los Alamos National Laboratory near Albuquerque, New Mexico is owned by the Department of Energy (DOE) and operated by the University of California. Since 1943, the laboratory's primary mission has been nuclear weapons research and development. It also does work in magnetic and inertial fusion, nuclear fission, nuclear safeguards and security, and laser isotope separation.

The laboratory emits radionuclides into the air through various stacks and other sources. Among the pollutants are tritium, isotopes of plutonium and uranium, and mixed fission products. The laboratory's own records indicate that it delivers the highest off-site radiation exposure dose to the public of any DOE facility. It has released more than 3.2 million curies of radioactivity into the atmosphere in the last decade alone. This is about 250,000 times the amount released in the accident at Three Mile Island.

In 1985, DOE and the University of California applied to the State of New Mexico under its EPA-approved Resource Conservation and Recovery Act program for a permit to operate a hazardous waste incinerator. (RCRA is described later in this chapter.) The state issued the permit, but attached conditions requiring assurances that wastes incinerated under the permit would contain no radionuclides, and requiring that stack gases from the incinerator be monitored for radiation. The New Mexico program contains no specific criteria for radionuclides, but it provides that a hazardous waste permit may be made subject to "any conditions necessary to protect human health and the environment."[112] DOE filed suit to have the conditions set aside, arguing that such radioactive materials were exempt from RCRA regulation, and that in any event the New Mexico rules were not specific enough to constitute "requirements" applicable to DOE under RCRA's federal facilities provision.[113] The court noted that the incinerator is sometimes used for hazardous wastes mixed with radionuclides, but it upheld the permit conditions on grounds that the permit applied only to unmixed, nonradioactive hazardous wastes. The court's decision in favor of the state was affirmed on appeal.[114]

The lower court concluded in the alternative that New Mexico's conditions on the Los Alamos incinerator permit could be upheld under the state's authority to regulate radioactive air emissions under the Clean Air Act. New Mexico could, the court noted, even set radionuclide emission standards more protective than those established by the EPA. However, the state has elected not to adopt hazardous air pollutant standards for radionuclides because, it insists, it lacks the staff and resources to enforce them.

In 1990, David Nochumson, an environmental engineer employed by the University of California, was assigned to manage the monitoring of radioactive air emissions at Los Alamos. He quickly determined that the laboratory failed to comply with a variety of EPA regulations for such emissions. When the laboratory refused to fund Nochumson's plan for achieving compliance with Clean Air Act requirements, and delayed notifying federal officials of its noncompliance, Nochumson contacted EPA officials informally on his own, and sought to publicize the violations in the laboratory's annual report to EPA on radionuclide emissions. His supervisors warned him to write only "positive factual statements" about the laboratory's radiation program in the future, then transferred him to other duties. Nochumson filed a harassment complaint with the Department of Labor under the Clean Air Act's whistleblower protection provision.[115] A Labor Department administrative judge ordered Nochumson reinstated and directed payment of back wages and damages for lost work and emotional distress.[116]

In fact, the laboratory's annual reports to EPA indicate that it has violated EPA's regulations for radionuclide emissions continuously since the regulations were published in March 1989. In November 1991, EPA sent the laboratory a notice of noncompliance, observing that 58 of 146 emissions from stacks there were not monitored for radioactive releases, and that the laboratory had not even identified all its sources of radionuclide emissions. EPA also found a violation of quality assurance programs required by the regulations. A second notice of noncompliance followed a year later. Internal documents at the laboratory reveal that the cost of bringing the facility into compliance will likely exceed $100 million dollars.

In September 1994, a local citizens' group and a Los Alamos resident who sometimes works at the laboratory filed suit in the federal district court to enjoin the continuing violations.[117] In particular, they complained that the laboratory had not adopted an adequate quality assurance program, as required by EPA regulations, and that it was not monitoring emissions from individual sources within the facility. DOE had sought to rely instead on ambient air quality sampling, a procedure for which the laboratory had sought but not been given approval by EPA. Moreover, the plaintiffs pointed out, EPA regulations call for corrections of any violations within ninety days after the effective date of the regulations, and the violations had persisted for more than four years. According to the plaintiffs, these failures make it impossible for members of the public to evaluate the health risks from airborne radiation. The suit is still pending at this writing.

The President may exempt any existing federal emission source from compliance if he determines that it is in the "paramount interest of the United States" to do so. He also may exempt any "weaponry, equipment, aircraft, vehicles, or other classes or categories of property which are owned or operated by the Armed Forces" for the same reason.[118] Any such exemption must be reported and explained to Congress.

Air pollutants tend to move freely across national borders. They may even affect the environment of the whole earth. The 1990 Clean Air Act amendments and other recent legislation address some of the global effects of local air pollution, as the following case study indicates.

Case Study: Defending Earth's Ozone Layer

A number of recent federal laws and international agreements are intended to slow the thinning of the earth's protective ozone layer. Depletion of stratospheric ozone allows more ultraviolet radiation to reach the earth's surface, causing increases in skin cancer, cataracts, and crop damage. The control measures are aimed particularly at the Defense Department and its contractors, which are the nation's largest consumers of ozone-depleting chemicals. These chemicals include chlorofluorocarbons (CFCs), halons, methyl chloroform, carbon tetrachloride, methyl bromide, and hydrochlorofluorocarbons (HCFCs). CFCs are used as solvents, especially in computer applications, and as refrigerants. Halons, which are far more destructive to the ozone layer, are used as fire-extinguishing agents.[119] During the Persian Gulf War, for example, the Navy and Air Force employed halons to purge the fuel tanks of fighter planes, so the tanks would not explode if hit by enemy ground fire. Halons are also used for fire suppression in enclosed spaces, such as command and control centers and inside armored vehicles.

The Montreal Protocol of 1987 called for an end to the production of CFCs before January 1, 2000.[120] The protocol was implemented in this country by 1990 amendments to the Clean Air Act.[121] In 1992, the protocol was amended in Copenhagen to halt the manufacture of CFCs by 1995, five years earlier than originally envisioned, to end production of halons by 1994 and HCFCs by 2030, and to add methyl bromide to the list of controlled substances.[122] A month later, EPA adopted regulations embracing those changes.[123]

Consistent with the Montreal Protocol, the Clean Air Act allows a limited exemption from the phase-out of halon production for aviation safety or for reasons of "national security."[124] Nevertheless, Congress has ordered the Pentagon to reduce unnecessary releases of CFC and halons into the atmosphere, and to mount a search for substitutes.[125] New defense contracts may not include specifications or standards requiring ozone-depleting substances unless a general or flag officer certifies that no suitable substitutes are available.[126] DOD now believes that by conserving existing supplies and developing alternatives it can probably meet its needs for the foreseeable future.[127] Training exercises using halons have been halted. The Defense Logistics Agency has formed a reserve bank of ozone depleting chemicals for "mission-critical" future uses, and the Air Force's Wright Laboratory is leading a multi-agency research effort to identify halon replacements. The Pentagon has also ordered each of its service branches to draw up new specifications for procurement contracts to reduce the use of ozone-depleting chemicals.[128]

The Clean Air Act exposes federal facilities to any "process and sanction whether enforced by Federal, State or local courts, or in any other manner."[129] Moreover, the act's citizen suit provision authorizes any person to bring suit against the United States or any federal agency for violations of emission standards or limitations, or of orders issued by the EPA Administrator or a state.[130] Sovereign immunity thus seems to be waived under both provisions, enabling states to maintain suits for enforcement against federal agencies, or to issue administrative orders or impose civil penalties for violations. Several lower courts have upheld state civil penalties under the federal facilities section of the Clean Air Act,[131] although the Supreme Court has decided that identical language in the Clean Water Act failed to waive sovereign immunity from state-imposed fines.[132]

Safe Drinking Water Act

Congress passed the Safe Drinking Water Act in 1974 to "assure that water supply systems serving the public meet minimum national standards for protection of public health."[133] The act employs a two-part strategy to ensure that water flowing from most kitchen faucets is fit to drink: (1) it imposes standards of purity for water supplied by public water systems, and (2) it seeks to safeguard groundwater sources that supply those systems. Applied to the nation's defense establishment, the act directly regulates water systems operated by defense installations. It also imposes significant constraints on a variety of national security activities that could contaminate America's groundwaters.

Overview of the Safe Drinking Water Act[134]

The Safe Drinking Water Act calls on EPA to establish standards of purity, called national primary and secondary drinking water regulations, for public water systems serving at least fifteen hookups or twenty-five individuals.[135] Smaller systems and users who depend on individual springs and wells to supply their water are not affected by these regulations, but they may benefit from the act's protection of some groundwaters, as described below.

The primary drinking water regulations are designed to protect human health.[136] They prescribe maximum contaminant levels (MCLs) for a number of different contaminants. Each MCL is set as close as is "feasible" to the "level at which no known or anticipated adverse effects on the health of persons occur and which allows an adequate margin of safety."[137] For example, under EPA rules a public water system must ensure that concentrations of trichloroethylene (TCE) do not exceed five parts per billion. TCE, a suspected carcinogen, is a common industrial solvent found in the groundwater beneath many military bases, where it has been used for cleaning aircraft engines and other equipment.[138] EPA has also set MCLs for radionuclides (radioactive elements) in drinking water.[139]

It is important to note that states with EPA-approved Safe Drinking Water

Act programs are free to adopt MCLs more stringent than those set by EPA, and that the stricter state standards bind federal facilities just as they do all others.[140] Equally important, the federally established MCLs (or their more restrictive state counterparts) are used by EPA as groundwater quality standards in the administration of other federal regulatory programs.[141]

The secondary drinking water regulations address the public welfare generally, rather than health alone. They provide standards for color and odor, and for contaminants such as pH, chloride, iron, and dissolved solids.[142] Unlike the primary regulations, enforcement of the secondary standards is left to the discretion of the states.[143]

Three separate Safe Drinking Water Act programs are designed to protect underground sources of drinking water: the Underground Injection Control (UIC) program, the sole source aquifer program, and the wellhead protection program. The UIC program[144] is aimed at the practice of disposing of untreated liquid wastes by pouring them into a hole in the ground, recalling the maxim "out of sight, out of mind." A permit is required for every "subsurface emplacement of fluids by well injection."[145] Underground injections of hazardous wastes are also covered by the Resource Conservation and Recovery Act (RCRA).[146] Under the UIC program, underground injection of wastes is forbidden if it is likely to "endanger drinking water sources" by causing them to violate maximum contaminant levels set out in the primary drinking water standards or otherwise to "adversely affect the health of persons."[147] RCRA categorically prohibits some hazardous waste injections either on the basis of their proximity to a current or potential drinking water source, or because of the character of the contaminant. For example, hazardous or radioactive wastes may not be injected into or above a geologic formation containing an underground source of drinking water within a quarter mile of the well bore.[148] The UIC program is administered by the EPA or by individual states with EPA approval. Program requirements are enforced through administrative orders and injunctions, as well as civil and criminal penalties.[149]

The sole source aquifer program[150] authorizes EPA to designate and protect any aquifer which is the "sole or principal drinking water source for [an] area and which, if contaminated, would create a significant hazard to public health."[151] Before it can issue a permit for a new underground injection well near a sole source aquifer, the EPA must find that operation of the new well will not threaten the purity of water in the aquifer.[152]

The Safe Drinking Water Act's wellhead protection program represents a third approach to protection of underground drinking water sources. A "wellhead protection area" is "the surface and subsurface area surrounding a water well or wellfield, supplying a public water system, through which contaminants are reasonably likely to move toward and reach such water well or wellfield." Within such an area the states, with EPA guidance, may impose regulations in the nature of zoning to control land uses that could release contaminants into the environment.[153]

Like the other major regulatory statutes, the Safe Drinking Water Act authorizes direct enforcement by EPA or by states with EPA-approved programs. It also enables any person to file suit in federal court to enjoin a violation, unless EPA, the Attorney General, or a state is already "diligently" prosecuting the case in court.[154] The act has a special provision for emergencies. If EPA discovers that a contaminant has entered or is likely to enter a public water supply or underground drinking water source, and the contaminant "may present an imminent and substantial endangerment to the health of persons," the Administrator may "take actions as he may deem necessary in order to protect the health of such persons."[155]

Federal Facility Compliance

More than 250 military installations in the United States operate public water systems regulated under the Safe Drinking Water Act. Twenty-one defense facilities overlie designated sole source aquifers.[156] The Safe Drinking Water Act provides that

> [e]ach Federal agency . . . engaged in any activity resulting, or which may result in, underground injection which endangers drinking water . . . shall be subject to, and comply with, all Federal, State, and local requirements, administrative authorities, and process and sanctions . . . in the same manner, and to the same extent, as any nongovernmental entity.[157]

This means that the Defense Department and the Department of Energy must obtain permits from the EPA or from states administering EPA-approved UIC programs before injecting any fluid wastes into the ground. EPA may waive the permit requirement if the President finds that the national security requires it, but no President has yet been called upon to do so.[158] Federal agencies also must comply with EPA-approved state wellhead protection programs.[159] Because such state programs may impose extensive constraints on uses of the land surface, they present an unusual opportunity for state and local governments to influence national defense planning.

In 1987, the Safe Drinking Water Act was applied in dramatic litigation affecting America's nuclear weapons program, as the following case study explains.

Case Study: Nuclear Weapons and Drinking Water

In 1982, Congress passed the Nuclear Waste Policy Act, directing the Department of Energy to build and operate permanent storage facilities for highly radioactive wastes from commercial nuclear power plants, military reactors, and weapons manufacturing.[160] The wastes are to be solidified, put into canisters, and placed in mines deep underground, in order to protect the environment from contamination by radioactive materials.[161] Two disposal sites are currently under consideration, one for civilian and military high-level waste at Yucca Mountain, Nevada, and the other for transuranic defense wastes at the Waste Isolation Pilot Project (WIPP) near Carlsbad, New Mexico. Both of the proposed sites and the controversy surrounding them are examined in detail at the end of chapter 4.

In 1985, EPA issued regulations for the Nuclear Waste Policy Act[162] establishing maximum cumulative totals of various radionuclides, such as uranium, that could be released from a "controlled area" over a period of ten thousand years. A controlled area could cover as much as one hundred square kilometers and extend up to five kilometers from a disposal site. No limits were set on contamination within such an area, which would become a sacrifice zone to help contain the radioactive material.

Outside a controlled area, radionuclide contamination of underground sources of drinking water was to be kept within limits prescribed by Safe Drinking Water Act regulations for at least one thousand years. The Nuclear Waste Policy Act regulations also set limits on individual human exposure to radiation for the same period. Those limits were intended to furnish additional protection from radioactive seepage that might enter the human food chain or drinking water supplies.

Several public interest organizations and four states challenged the new EPA regulations in court.[163] The First Circuit Court of Appeals decided that in adopting regulations for the Nuclear Waste Policy Act, EPA could not ignore its obligations under the Safe Drinking Water Act. The latter act provides that EPA has a duty to ensure that underground sources of drinking water will not be "endangered" by any underground injection. This duty was implicitly waived by Congress in the Nuclear Waste Policy Act only for the "controlled area" surrounding a repository site, the court said. Otherwise, the disposal of highly radioactive wastes in a deep geological formation "likely" constitutes underground injection of "liquids" subject to the Safe Drinking Water Act, even if those wastes are solidified and containerized. That is so, said the court, because

> [t]he dangerous component of this waste, i.e., the radiation, regardless of whatever "form or state" it is emitted from, will flow or move, thus

> having the capacity to do harm to drinking water sources far distant from the original site as more conventional injected fluids would do. . . .
>
> [Congress'] overall intent [was] to protect future supplies of drinking water against contamination. Unusable ground water is unusable ground water no matter whether the original source of the pollution arrived in a loose, free form manner, or in containers injected into the ground.[164]

It was not enough, the court held, to protect underground sources of drinking water outside the "controlled areas" for only(!) one thousand years. Nor was it acceptable to allow nearby residents to receive individual radiation doses of up to twenty-five millirems per year, far more than the four-millirem limit set by EPA in regulations for the Safe Drinking Water Act. These regulatory anomalies made EPA's rules under the Nuclear Waste Policy Act arbitrary and capricious, according to the court.

The court set aside two sections of the regulations that dealt with the protection of individuals and groundwater, and remanded the rest to EPA for revision. While other Safe Drinking Water Act rules might prevent the actual use of contaminated water for drinking, the court observed, they could not restore the water to its original quality. "Once [radioactive waste] is placed in a repository, the situation may well be irreversible: there may be no feasible way, years later, to arrest ongoing contamination of surrounding water supplies."[165]

In 1992, Congress reinstated all of the earlier regulations except the two sections specifically vacated by the court.[166] In late 1993, EPA issued amended regulations to replace those two sections.[167] The amendments extend from one thousand to ten thousand years individual human exposure limits outside the controlled areas. Those exposure limits are lowered from twenty-five to fifteen millirems of radiation per year, about the equivalent of two chest x-rays, but still higher than levels established under the Safe Drinking Water Act. The amendments also provide that radioactive pollution of offsite underground drinking water sources must not exceed Safe Drinking Water Act maximum contaminant levels for ten thousand years. However, the new regulations do not apply to a disposal above or into a geologic formation that within one-quarter mile contains an underground source of drinking water (a Class IV well, in Safe Drinking Water Act parlance). Neither do they apply to the proposed Yucca Mountain repository in Nevada, for which rules are to be issued in 1995.[168]

EPA insists that despite the contrary suggestion of the First Circuit Court of Appeals, the disposals addressed by its new regulations are not underground injections of fluids subject to regulation under the Safe Drinking Water Act. If EPA is correct, the Department of Energy will not have to obtain UIC permits for the repositories. That would make nonfederal oversight and enforcement of compliance more difficult, since the Safe Drinking Water Act's reporting and citizen suit provisions are not replicated in the Nuclear Waste Policy Act. Nor would states with EPA-approved UIC programs be able to impose environmental standards more stringent than the federal standards, a serious point of contention between the federal government and the States of Nevada and New Mexico. In addition, EPA regulations under the Safe Drinking Water Act that prohibit new Class IV wells (wells that inject hazardous or radioactive wastes within one-quarter mile of an underground drinking water source)[169] would not apply. Nevertheless, each of the planned radioactive waste repositories will presumably be subject to similar requirements under the Resource Conservation and Recovery Act (RCRA), described in the next section.

Citizens and states are expressly authorized to bring a civil suit against the United States for violation of any Safe Drinking Water Act requirement.[170] Moreover, as under the Clean Air Act and Clean Water Act, federal agencies are subject to "any process or sanction, whether enforced in Federal, State, or local courts or in any other manner . . . notwithstanding any immunity of such agencies, under any law or rule of law."[171] Whether this language provides a waiver of sovereign immunity, enabling states to enforce administrative orders or impose civil penalties for violations, has not yet been resolved in litigation applying the Safe Drinking Water Act.

Resource Conservation and Recovery Act (RCRA)

The Resource Conservation and Recovery Act[172] (universally known by its acronym RCRA) is the federal statute aimed at protecting human health and the environment from the immense volume of solid wastes generated by our society. "Solid waste" for this purpose means "discarded material," and includes not only solids but also liquids and contained gases.[173] RCRA includes two broad regulatory programs, one for hazardous wastes and another for everything else, as well as a special program to regulate underground storage tanks.

Federal agencies are generally subject to the same rules for handling solid wastes that govern everybody else. Both DOD and DOE produce a lot of ordinary garbage that is routinely recycled, incinerated, or sent to sanitary landfills. They also generate prodigious amounts of hazardous wastes in a wide variety of activities—from cleaning aircraft engines to manufacturing nuclear weapons.[174]

In this section, we examine the use of RCRA to prevent the release of both hazardous and nonhazardous wastes into the environment. We begin with a brief overview of the statute, then examine its application to defense facilities. In the next chapter, we see how provisions of the same statute are used to clean up hazardous materials that have already been spilled or disposed of improperly.

Overview of RCRA[175]

The disposal of most nonhazardous solid waste is regulated by the states under Subtitle D of RCRA. That part of the law determines what happens, for example, to ordinary household garbage from municipalities and military bases alike. EPA regulations furnish guidelines for state planning and for the design and operation of "sanitary landfills." Federal technical assistance and funding are available to states with EPA-approved regulatory programs that conform to the guidelines.[176]

The disposal of nonhazardous wastes in sanitary landfills may nevertheless pose enormous and lasting threats to the environment, especially since such wastes typically include significant quantities of materials that by themselves would be regarded as hazardous. That is why EPA regulations now call for the installation of plastic liners, monitoring of groundwater quality, and eventual closure of each landfill, following procedures designed to minimize the risk of releases into the surrounding environment.[177] While this part of RCRA places primary reliance on the states to prevent open dumping of wastes, EPA is authorized to enforce its regulations directly if a state fails to do so.[178]

Other RCRA provisions are aimed at reducing the need for disposal of solid wastes by discouraging their production in the first place. Generators of hazardous wastes must adopt programs to reduce their quantity and toxicity to the degree "economically practicable." And federal agencies must follow EPA guidelines in procuring products that contain the highest "practicable" percentage of "recovered" materials, such as waste paper, boxboard, old newspapers, and manufacturing residues.[179]

Subtitle C of RCRA[180] governs the day-to-day management of hazardous wastes, as well as their cleanup. According to the statute, wastes are "hazardous" if they pose a significant threat to human health or the environment when improperly handled.[181] Any waste is subject to regulation under this part of RCRA if it appears on a list of hazardous wastes prepared by EPA or if it meets criteria for identifying unlisted waste as hazardous based on its ignitability, corrosiveness, reactivity, or toxicity.[182] Both the list, now including hundreds of substances, and the criteria are set forth in EPA regulations.[183] Nonhazardous waste generally becomes hazardous when it is combined with a listed hazardous waste.[184] By definition, hazardous waste does not include "source, special nuclear, or byproduct material" regulated under the Atomic Energy Act,[185] although such radioactive materials are subject to RCRA regulation if they are "mixed" with other hazardous wastes.[186]

Wastes listed by EPA as hazardous or satisfying the criteria in EPA regulations must be managed from "cradle to grave," that is, from their creation to their final safe disposal. Generators and transporters of hazardous wastes have to identify and track those wastes utilizing a system of manifests reported to EPA. Owners and operators of facilities that treat, store, or dispose of hazardous wastes must apply for a permit from EPA or from a state authorized by EPA to run its own hazardous waste program.[187] However, if an existing facility qualifies for "interim status," it may continue to operate until a final permit is issued.[188]

RCRA provides that no permit can be issued for a disposal facility unless the owner or operator demonstrates that the hazardous waste disposed of there will not escape into the surrounding environment for as long as it remains hazardous.[189] EPA regulations set forth operating criteria, as well as requirements for facility housekeeping and contingency planning.[190] For example, land disposal facilities must carefully monitor nearby groundwater to detect any contamination from releases. If a hazardous constituent is found in the groundwater, it generally must not exceed the background level for that substance or a maximum contaminant level (MCL) established for it under the Safe Drinking Water Act.[191] Permitted and interim status facilities must be inspected at least every other year, and the reports of those inspections must be made available to the public.[192]

Special rules apply to the construction and maintenance of underground storage tanks.[193] The disposal of hazardous wastes in wells is regulated under the Safe Drinking Water Act UIC program.[194] EPA has also developed special requirements for the destruction of hazardous wastes in incinerators.[195]

RCRA requires hazardous waste facilities to clean up any releases that occur at a site.[196] Each permitted facility also must have a "closure plan" that would enable it to shut down safely at any point during its operation.[197] These cleanup and closure requirements are spelled out in greater detail in chapter 4.

As with the statutes described earlier in this chapter, RCRA's requirements may be enforced directly through administrative compliance orders and civil penalties, judicial injunctions, and criminal sanctions. Private citizens and states also may bring suit against any "person," including a federal government agency, for a RCRA violation, or may sue EPA for failing to perform a nondiscretionary duty under the act.[198] In an emergency, EPA may issue an order or bring suit for an injunction against any person whose handling of either hazardous or nonhazardous waste could "present an imminent and substantial endangerment to health or the environment," regardless of the location of the waste. A state or private citizen also may initiate a suit in federal court to abate such an imminent hazard, without waiting for EPA to act.[199]

Federal Facility Compliance

RCRA employs sweeping language to ensure federal facility compliance with "all Federal, State, interstate, and local requirements, both substantive and procedural," concerning solid waste management and disposal "in the same manner, and to the same extent, as any person is subject to such requirements."[200] Thus, if an Army weapons depot uses trichloroethylene (TCE), a listed hazardous substance, as a degreaser to clean equipment, and it sends TCE contaminated wastes to a commercial disposal facility off-base, the shipment must be accompanied by a manifest, one copy of which goes to EPA and becomes a public record. Likewise, when DOE stores high-level nuclear wastes mixed with hazardous chemicals used in processing spent fuel rods, it must obtain a RCRA permit from EPA. The President may exempt a federal facility from compliance "in the paramount interest of the United States."[201]

Although the Departments of Defense and Energy produce immense quantities of hazardous wastes, they have not always been eager to comply with RCRA's requirements. As recently as 1988, an EPA survey of federal facilities found that the Navy had "one of the worst compliance records."[202] For its part, DOE has admitted that throughout the 1980s it followed a deliberate policy of avoiding compliance at its nuclear weapons complex, and it instructed its contractors to do the same.[203] Today both agencies are working hard to come into compliance. For example, the Army reports that 96 percent of RCRA violations at its facilities can be cured by administrative or procedural corrections, and that such violations are being reduced by increased staffing and improved training.[204] Both the Pentagon and DOE are also trying to ease their regulatory burden by adopting new operating procedures that generate less waste and recycle as much as possible.

The following case study describes a citizen suit to enforce the provisions of RCRA against a key facility in the Department of Energy (DOE) nuclear weapons complex. It illustrates the reluctance some agencies have felt to comply with environmental regulations that might somehow hamper their defense missions. It also shows the difficulty both Congress and the courts have experienced in trying to harmonize national security with environmental protection.

Case Study: Toxic Fish at Oak Ridge

"State-of-the-art 1948" was one expert's description of DOE's waste management practices at its Oak Ridge, Tennessee Y-12 plant during the early 1980s.

Throughout the Cold War, Y-12 processed and recycled highly enriched uranium and other materials for nuclear weapons.[205] From 1950 to 1963, the plant dumped or spilled more than two million pounds of mercury at the plant site. The mercury was used to purify a form of lithium for hydrogen bombs. In addition, four waste disposal ponds leaked nearly five million gallons of metal plating wastes, acids, and solvents into the local groundwater each year. Fish consumed by residents were contaminated with mercury well above federal limits. To make matters worse, dirt taken from a contaminated flood plain near the plant was used in the early 1980s for construction projects in and around the town of Oak Ridge.[206]

The Legal Environmental Assistance Foundation and the Natural Resources Defense Council (NRDC) filed suit in 1983 to force DOE to obtain a permit for the treatment, storage, or disposal of hazardous waste under RCRA.[207] DOE claimed complete exemption from RCRA at the Y-12 Plant, based on the following provision of the statute:

> Nothing [in RCRA] shall be construed to apply to (or to authorize any State, interstate, or local authority to regulate) any activity or substance which is subject to . . . the Atomic Energy Act of 1954 (42 U.S.C. 2011 and following) except to the extent that such application (or regulation) is not inconsistent with the requirements of such Act.[208]

DOE argued that the Atomic Energy Act gave it, not EPA or the states, the authority to set standards for waste disposal. Unlike RCRA, it insisted, the Atomic Energy Act did not permit state regulation of DOE activities or allow the dissemination of information about nuclear weapons and materials.

The court disagreed, and ordered DOE to apply for a RCRA permit. It noted that DOE facilities are in fact subject to state and local regulation under a variety of other environmental statutes, such as the Clean Air Act and the Clean Water Act. Furthermore, the court pointed out, application of RCRA is precluded only to the extent that it is inconsistent with the Atomic Energy Act. Only the regulation of nuclear wastes is expressly precluded by RCRA.[209] The two statutes could thus be harmonized, the court decided, by subjecting nonradioactive wastes at DOE facilities to RCRA regulation. If the security of nuclear material data were threatened, said the court, the President could exempt Y-12 from RCRA. However, where "DOE has not applied for a Presidential exemption, national security considerations should not be considered by the Court."[210]

In spite of its finding that DOE was in violation of RCRA (as well as the Clean Water Act), and despite DOE's refusal to apply for a presidential exemption, the court seemed to create its own judicial national security exemption by allowing the plant to continue operating while DOE applied for a permit. It did so, the court said, because the "Y-12 Plant is a unique and essential element of this nation's system of nuclear defense," and because DOE had already taken or had agreed to take "steps that will reduce environmental harm" caused by the statutory violations.[211]

DOE decided not to appeal the decision in the *Y-12* case, but it claimed exemption from RCRA for chemical wastes mixed with radioactive materials ("mixed wastes").[212] The Department was particularly concerned about having to comply with RCRA's storage and disposal requirements at its huge tank farms at Hanford, Washington[213] and Savannah River, South Carolina, which contain high-level radioactive wastes and toxic chemicals from chemical reprocessing of spent nuclear fuel. However, in 1986 EPA determined that mixed wastes are subject to RCRA regulation. The following year, after heavy criticism from states, citizen groups, and members of Congress, DOE finally agreed.[214]

More recently, DOE claimed that when it stored hazardous chemical waste laced with plutonium at its Rocky Flats plant in Colorado, it could avoid complying with RCRA storage rules because it intended to recover the plutonium for reuse. A federal district court decided otherwise:

> Once the plutonium is separated from the hazardous waste, the plutonium itself is clearly no longer subject to RCRA. . . . [B]ut plutonium *mixed* with this waste is itself hazardous waste. . . . [T]he materials at issue are part of the waste problem that Congress intended to regulate under RCRA.[215]

The court ordered DOE to obtain a RCRA permit for storage of the mixed wastes or shut down any operations that would generate more of them.[216]

In 1991 DOE attempted to open its Waste Isolation Pilot Plant (WIPP) in New Mexico for geologic disposal of nuclear and hazardous wastes without a RCRA permit. The Environmental Defense Fund, the Natural Resources Defense Council, and two New Mexico environmental groups sued to enjoin the opening. DOE argued that WIPP qualified for RCRA interim status, so that it could proceed without a permit. But the federal district court permanently enjoined the unauthorized opening.[217] The court of appeals affirmed the permanent injunction, not on RCRA grounds, but because the Secretary of the Interior had violated the Federal Land Policy and Management Act (FLPMA)[218] by failing to properly "withdraw" the WIPP site for the purpose of depositing radioactive waste.[219] The WIPP project is described in greater detail in chapter 4.

The Department of Energy has never applied for a RCRA permit or a no-migration variance for any of its underground tests of nuclear weapons in Nevada. Those tests nevertheless invariably leave behind quantities of lead and other hazardous materials from the explosive devices and diagnostic packages, along with radionuclides. Nor has DOE sought to conform to Underground Injection Control program requirements under the Safe Drinking Water Act. It has, however, apparently reported each test shot as a "release" of a "hazardous substance" under the Comprehensive Environmental Response, Compensation, and Liability Act (CERCLA). DOE's omissions have not been challenged in court.

When hazardous wastes are incinerated, EPA's RCRA regulations for incinerators provide that 99.99 percent of all organic constituents must be eliminated.[220] Despite such precautions, persons living near the incinerators have sometimes felt apprehensive about the threat of air pollution, as the following case study demonstrates, especially when the incinerators are to be used for the destruction of chemical weapons.

Case Study: Regulating the Disposal of Chemical Weapons[221]

In a very controversial 1985 vote, Congress approved the production of a new generation of binary chemical weapons, but not before ordering the destruction of all existing chemical weapons in the United States arsenal by 1994.[222] Many of the existing weapons are decades old, some dating from World War II, and in a dangerous state of decay. The deadline has now slipped to 2004,[223] although even that date may be too optimistic.

Chemical weapons are stored at eight Army installations around the country: Tooele Army Depot, Utah; Pine Bluff Arsenal, Arkansas; Umatilla Depot, Oregon; Pueblo Depot, Colorado; Anniston Army Depot, Alabama; Aberdeen Proving Ground, Maryland; Newport Army Ammunition Plant, Indiana; and Lexington–Blue Grass Army Depot, Kentucky. Neighbors at each of the sites are eager to see the weapons removed, but they are concerned about the Army's plans to destroy the weapons in high-temperature incinerators at the local sites.

The Army insists that it can easily meet EPA's proposed safety standards for civilian incinerators, although prototypes at Tooele and at Johnston Atoll in the Pacific have experienced repeated breakdowns. The neighbors are worried about releases of combustion products that could include toxic furans and dioxins, and about mechanical failures or human errors that could result in dispersal of the chemical warfare agents.[224] Their fears were fanned by Army plans at one site to outfit local schools with special ventilation systems, provide children with protective clothing, and install outdoor warning sirens.[225] They also worry that once the chemical weapons are destroyed, the incinerators will continue to be used for the disposal of other kinds of hazardous wastes. For its part, the Army is reluctant to transport the aging and often highly unstable weapons for destruction at other sites, and it would like to avoid the objections of cities and states along the transportation routes.[226]

In 1986, the Army published a draft programmatic environmental impact statement (PEIS), as required by NEPA, for the entire chemical weapons disposal program. The following year, Congress directed the Army to prepare a final PEIS containing a thorough evaluation of alternative disposal technologies.[227] When the PEIS was completed in 1988, it was criticized for relying entirely on incineration as a method of disposal, for failing to fully evaluate data from local storage sites, and for including information developed after the close of the public comment period.[228]

Opponents of the incinerators argue that the chemicals in the weapons should be reprocessed and utilized for fertilizers, semiconductors, and other valuable products. In the alternative, they say, low-temperature chemical detoxification or biological reduction would avoid the risks associated with incineration.[229] Congress responded in 1992 by directing the Army to look into alternatives to incineration recommended by the National Research Council, an investigative arm of the National Academy of Sciences, considering safety, environmental protection, and cost-effectiveness.[230]

The National Research Council's report, published in 1994, recommended that the Army further investigate several alternative technologies, while at the same time continuing with its "baseline" incinerator program, in order to avoid delays that might increase the cumulative risk to the public from extended storage of the chemical agents.[231] In its response to the report, the Army proposed to undertake further research, but concluded that no alternative technology would be more cost-effective than incineration or able to meet the congressional deadline of 2004.[232] Whether any alternative would be safe can only be determined after additional research. Programs utilizing low-temperature chemical neutralization and biological treatment are being developed as backups to incineration in case public opposition causes delays at individual sites. Although the National Research Council urged that both comprehensive and site-specific risk assessments be updated, the Army has so far prepared environmental impact statements for only three local sites, and it currently has no plans to revise its earlier environmental analyses.

In the midst of this controversy, the United States signed the Chemical Weapons Convention[233] in 1993. After the convention's expected effective date in 1995, each signatory (the U.S. and Russia are the only states to admit possession of such weapons) will be required to destroy all chemical warfare agents within ten years, although a five-year extension is possible under exceptional circumstances. In disposing of the weapons, states are bound to "assign the highest priority to ensuring the safety of people and to protecting the environment."[234] Otherwise, each state may choose its own methods of destruction and apply its own domestic environmental laws, except that land burial, open-air burning, and dumping in any body of water are expressly prohibited by the treaty.[235] Some members of the Senate have expressed their reluctance to approve the convention until they are assured that the Russians have effective means of safely destroying their stockpiles. Such assurances are not likely to be given until the United States itself has developed the technology needed to do the job. While the 1985 legislation orders only the destruction of active chemical weapons stockpiles, the convention calls for elimination of obsolete weapons and production facilities as well.

Before chemical weapons can be incinerated at each site, the Army will have to obtain a RCRA permit from the state or EPA. Several states have recently adopted more restrictive environmental standards for the incinerators. Whether the Army will be able to satisfy all requirements for a permit at each site remains to be determined. The only thing certain in all this is that whatever method is chosen, it will be controversial, and it will involve some environmental risk.

As with the other environmental statutes, the Justice Department for years invoked its "unitary executive" theory to limit EPA's RCRA enforcement efforts against federal facilities to negotiated compliance agreements, consent orders, and memoranda of understanding.[236] It also insisted, and the Supreme Court agreed, that in enacting RCRA Congress did not intend to waive the government's sovereign immunity from state-imposed punative penalties for past RCRA violations by federal agencies.[237] However, in 1992 Congress passed the Federal Facility Compliance Act, authorizing EPA to proceed against federal facilities just as it does against other RCRA violators.[238] EPA invoked that authority for the first time in 1993, when it proposed RCRA penalties against the Navy for improper storage of hazardous wastes at the El Centro Naval Air Station.[239] Shortly thereafter, EPA issued a compliance order under RCRA's substantial endangerment provision against Reese Air Force Base in Texas, to abate threats to a nearby community from releases of trichloroethylene.[240] The order includes a strict compliance schedule and a directive to reimburse EPA for response costs.

The 1992 legislation also allows states to issue administrative orders and levy civil penalties against federal facilities for past violations.[241] The new provision was first invoked by the State of Washington in 1993, when it imposed $100,000 in fines and issued an enforcement order against DOE for failing to adequately test and label two thousand drums of hazardous wastes at its Hanford nuclear weapons plant.[242] Even before 1992, states as well as private citizens could obtain court orders enjoining federal agencies to comply with RCRA.[243] The subject of enforcement generally, and the effect of the Federal Facility Compliance Act in particular, are addressed in more detail in chapter 7.

Other Regulatory Statutes

Instead of limiting the impact of an activity on a particular environmental medium, such as air, land, or groundwater, several statutes are aimed at specific environmental threats, such as noise, pesticides, and toxic chemicals. Like the other regulatory statutes, they apply to activities of the military and the Department of Energy, so they have serious implications for defense preparedness. They also may provide important information and opportunities for public involvement in national security decisions that affect public health and the environment. Four of the most important of these statutes are briefly described below. Others are listed in Appendix A.

Toxic Substances Control Act (TSCA)

The Toxic Substances Control Act[244] requires EPA to establish a registry of all chemicals manufactured or processed in the United States, except those already regulated under another statute, such as the Federal Insecticide, Fungicide and Rodenticide Act (FIFRA).[245] If EPA finds that a particular chemical might present an unreasonable risk of injury to health or the environment, it can prohibit or restrict the manufacture, processing, shipment, use, or disposal of that chemical. This authority has been used to impose restrictions on CFCs, asbestos, PCBs, and certain chemicals used in metal working.[246]

No new chemical substance may be produced without first notifying EPA. EPA then may ban or limit the chemical's manufacture, or require testing to prove its safety; the burden of proof rests on the manufacturer to justify any risk posed by the introduction of that chemical. EPA may bring suit in federal court to enjoin the production, use, or distribution of any chemical that represents an imminent hazard.[247]

A special provision of TSCA bans the manufacture and nearly all uses of PCBs, and calls for regulation of their disposal.[248] This measure is of special concern to the military services, many of whose bases are contaminated with PCB wastes. For example, at the Twin Cities Army Ammunition Plant in Minnesota, some fourteen hundred cubic yards of soils laced with PCBs had to be dug up and decontaminated in a high-temperature incinerator. Efforts are underway at the Defense Department to develop a safe chemical process for destroying PCBs.[249]

Unlike most of the other regulatory statutes, administration of TSCA is the sole responsibility of EPA, although states may receive grants for the development of supplemental programs.[250] Nevertheless, the act includes a citizen suit provision that allows judicial enforcement by states and others.[251] Compliance with various provisions of the act may be waived "in the interest of national defense."[252]

Federal Insecticide, Fungicide, and Rodenticide Act (FIFRA)

The Federal Insecticide, Fungicide, and Rodenticide Act[253] requires that all pesticides sold or distributed in commerce be registered with the Environmental Protection Agency. EPA has the authority to restrict applications of pesticides that could have "unreasonable adverse effects on the environment."[254] Whether the risk to "man and the environment" is unreasonable depends upon a balancing of economic, social and environmental costs against benefits from using a particular pesticide.[255]

Applicators must be certified by EPA or by a state with an EPA-approved program as competent to handle and use pesticides.[256] Otherwise, the only practical constraint on application of a pesticide, unless its use is altogether forbidden by EPA, comes from the label affixed to it by the manufacturer; it is unlawful "to use any registered pesticide in a manner inconsistent with its labeling."[257] The act sets out both civil and criminal penalties for violations.[258]

Because they maintain enormous inventories of land, buildings, and equipment, the Defense Department and DOE are among the largest domestic consumers of pesticides. DOD has a special Armed Forces Pest Management Board that oversees all pesticides in the military supply system and trains its applicators.[259] Because of continuing criticism over use of the defoliant Agent Orange during the Vietnam War, as well as United States obligations under the Environmental Modification Convention of 1977, current Army regulations specify that if herbicides are employed in combat, the commander will comply with EPA regulations or manufacturer label requirements concerning droplet size, concentration, and rate of application, in accordance with FIFRA.[260]

Noise Control Act

The Noise Control Act[261] was created "to promote an environment for all Americans free from noise that jeopardizes their health or welfare." EPA must develop criteria for noise levels in various settings to achieve this goal. The act also directs EPA to publish noise emission standards for construction and transportation equipment, motors and engines, and electrical and electronic equipment that constitute major sources of noise. However, some of the noisiest products of all—aircraft and military weapons—are expressly exempted from the provisions of the act.[262] Thus far, EPA has only developed standards for construction equipment and vehicles.[263] Manufacture, sale, or importation of products that fail to meet the standards is unlawful and may be enjoined by citizen suit or punished criminally.[264]

The Noise Control Act generally applies to federal agencies, including the Defense Department and DOE, the same way it does to everybody else. The President can exempt any activity, facility, or noise emission source from compliance if he finds that it is in the "paramount interest of the United States" to do so.[265] In addition, the act permits the Administrator of EPA to waive the prohibition on manufacture, sale, or import of any product not conforming to the noise emissions standards "for reasons of national security . . . upon such terms and conditions as [s]he may find necessary to protect the public health and welfare."[266] The Administrator has chosen to exercise that authority by adopting a blanket exemption for all products made to specifications developed by a "national security agency" and purchased by that agency,[267] although there is no apparent nexus between the exemption and protection of the public health and welfare.

Each federal agency must "comply with Federal, State, interstate, and local requirements respecting control and abatement of environmental noise to the same extent that any person is subject to such requirements."[268] However, when the Governor of Puerto Rico brought suit under the Commonwealth's criminal nuisance statute to curtail the noise from explosions, gunfire, and jet aircraft at the Navy's Vieques Island training facility, a federal court of appeals ruled that the Puerto Rico law lacked standards specific enough to qualify as "requirements" within the meaning of the federal statute.[269]

The act directs all federal agencies to implement its policy "to the fullest extent consistent with their authority under [other] Federal laws administered by them."[270] The scope of this provision was tested when property owners near Luke Air Force Base in Arizona complained about noise from the operation of F-15 jet fighter planes.[271] The court decided that the Federal Aviation Administration (FAA), which generally regulates aircraft traffic in the United States, had no authority to control noise pollution from military flights.

Despite the broad exemptions and limitations just described, the Pentagon is not insensitive to the impact of its operations on its neighbors. By far the greatest cause for concern is aircraft noise. However, the Air Force has not responded to this concern by purchasing quieter airplanes or altering its schedules and flight patterns. Instead, it has developed a program for mapping noise levels from takeoffs and landings at each of its air bases, then turning that information over to local officials, whom it hopes will employ municipal land use planning controls to keep homes, schools, and churches away from the areas of greatest impact. Home mortgage guarantees from HUD and the

Veteran's Administration may be unavailable for new construction in the noisiest areas.[272]

Hazardous Materials Transportation Act

Pursuant to the Hazardous Materials Transportation Act,[273] the Secretary of Transportation may find that materials such as explosives, radioactive materials, flammable liquids, and oxidizing agents pose an unreasonable risk to health and safety or property when they are transported in a particular quantity or form. If so, the Secretary must issue regulations for the transportation of those materials.[274]

The Secretary's regulations generally preempt inconsistent state or local rules.[275] In the 1970s, for example, the City of New York adopted an ordinance to prevent trucks carrying spent nuclear fuel from reactors at DOE's Brookhaven Laboratory on Long Island from driving through the city. When the Department of Transportation promulgated regulations for the highway shipment of radioactive materials, the city was forced to seek a determination that its ordinance was not preempted by the federal requirements.[276] Nevertheless, states may designate transportation routes for hazardous shipments, applying federal standards, and taking into account such factors as population density, type and quantities of materials, and emergency response capabilities.[277]

Because the regulations apply to federal agencies just as they do to other shippers, they affect the day-to-day operations of the Defense Department and DOE.[278] For example, the Air Force uses private contractors to ship the rocket fuels nitrogen tetroxide and hydrazine by truck from the manufacturer in Vicksburg, Mississippi to defense installations in California, Colorado, New Mexico, and Florida. Both substances are exceptionally dangerous when released into the environment.[279] The Air Force and its transporters must comply with regulations limiting the size and prescribing the packaging of shipments,[280] and follow routes designated by states or the Department of Transportation.

Despite such precautions, the risk to the public from defense-related shipments can be expected to increase substantially in coming years. The ongoing cleanup of Defense Department and DOE facilities, described in the next chapter, will produce large quantities of hazardous and radioactive wastes, many of which will have to be transported by rail or highway to remote sites for treatment, storage, and disposal.

Applying Environmental Regulations Abroad

The United States military has hundreds of operations in foreign countries and territories. Like their domestic counterparts, these operations affect their surrounding environments in various ways. At several bases in Germany, for example, underground sources of drinking water have been contaminated with spilled jet fuel and trichloroethylene from U.S. military operations.[281] In the Philippines, the U.S. Navy has poured "tons of toxic chemicals into Subic Bay."[282] Elsewhere, local citizens complain that U.S. military operations monopolize fertile farmlands in Guam, threaten bird sanctuaries in Japan, and fill the air with jet noise and exhaust in Germany.[283]

A 1991 General Accounting Office report found that the Defense Department had made only limited progress

> in improving its management of hazardous waste overseas. Guidance has not been issued to clarify applicability of U.S. laws when host country hazardous waste laws either do not exist or are not as stringent as U.S. laws. Most overseas bases GAO visited did not have adequate hazardous waste management plans. Hazardous waste management training did not meet the Department's requirements, and the Department's oversight of activities that generated hazardous waste was still minimal.[284]

Among other concerns, these failures could be costly, according to the GAO: at the time of the report the Army had received eighteen host country claims for damages allegedly resulting from the mishandling, storage, or improper disposal of hazardous waste.

It has long been assumed that United States environmental laws which regulate military activities in this country have no application abroad. The controversy surrounding application of NEPA to defense activities overseas was examined in the last chapter. As a general proposition, courts have adopted a presumption that domestic laws do not apply outside our borders unless Congress makes clear its intent that they should apply extraterritorially.[285] Nevertheless, in some recent decisions the courts seem to have relaxed that presumption to some degree. For example, in one case the court found no congressional intent to apply the Resource Conservation and Recovery Act (RCRA) to extraterritorial disposal of hazardous wastes by private parties, but not before carefully reviewing the statute's legislative history.[286] In another, the court found evidence of congressional intent to apply the Endangered Species Act abroad in language resembling that in NEPA, as well as in the legislative history. The same court noted that, like NEPA, "the [Endangered Species] Act is directed at the actions of federal agencies, and not at the actions of sovereign nations. Congress may decide that its concern for foreign relations outweighs its concerns for foreign wildlife; we, however, will not make such a decision on its behalf."[287] In a third case, the court found that NEPA applied to a U.S. decision to incinerate wastes in Antarctica because "the presumption against the extraterritorial application of statutes . . . does not apply where the conduct regulated by the statute occurs primarily, if not exclusively, in the United States, and the alleged extraterritorial effect of the statue will be felt in . . . a continent without a sovereign."[288]

In 1978, purporting to act independently of any congressional mandate, President Carter ordered all federal agencies responsible for construction or operations abroad to comply with pollution controls applicable in each host country.[289] Unfortunately, host country laws, if any existed, were often far more lax than their United States counterparts. In 1990, Congress directed the Pentagon to "develop a policy for determining applicable environmental requirements for military installations located outside the United States" in order to protect "the health and safety of military and civilian personnel assigned to such installations."[290] There was no mention of protection for the local environment or citizens of other countries. Two years later, the Pentagon published its "Overseas Environmental Baseline Guidance Document," directing military personnel at each overseas location to develop environmental standards based on DOD "suggested criteria" for pollutants such as air emissions, drinking water contaminants, hazardous wastes, noise, and pesticides.[291] These criteria resemble but do not replicate the requirements of domestic environmental laws and EPA regulations. Standards for each foreign facility must be at least as stringent as DOD's suggested criteria or host country laws, whichever provides greater protection for the environment, and must be consistent with status of forces agreements and other international requirements. Such standards may be waived by the Defense Department, however, if compliance "would seriously impair [a facility's] operations, adversely affect relations with the host nation or require substantial expenditure of funds not available for such purpose."[292] For example, a waiver might be sought to avoid ruffling local political feathers when a wastewater treatment plant built to DOD standards would be more effective than comparable host country facilities. The directive is careful to note that it does not govern cleanup or remedial actions arising from DOD's past activities.[293]

The Pentagon's new policy reflects a serious commitment to protect the environments of United States allies, as well as the global commons, from the effects of military operations abroad. Nevertheless, even in unexceptional circumstances it allows for different performance standards than the ones that govern DOD activities at home. This double standard is not justified by either military necessity or diplomatic concerns, since DOD reserves the right to waive the rules for almost any reason. The policy also includes no mandate for monitoring and public reporting, and no provision for citizen enforcement, leaving our overseas neighbors without the environmental protections that we in this country now take for granted.

Defense Programs for Compliance

Officials at both the Defense Department and the Department of Energy have launched far-reaching programs to bring their agencies into full compliance with the regulatory environmental laws. These initiatives have been spurred by pressure from Congress, state and local governments, and members of the public, all aided by improved access to information about agency operations. But the most important changes have come from within each agency, where a new generation of officials seems determined to protect the environment while performing their defense missions, and determined as well to do what the law requires.

Each year's progress toward achieving compliance is traced in public reports from DOD and DOE. The Defense Department produces an annual "Report on Environmental Compliance," briefly describing environmental activities within each DOD component.[294] The report also sets out personnel and funding requirements at each defense installation for the coming five years, and analyzes the effect of compliance with environmental laws on military operations and mission capabilities. The 1994 report noted that while compliance has made certain training and other operations more difficult and expensive, it had not caused any reductions in unit readiness.[295] According to the same report, the Pentagon is incorporating pollution prevention principles into military planning, operations, and training at all levels, in order to reduce compliance costs and disruption of defense missions.

For several years, the Department of Energy published an annual "Environmental Restoration and Waste Management Five-Year Plan."[296] Each plan described the previous year's successes and failures in achieving major environmental milestones at DOE facilities, along with prospects and projected costs for coming into compliance. For example, DOE's 1993 report revealed that its Rocky Flats nuclear weapons plant had implemented a required surface water management plan but had failed to obtain a permit for a new sanitary landfill.[297] Each DOE five-year plan included budget requests for environmental activities during the coming year. Beginning in 1995, the five-year plan will be replaced by a "Baseline Environmental Management Report."

The expense of complying with the environmental laws is substantial, yet it represents only a small part of each year's defense budget. The Department of Defense expects to spend between $2.5 and $3 billion each year between 1995 and 1999 on environmental compliance in this country, and another $200 million or so at military installations abroad.[298] These amounts do not include expenditures for research and development or for cleanup of contaminated military bases. The Department of Energy devoted nearly half of its $6.3 billion environmental budget for 1995 to what it calls "waste management and corrective activities."[299] These activities include the development of means to properly treat, store, and dispose of more than a million cubic meters of radioactive wastes at sites around the country. The largest item in the 1995 budget is more than $600 million for reducing the risks posed by underground waste storage tanks at DOE's Hanford Reservation.[300]

The cost of compliance with environmental requirements is always an issue, as environmental programs within each agency compete with more traditional national security missions for a slice of the shrinking budget pie. There is even a spirited competition among environmental programs. Congress has addressed this problem in part for the Defense Department by creating a special budget account for environmental restoration at closing military bases, and another account for cleanups at operating installations and formerly used defense sites. However, ongoing compliance efforts are paid for out of current operating expenses, along with other military activities.[301] DOE has tried to reconcile competing demands for limited resources by setting priorities within each of its major environmental programs, as follows: protection of workers, the public, and the environment; performance of interagency and other compliance agreements; compliance with environmental laws, such as the Clean Water Act; and what DOE calls "desirable" activities, such as research and development.[302] However, no statutory authority permits DOE to so pick and choose among its regulatory obligations.

Research and development are not merely desirable but essential to achieving full compliance. The technology needed to comply with some requirements, especially those concerned with cleanup of contaminated sites, simply does not yet exist, or in some cases it is not affordable. Both DOE and the Pentagon actually have robust research and development programs, each drawing heavily on university and private industry resources. The Defense Department will spend almost $300 million on environmental research and development in 1995; DOE's research budget for the same period is about $425 million. Some research is driven by statutory mandates. For example, the Federal Facility Compliance Act of 1992 directed DOE to draw up a National Compliance Plan to help develop site-specific programs for the disposal of its immense inventory of "mixed waste."[303]

The Defense Department and the Department of Energy spend billions of dollars each year for the purchase of goods and services from private contractors. These contractors, like the government agencies that employ them, create enormous impacts on the environment. The agencies have begun to try to regulate these impacts by writing environmental performance standards into procurement contracts and monitoring contractor behavior more carefully. To increase the environmental sensitivity of military purchases, the Pentagon has recently appointed its Deputy Under-Secretary for Environmental Security to membership on the Defense Acquisition Board.

In general, defense contractors pay their own environmental costs if they operate under fixed-price agreements for goods or services.[304] However, under so-called "cost-reimbursable" contracts, compliance and other environmental expenses may be passed along to the government as overhead or as "reasonable" direct costs. Federal regulations say such costs are reasonable if, among other considerations, they are "generally recognized as ordinary and necessary" and reflect "sound business practices."[305] Unfortunately, such regulatory restrictions have not always prevented defense agencies from picking up the tab for their contractors' misbehavior. In one celebrated case, the California firm of Aerojet-General, which manufactures rocket motors for the MX and other missiles, dumped hundreds of thousands of gallons of toxic wastes into unlined pits. Suits by state and federal regulators resulted in Aerojet-General's agreement to clean up its contaminated site at a cost of at least $62 million. However, the Air Force is contributing up to $37 million as a reimbursable cost of doing business, in part because Aerojet-General reportedly followed government guidance in its operations.[306]

Contractors may be indemnified for their environmental costs to "facilitate the National defense," when the subject matter of the contract involves "unusually hazardous or nuclear risks." However, penalties imposed on defense contractors for environmental law violations cannot be reimbursed by the government unless they result from compliance with specific contract terms or conditions, or written instructions from the responsible contracting officer.[307] In one recent case, DOE was forced to pay a $78 million settlement of claims against its contractor operator at the Fernald Feed Materials plant in Ohio for discharging millions of pounds of uranium into the environment, because its contract required indemnification.[308] According to federal acquisition regulations, new contracts may not be awarded to firms that violate the Clean Air Act or Clean Water Act, although the head of an agency may exempt a firm from this prohibition if he finds that it is "in the paramount interest of the United States to do so."[309] A firm also may be debarred or suspended from obtaining government contracts if it is found guilty of any offense "indicating a lack of business integrity or business honesty,"[310] such as a violation of RCRA.

The agencies have adopted programs to draw members of the public and other affected constituencies into the process of making decisions that have environmental implications. The Department of Energy has organized a national Stakeholders' Forum that includes representatives of labor unions, business, state and local governments, and environmental organizations. The purpose of the Forum is to advise DOE about compliance and other environmental policy issues, and to help in the development of the annual five-year plan for environmental management. A State and Tribal Government Working Group has similar responsibilities. Representative working groups and advisory panels have also been established at several DOE nuclear weapons facilities to consult with federal officials about environmental cleanup operations and future uses of government lands. A separate environmental management advisory committee has the task of addressing DOE's overall plan for cleaning up the entire weapons complex. Public workshops on a variety of topics have helped provide information about Department operations.[311] Perhaps the most important development, however, has been Energy Secretary Hazel O'Leary's public release of millions of pages of previously classified documents describing DOE operations throughout the Cold War.[312]

The Defense Department is developing its own outreach programs to foster community involvement. In planning its cleanup of Moffett Field Naval Air Station in Sunnyvale, California, for instance, the Navy set up a technical review committee that included members of the public and held public meetings to solicit comments on draft plans.[313]

Both the Defense Department and the Department of Energy conduct regular environmental audits of their facilities to test compliance with statutory and regulatory requirements.[314] Each agency, and each DOD component, has an inspector general who oversees agency policies and operations and prepares public reports of her findings. A recent audit report by the Defense Department Inspector General, for example, concluded that the Navy had paid higher award fees than contractors were entitled to receive under cost-reimbursement contracts for cleaning up contaminated sites, thereby removing the contractors' incentive for good performance.[315]

Each agency is training its personnel to better understand their environmental responsibilities. The Army has adopted "a vigorous program to incorporate environmental . . . awareness into the fabric of the Army's training mission."[316] Under its Army Environmental Training Master Plan, all soldier and civilian employees are to receive some environmental instruction at various stages in their military careers. The Navy and Air Force have similar programs. All three service branches have created special environmental leadership courses for high-ranking officers, as well as programs aimed at particular compliance issues, such as the 1990 Clean Air Act amendments. Each branch offers detailed program guidance through publications such as the Navy's "Environmental and Natural Resources Program Manual." In addition, the Army's Environmental Policy Institute and the Air Force's Environmental Center of Excellence provide expertise and support on remediation, compliance, and environmental planning.

Finally, defense agency and contractor employees are being encouraged to help monitor compliance and report violations. According to Energy Secretary O'Leary, whistleblowers "should be celebrated, not punished. . . . We want an environment where employees feel safe to voice their concerns. We have zero tolerance for reprisals. It's as simple as that."[317]

These defense programs offer no guarantee that important environmental values will be protected, or even that the laws designed to protect those values will be observed. Still, they represent an important public commitment to carry out the policies reflected in those laws. The Army Corps of Engineers publishes a widely used Commander's Guide to Environmental Management, which begins with this injunction from the Chief of Engineers:

> As a commander, you are entrusted with the stewardship of the land, air, water, and natural and cultural resources associated with performing your military mission. These resources must be carefully managed to serve both the Army's and the nation's short and long-term needs. Today, environmental considerations must be a part of your decisions.[318]

This directive is only part of a growing body of evidence that managers of all the nation's defense programs have embraced a new environmental ethic. With better oversight by EPA and by state and local governments, along with support from informed citizens, we can expect steady improvement in compliance with the regulatory environmental laws, even as we maintain our military preparedness in a dangerous world.

Chapter 4

Dangerous Legacy: Cleaning Up after the Cold War

More than four decades of Cold War have left soil, groundwater, and buildings at defense facilities from coast to coast contaminated with hazardous and radioactive wastes.[1] Whether the nation was rendered more secure or less so by the activities that produced these wastes is now beside the point. The important thing today is to clean up the contamination and safely dispose of the wastes, which pose a continuing threat to human health and the environment. By most estimates, the cleanup is going to cost more than $200 billion, maybe a great deal more. The Department of Energy (DOE) cleanup budget for fiscal year 1995 comes to $6.3 billion. In the same year, the Department of Defense (DOD) will spend another $2.2 billion for environmental remediation at active and formerly used military installations, and $500 million more at closing bases.

At the end of 1993, DOD counted 19,694 contaminated sites at 1,722 active facilities nationwide.[2] For example, investigations at Norton Air Force Base in California have revealed accumulations of waste oil and fuels, solvents, paint strippers, acid plating solutions, and heavy metals at twenty-two locations. The Army's Fort Riley in Kansas is contaminated with tetrachloroethane, mercury, pesticides, acetone, methylene chloride, and carbon tetrachloride. Cost estimates for investigation and cleanup at all DOD sites range from $25 to $42 billion dollars.[3] It will take at least thirty years to complete the work, probably much longer.[4]

Measured by the cost and complexity of cleaning up, the overall danger to human health and the environment may be far greater from contamination at DOE's nuclear weapons complex. Radioactive or hazardous chemical wastes, or both, are found at 137 DOE installations in thirty-four states and territories.[5] At the Fernald, Ohio, Feed Materials Plant, for example, DOE fabricated uranium metal for use in nuclear production reactors. Merely planning for environmental restoration at the plant has turned out to be so difficult that it will be 1998 before actual cleanup begins at all of Fernald's operating units. By that time, the annual bill for Fernald alone will be about half a billion dollars. DOE officials now say that restoration of the entire weapons complex is going to cost between $200 and 300 billion or even as much as $1 trillion.[6]

Concerned about the political difficulty of sustaining public support for such enormous budget commitments over several decades, the Environmental Protection Agency organized the Federal Facilities Environmental Restoration Dialogue Committee in 1992.[7] The Committee, which includes DOD and DOE officials, federal and state environmental regulators, and representatives of environmental and pubic interest organizations, has recommended new procedures to expand public participation and set priorities among federal sites awaiting cleanup. Up to now, Congress has shown its support by voting funds for environmental restoration while cutting the Pentagon budget for weapons.[8]

Some defense facilities are so contaminated that it would be prohibitively expensive to try to clean them up using current technology. The technology needed to restore other areas simply does not yet exist. These "sacrifice areas" will have to be fenced off and clearly marked to prevent public access for the foreseeable future, or until new technologies are developed.[9] At the same time it will be necessary to thwart any migration of wastes during the hundreds or thousands of years that they remain extremely dangerous. Whether a foolproof program can be developed to protect present and future generations from these threats remains to be seen.

Where cleanup is possible, the process begins with a survey of the site, analysis of contaminants, review of historical records, and interviews with personnel who might know how the property was used. When hazardous or radioactive materials are discovered, an assessment is made of risks to human health and the environment, and a remediation plan tailored to the site is developed in compliance with federal, state, and local laws, as well as agency regulations. The cleanup itself involves treatment of the waste in situ or removal and treatment elsewhere, followed by disposal and monitoring to ensure that any future releases are detected, at least for a while.

Two federal environmental statutes are especially important here. The Resource Conservation and Recovery Act (RCRA)[10] governs not only the day-to-day management of hazardous wastes described in chapter 3, but also the cleanup of such wastes when they have been deliberately or inadvertently released into the environment. The Comprehensive Environmental Response, Compensation, and Liability Act (CERCLA)[11] likewise provides for the cleanup of sites contaminated with hazardous substances. In this chapter we first briefly examine the substantive provisions of these two laws, then see how they are being used to restore defense facilities across the country. Later in the chapter we examine special programs established at the Defense Department and DOE to help carry out the required cleanup. Finally, we review the search for suitable sites and safe techniques for permanent disposal of nuclear weapons wastes.

Resource Conservation and Recovery Act (RCRA)[12]

Overview of RCRA's Cleanup Provisions[13]

The Resource Conservation and Recovery Act, like Janus, looks forward and backward. As we saw in the previous chapter, RCRA applies prospectively to regulate the handling of hazardous wastes, in order to prevent their release into the environment. RCRA also has a retrospective aspect: it requires "corrective actions" to clean up and control hazardous materials already released into the environment. In this application, RCRA overlaps the coverage of CERCLA, which also provides for cleanup of hazardous substances. Both laws may be applicable to the cleanup of a single hazardous waste site, sometimes creating confusion about which law controls. When EPA elects to follow CERCLA rather than RCRA procedures at a given site, a state with RCRA corrective action authority may order its own concurrent RCRA cleanup, in order to apply stricter state standards or simply maintain a greater measure of control over the process.[14]

Anyone wishing to treat, store, or dispose of hazardous wastes must apply for a RCRA permit.[15] Before a permit will be issued, the applicant must provide assurance that any existing hazardous waste contamination at the site will be cleaned up, no matter when the hazardous wastes were brought to the site or when they were released into the environment.[16] The obligation to take "corrective action" continues for releases that occur during the operation of a permitted or interim status facility, and following its closure or abandonment.[17] If hazardous waste migrates beyond the boundary of a facility, corrective action must be taken "where necessary to protect human health and the environment."[18]

When cleanup is required, hazardous contaminants must be removed from the site or treated in place. Contaminated groundwater generally must be restored to background levels for each contaminant, to maximum contaminant levels (MCLs) developed under the Safe Drinking Water Act, or to some alternative level set by the EPA Regional Administrator.[19]

Before a facility can obtain a RCRA permit for treatment, storage, or disposal of hazardous waste, it must have a closure plan that would enable it to be safely closed and cleaned up at any time.[20] A closure plan includes detailed steps for either removing any contaminants remaining at the site or capping them with an impermeable cover, as well as arrangements for monitoring the site for at least thirty years.

RCRA's corrective action requirements may be applied and enforced by states with programs specifically approved for that purpose by EPA. Otherwise, RCRA's cleanup provisions are administered entirely by EPA.[21] While forty-six states have EPA-approved RCRA permit programs, only five are authorized to administer their own corrective action programs, and no more than a dozen others are expected to apply for such authority.[22]

As with the other major environmental laws, RCRA requirements may be enforced through administrative compliance orders, civil actions for injunctions, civil penalties, and criminal sanctions.[23] EPA also may issue an administrative order or bring suit for an injunction against anyone responsible for hazardous waste that presents "an imminent and substantial endangerment to health or the environment," to require them to "take such action as may be necessary."[24] This provision represents an especially powerful tool for environmental protection, since it applies not only at RCRA-permitted facilities, but wherever hazardous wastes are discovered in the environment. If EPA fails to act, anyone may sue for abatement of an imminent hazard.[25]

RCRA Cleanups at Federal Facilities

By its terms, RCRA generally applies to federal facilities just as it does to commercial facilities.[26] That means RCRA cleanup provisions are applicable to DOD and DOE, as well as to their contractors. In fact, RCRA imposes requirements on federal agencies that are more stringent in several particulars. For example, EPA is required to conduct a thorough inspection of each federal facility every year,[27] instead of every two years as with private facilities, and to make its inspection reports public. In addition, every two years each federal agency must compile an inventory of sites on property it owns or operates, or has ever owned or operated, where hazardous wastes have ever been stored, treated, or disposed of.[28] The inventory must describe the amount, nature, and toxicity of wastes at each site, as well as efforts to clean it up. Given the need to review operational records stretching back several decades in many cases, the practical difficulty of simply locating some contaminated sites is apparent. Evaluating the current condition of those sites may be no less daunting. Once these inventories are compiled, however, they must be made available to the public.[29]

Until 1992, when Congress passed the Federal Facility Compliance Act, amending RCRA,[30] EPA did not enforce RCRA's cleanup requirements against federal facilities the same way it did against private ones. The 1992 legislation expressly authorizes administrative and judicial enforcement of RCRA's provisions by EPA against federal facilities. It also permits states to levy civil penalties against federal agencies for past violations. The act does not address the waiver of sovereign immunity for cleanup costs and resource damages imposed under state law. One court has held that the Navy was not responsible for such charges when hazardous wastes sent by it to a commercial disposal facility were spilled.[31]

EPA has already begun using its new powers against DOD and DOE facilities. In 1993, EPA issued its first administrative compliance order against a federal facility under RCRA's imminent endangerment provision to compel the cleanup of contamination at Reese Air Force Base in Texas. The Air Force was directed to adhere to a strict cleanup schedule and to reimburse EPA for its investigative costs.[32] Whether EPA will now hail the defense agencies into court for violations remains to be seen.

As with many of the other federal environmental laws, private citizens and states may obtain court orders enjoining federal agencies to comply with RCRA. In addition, both EPA and states with corrective action authority can use their full enforcement powers under RCRA against commercial contractors operating federal facilities. Individual federal employees, as well as commercial defense contractors and their employees, may be held criminally liable for RCRA violations.[33] Questions of enforcement are addressed in greater detail in Chapter 7.

Comprehensive Environmental Response, Compensation, and Liability Act (CERCLA)

In 1980, following widespread publicity about toxic waste dumping at Love Canal, Congress enacted the Comprehensive Environmental Response, Compensation, and Liability Act (CERCLA),[34] commonly referred to as the Superfund law. Six years later, Congress strengthened CERCLA with the passage of the Superfund Amendments and Reauthorization Act (SARA).[35] This legislation provides for the systematic identification and cleanup of sites contaminated with hazardous substances, and it fixes liability for cleanup costs. It applies to government and nongovernment owners and operators alike.

CERCLA and RCRA are often applicable at the same site. While the ultimate goals of the two statutes are exactly the same—protection of human health and the environment—the procedures and nomenclature are quite different. Moreover, as we see below, the decision to restore a site under the authority of one statute rather than the other may affect the degree to which the site is finally cleaned up, as well as the ability of states, local governments, and citizens to influence the outcome.

Overview of CERCLA[36]

The CERCLA cleanup process is activated by the "release" of a "hazardous substance" into the environment. Release means almost any kind of discharge or disposal, accidental or deliberate.[37] A hazardous substance is defined by CERCLA as any material deemed hazardous or toxic by RCRA, the Clean Water Act, the Clean Air Act, or the Toxic Substances Control Act, or separately characterized by EPA as presenting a substantial danger to public health or welfare or the environment.[38] To promote prompt cleanup and avoid wider dispersal into the environment, the release of any hazardous substance must be reported to the National Response Center immediately by the person in charge of the facility or vessel from which the release occurs. To help locate sites contaminated in the past, anyone who was ever in charge of a facility where hazardous substances were treated, stored, or disposed of must report that information to EPA, unless the facility currently holds a RCRA permit or enjoys interim status under that statute.[39] Failure to report can lead to civil or criminal penalties.[40]

Once the release of a hazardous substance is discovered, even if the release occurred long ago, or once there is a substantial threat of a release, EPA is authorized to begin removal or remedial actions to clean it up, or take other measures to reduce the risk of harm.[41] CERCLA defines "removal" as a short-term measure to minimize risk; "remedial" actions are intended to effect more nearly permanent solutions.[42]

Cleanup procedures are set forth in a National Contingency Plan.[43] The Plan includes methods and criteria for locating and investigating contaminated facilities, as well as guidelines for coordinating response actions by federal, state, and local governments. The process begins at a particular site with a preliminary assessment (PA) based on records searches, a visual site inspection, and interviews with facility personnel. If further investigation is indicated, a site investigation (SI) will be conducted to collect physical samples and determine whether hazardous waste is being released into the environment. CERCLA authorizes the government to take immediate removal actions to stabilize a site and prevent the spread of contaminants if it discovers an "imminent and substantial endangerment" to human health or welfare or the environment.[44]

If an immediate removal action is not required, information from the PA/SI is used to set priorities among sites based on their relative danger to public health, welfare, or the environment.[45] Sites that pose the greatest risk are placed on the National Priorities List (NPL) for the earliest possible cleanup, although listing is not a precondition for remediation.[46] The next step in the cleanup process is a remedial investigation (RI), to better identify the sources and extent of contamination, and a feasibility study (FS) to evaluate specific alternatives for cleaning it up. The RI and FS are often performed together. Once a cleanup plan is selected, it is documented in a public record of decision (ROD).

Cleanup actions are supposed to protect human health and the environment, be cost effective, and be permanent insofar as possible. Preference is given to actions that significantly reduce the volume, toxicity, or mobility of contaminants.[47] Contaminants must be removed to a degree that "assures protection of human health and the environment," utilizing procedures that are "relevant and appropriate under the circumstances."[48] This welter of ambiguous and often conflicting commands has been widely criticized and is a principal target for reform.[49]

When contaminants are not entirely removed from a site, because of the expense or the absence of suitable technology, CERCLA calls for cleanup to at least the degree required by other federal environmental laws or applicable state laws. The term used in the statute is "applicable or relevant and appropriate requirements," or ARARs. For example, when a cleanup is completed, groundwaters beneath the site must not exceed maximum contaminant level goals set under the Safe Drinking Water Act, and surface waters must conform to water quality standards established under the Clean Water Act.[50] However, EPA has discretion to waive a particular ARAR if, among other considerations, compliance is "technically impracticable from an engineering perspective," or the level of cleanup chosen is "equivalent" to that required by the ARAR.[51]

The state in which a CERCLA site is located must be given a "substantial and meaningful" role in developing a plan for remedial action.[52] Members of the public also must have an opportunity to participate in a public hearing and comment on the plan in writing. The final plan must be made public, along with any subsequent actions to enforce the terms of the plan or to enter into settlements that differ from what the plan prescribes.[53] If EPA adopts a plan that does not require attainment of a particular ARAR, the state may test the waiver in court on grounds that it was not supported by substantial evidence. If it loses, the state can still insist on a more stringent cleanup standard, providing it is willing to pay the additional cleanup costs.[54] Thus, EPA may shift the economic burden of a thoroughgoing cleanup from the federal budget to the state.

As of mid-1994, there were 1,232 sites on the NPL.[55] However, despite the importance attached to restoring them, only about 200 NPL sites have been completely cleaned up since CERCLA was first enacted. Some critics of the current law believe that the slow pace of cleanups is attributable to unnecessarily strict standards embodied in the statute or applied to individual sites. They argue that more sites could be restored more quickly without compromising the goals of the law if each one were cleaned up no more than necessary for its immediate intended use. Thus, the soil of a contaminated tract destined for use as a paved parking lot would not be cleaned up to the same degree as if a day care center were going to be built there. But opponents of site-specific, use-based standards worry that contaminants left in place could leach into underlying aquifers or be blown onto neighboring lands. They also point out that if the site were someday needed for a different kind of use, a new owner might not be able to pay for additional cleanup or might even be unaware of the danger.

Another reason so many sites remain on the National Priorities List is the elaborate planning and evaluation required before actual cleanup can begin. Fiscal year 1993 was the first year DOD spent more on actual cleanup than it did on studies. In the spirit of the carpenter's maxim that counsels "measure twice, cut once," environmental officials have taken time and care (and spent money) to be certain that the remedies they select will be cost-effective and will not actually make matters worse. Yet in the view of one top Defense Department official, "We ought to spend less time and money on remedial design, and get busy cleaning up." Another noted that we have spent about one-third of the Pentagon's environmental budget on studies. "We approach every site," he said, "as if it was the first one" ever cleaned up.[56] In response to this last concern, both the Pentagon and the Department of Energy are working to develop generic cleanup plans for sites with the most common contaminants, in order to speed the process and reduce costs.

The CERCLA remedial process is expensive. EPA estimates the average cleanup cost for each NPL site at more than $30 million.[57] Investigation and cleanup costs incurred by the federal government may be charged to "responsible parties," including any individual, company, or government entity that caused the release, or any past or present owner of a site, or a generator or transporter of the released contaminant.[58] Liability is strict, joint, and several. Defenses to liability are limited to acts of God, of war, and of independent third parties.[59] Liability can also be avoided if the release was authorized by a valid federal permit.[60]

Responsible parties, including the federal government, may seek contribution for the costs of cleanup from other responsible parties, or may have to reimburse others for such costs. For example, the Air Force was forced to contribute to the cost of cleaning up a county landfill in which it had dumped hazardous wastes.[61] By the same token, the Navy was able to recover cleanup costs from a chemical company that contaminated land later taken by eminent domain for military use.[62]

In some instances, the government may have to bear at least part of cleanup costs assessed against its contractors, either because the government is characterized as a responsible party, or because of contract provisions holding the contractor harmless from such claims. In a 1993 report, the Defense Department found that there was no explicit provision for such payments in existing laws or regulations. Nevertheless, it said such costs might be paid as part of a contractor's "ordinary and necessary business overhead" if the contractor had complied with all applicable environmental laws and regulations and had acted promptly to minimize damages and costs.[63] The Defense Department's Inspector General has criticized DOD for failing to estimate its actual and potential liabilities for such costs.[64] In one remarkable case, the federal government was held jointly liable for cleanup costs at an NPL site contaminated during World War II by a company that manufactured rayon for the war effort under close government supervision.[65]

In practice, the federal government does the actual cleanup at nonfederal sites only if parties responsible for the contamination cannot be located or are insolvent, or if they refuse to clean up or cannot be trusted to do the job properly.[66] The prevailing view is that responsible parties should do the work whenever possible, since they have the greatest incentive to control costs.

To make certain that financing is available to clean up all contaminated sites, Congress created a "Superfund" of up to $13.6 billion for cleanups from October 1986 through September 30, 1994.[67] Superfund is financed by taxes on petroleum and chemical products, an "environmental tax" on corporations, reimbursements for government cleanup expenses from responsible parties, penalties and damages, and a total of up to $1.7 billion from general congressional appropriations.[68] Money from the Superfund is used to initiate cleanup actions, and later to conduct full remedial actions if responsible parties with sufficient resources cannot be found.[69]

The federal government, a state, or an Indian tribe may recover damages from responsible parties for injuries to natural resources.[70] The federal government itself may be liable to a state for such damages growing out of national defense activities. But when the town of Bedford, Massachusetts, tried to recover damages from the Air Force and the Navy for polluting its principal source of drinking water, the court held that the federal government's waiver of sovereign immunity for suits by a "state" did not extend to claims by a municipality.[71]

As with the major regulatory environmental laws, any person or a state may file suit in federal court to compel compliance with CERCLA or with any regulation, order, or agreement entered into pursuant to it, or to require a government official to perform some nondiscretionary duty.[72] There is one extremely important exception: no federal court may hear a "challenge" to a CERCLA response action once it has been adopted by EPA or another federal agency, other than to test the waiver of an ARAR, as described above.[73] Congress adopted this exception to prevent responsible parties from filing suits designed to delay costly cleanups. However, it has been invoked to frustrate efforts by public interest plaintiffs and states wishing to compel federal agency compliance with various federal and state requirements, orders, and interagency agreements once a cleanup is underway. Thus, two recent citizen suits for Clean Water Act and RCRA violations, one at the Twin Cities Army Ammunition Plant, the other at DOE's Hanford facility, were barred by ongoing CERCLA cleanups at those facilities.[74]

The 1986 legislation amending CERCLA included a free-standing measure called the Emergency Planning and Community Right-to-Know Act, also known as SARA Title III.[75] It requires facilities that handle hazardous chemicals to provide descriptions of those chemicals and their locations to state and local community planners, whose job it is to provide for emergency responses in the event of an unplanned release. Leaks and spills must be reported immediately to local response authorities. Routine releases of these chemicals into the air, water, or land, or their shipment off-site for treatment and disposal, must be reported to EPA, which compiles the data in an annual Toxic Release Inventory. All these planning and reporting requirements are made applicable to federal facilities by a 1993 executive order. The reports give state regulators and members of the public a new tool for monitoring the environmental performance of defense facilities.

CERCLA Cleanup of Federal Facilities

Like RCRA, CERCLA applies equally to federal and nonfederal hazardous waste sites for most purposes:

> Each department, agency, and instrumentality of the United States (including the executive, legislative, and judicial branches of the government) shall be subject to, and comply with, this Chapter in the same manner and to the same extent, both procedurally and substantively, as any nongovernmental entity.[76]

By executive order, individual agencies responsible for contaminated sites, rather than EPA, are directed to carry out cleanups, except in an emergency. These agencies must nevertheless act in accordance with EPA practices and criteria.[77] Unfortunately, EPA has not yet promulgated provisions in its National Contingency Plan to address the special concerns of federal facility cleanups.[78] The President may exempt a Defense Department or DOE site from CERCLA's provisions in order to protect the "national security interests of the United States,"[79] although no President has yet found it necessary to do so.

CERCLA calls on EPA to establish a special Federal Agency Hazardous Waste Compliance Docket. The Docket is a public compilation of inventories that each agency is required by RCRA to prepare to identify all contaminated properties under its control.[80] The amount, nature, and toxicity of hazardous wastes at each site are described, as well as any efforts to clean it up, although properly classified information need not be revealed.[81] EPA must publicize additions to the Docket every six months.

Once a federal property is placed on the Docket, EPA is supposed to make a prompt preliminary assessment to decide whether it should be listed on the NPL.[82] If a site is added to the list, the agency proprietor must begin a remedial investigation and feasibility study (RI/FS) within six months. After completion of the investigation and planning, the agency must enter into a legally binding agreement with EPA covering details of the cleanup, following public notice and opportunity for public hearing and comments.[83] This interagency agreement (IAG) includes schedules for completion and arrangements for long-term operation and maintenance of a facility if necessary.[84] Actual on-site remediation must begin within fifteen months. EPA practice is to bring state governments into the planning immediately after listing and make them parties to interagency agreements whenever possible.[85] For example, the State of Washington is a party to a 1989 agreement with the Department of Energy and EPA to clean up DOE's Hanford Reservation in that state. CERCLA gives EPA the final say in any controverted cleanup decision.[86] Nevertheless, the terms of an interagency agreement may be enforced by citizen suit[87] or by imposition of civil penalties.[88]

CERCLA's federal facilities section provides that "[s]tate laws concerning removal and remedial action" apply to federal facilities not included on the NPL.[89] This provision is extremely important, since most federal sites are not on the NPL. In one reported case, it was applied to allow the Commonwealth of Pennsylvania to regulate the cleanup of a contaminated drainageway at the Navy Ships Parts Control Center that was not on the NPL.[90] However, the provision has been held to apply only to facilities currently owned by the federal government, apparently ruling out claims based on state law for cleanup costs at military bases already transferred into nonfederal ownership.[91]

On the basis of the same provision, it has been argued that state laws do *not* apply to properties *listed* on the NPL.[92] However, two other CERCLA sections indicate that states retain wide-ranging authority in the cleanup of any site. One declares that nothing in CERCLA is to be "construed or interpreted as preempting any State from imposing any additional liability or requirements with respect to the release of hazardous substances."[93] The other provides that CERCLA does not "alter or modify in any way the obligation or liabilities of any person under other Federal or State law, including common law, with respect to releases of hazardous substances or other pollutants or contaminants."[94] We saw earlier that ARARs for a site may include state standards more stringent than the relevant federal ones, although states have only limited ability to contest the waiver of such ARARs. We also know that once a CERCLA response action is set in motion, it is difficult to challenge in federal court. Otherwise, the balance of power between state and federal authorities at an NPL site remains to be worked out, at least within the context of a CERCLA response action.

EPA has adopted a policy of listing a federal site on the NPL if it meets the usual criteria for listing, even though the site can be satisfactorily cleaned up by a RCRA corrective action.[95] Nevertheless, EPA, or another agency cleaning up its own site, enjoys considerable discretion in invoking the authority of one statute rather than the other to conduct the actual cleanup.[96] Not surprisingly, some federal program managers would prefer to remain in total control of cleanups at their facilities, free from interference by state or local officials. That concern may influence their decisions about whether to conduct a cleanup under the authority of RCRA or CERCLA, or possibly the Toxic Substances Control Act. One Defense Department manual cautions that the "use of RCRA Corrective Action as a response process can seriously jeopardize the Air Force's authority (as the CERCLA lead agency) to select remedies and CERCLA removal authority to take early removal actions to control and reduce the environmental risks posed by contamination at a site."[97] The quoted language presumably refers to the fact that a RCRA cleanup provides no opportunity for federal waiver of otherwise applicable cleanup standards, as CERCLA does, and that a state with RCRA corrective action authority is entitled to supervise the cleanup. The manual neglects to mention that a state may be far more eager than the federal agency proprietor to clean up a site promptly and thoroughly. It also fails to note that states have been frustrated by EPA's unwillingness to seek administrative and judicial orders against other federal agencies for enforcement of cleanup requirements.

Can a state somehow take the initiative and preempt the federal agency's choice, or at least carve out a significant role for itself in the cleanup?[98] That is the question raised and answered in the following case study.

Case Study: Struggle for Control at Basin F

The Rocky Mountain Arsenal is a twenty-seven-square mile Army installation northeast of Denver, Colorado. Beginning in World War II, the arsenal manufactured chemical and incendiary weapons. From 1946 to 1982, the Army also leased part of the premises to companies that produced a variety of pesticides. Millions of gallons of liquid wastes from these processes, along with solid wastes and munitions, were disposed of in unlined natural depressions, until contaminated groundwater water from the arsenal began killing nearby crops and livestock. In 1956, the Army constructed a 92.7-acre asphalt-lined impoundment known as "Basin F" to hold contaminated liquid wastes. Nevertheless, thousands of migratory birds died when they landed in the impoundment, and the toxic liquids soon began leaking into the underlying soil and groundwater.

In 1982, the Army, the State of Colorado, EPA, and the Shell Corporation entered into a memorandum of agreement acknowledging the applicability of both RCRA and CERCLA to cleanup of the arsenal generally. The parties also agreed that Basin F would be cleaned up pursuant to RCRA. In June 1983, the Army submitted a RCRA closure plan for Basin F that was rejected by EPA.

In 1984, EPA authorized Colorado to administer and enforce its own state RCRA program.[99] When the Army submitted the same RCRA closure plan for Basin F to Colorado that EPA had earlier rejected, it was turned down by the state. Two years later, Colorado issued its own closure plan for Basin F pursuant to its state RCRA program. The Army then announced that because it had by that time begun a CERCLA remedial action, it did not need to comply with Colorado's plan.

Toward the end of 1986, Colorado filed suit against the Army for violations of the state's RCRA closure plan. The Army argued that its ongoing CERCLA remediation preempted Colorado's RCRA cleanup of Basin F. In 1989, a federal district court ruled in favor of the state.[100] EPA's monitoring of the cleanup under CERCLA did not serve as an "appropriate or effective check on the Army's efforts," the court declared, and Colorado's involvement "would guarantee the salutary effect of a truly adversary proceeding that would be more likely, in the long run, to achieve a thorough cleanup."[101]

Meanwhile, between May and December 1988 the Army removed eight million gallons of contaminated liquids from Basin F to three storage tanks and a double-lined holding pond on-site. It also excavated five hundred thousand cubic yards of contaminated solid material from Basin F and stockpiled it in a sixteen-acre waste pile near the basin. During this process, arsenal neighbors complained of strong odors and various adverse health effects, including headaches, rashes, nosebleeds, and respiratory problems. Their claims for damages were later rebuffed in separate litigation.[102]

When EPA added Basin F to the National Priorities List in 1989,[103] the Army filed a new suit to halt Colorado's enforcement of its RCRA plan. The court decided that placement of the site on the NPL prevented the application of state laws, and it ruled that it lacked jurisdiction to hear Colorado's "challenge" to the Army's continuing CERCLA response.[104]

On appeal, the 10th Circuit Court reversed, holding that the State of Colorado may carry out its own independent RCRA closure plan concurrently with the Army's CERCLA remedial action.[105] Two provisions of CERCLA, the court pointed out, guarantee the right of states to enforce their own cleanup remedies without interfering with the federal CERCLA cleanup.[106] CERCLA also states that nothing in its federal facilities provision shall "affect or impair" the obligations of an agency to comply with any requirements of RCRA.[107] Finally, the court ruled that Colorado's suit to enforce its own RCRA program did not constitute a "challenge" to the Army's CERCLA response action, which the court would have had no authority to entertain. Listing of a site on the NPL, said the court, is for this purpose irrelevant. Further review of the case was refused by the U.S. Supreme Court in 1994.[108]

The dispute between the Army and the State of Colorado is far from over, however. The most contentious remaining issue is selection of an appropriate standard for cleanup of the toxic chemical diisopropyl methyl phosphonate (DIMP), a manufacturing byproduct of Sarin nerve gas, which has appeared in more than a hundred wells near the arsenal. DIMP is found free in the environment nowhere else in the United States. Colorado's standard for DIMP is eight parts per billion (p.p.b.), while EPA's maximum contaminant level goal for the same substance under the Safe Drinking Water Act is 600 p.p.b. The Army has filed yet another suit seeking to set aside Colorado's standard for DIMP, alleging that the standard is not supported by "any reliable toxicological data." That suit is still pending at this writing.[109]

Each federal agency must report annually to Congress and the affected states on its progress in cleaning up sites both on and off the NPL, explaining any postponements or failures.[110] By mid-1994, one hundred fifty federal facilities had been listed on the NPL, but none had been completely cleaned up. Almost all are defense-related.[111]

Superfund monies generally are not available for CERCLA cleanups of federal facilities.[112] Congress has established a Defense Environmental Restoration Account to finance cleanups at active DOD facilities.[113] Cleanups at closing military bases are paid for out of special accounts established in base closure legislation in 1988 and 1990.[114] The Department of Energy pays for cleanups from funds appropriated for specific program activities. Within each agency a competition has developed for limited cleanup funding. Base commanders and facility managers have learned that one way to get the money they need is to be placed on the National Priorities List. It should be noted that even though agencies enter into multi-year interagency cleanup agreements with EPA and state and local governments, long-term financing commitments must be conditioned upon the availability of appropriated funds in future years.[115]

In general, EPA has adhered to the "unitary executive" theory in CERCLA cases by refusing to issue administrative orders or file suit against other federal agencies for violations. Federal agency noncompliance with IAGs is addressed through the assessment of "stipulated penalties."[116] However, EPA has agreed to issue administrative orders to federal agencies pursuant to CERCLA's "imminent endangerment" section.[117] Further complicating enforcement efforts, as noted above, once a CERCLA response action is underway, the agency proprietor's performance is nearly impossible to attack in court. In addition, a federal appeals court recently held that the Navy was not exposed to state civil penalties for CERCLA violations at its Kittery, Maine, shipyard.[118]

Congress is expected to consider legislation to reauthorize CERCLA in 1995. The new legislation will most likely include amendments that would increase community involvement, create an expanded role for states in the cleanup of federal facilities, and permit the selection of cleanup methods to be based on expected future uses of the land.

Cleaning Up America's Military Bases[119]

For almost a half-century of Cold War, the United States prepared for a hot war. During that period, the Department of Defense (DOD) manufactured or bought weapons and equipment, and trained relentlessly. In the process, DOD and its contractors generated an immense quantity of dangerous wastes. Some of these wastes spilled into the environment when pipelines leaked or storage tanks ruptured. Some were deliberately dumped in unlined pits or landfills, injected into wells, burned in the open air, or left in containers that are now corroded and leaking. The environmental impact of these actions, perfectly legal throughout much of this period, is enormous. So is the cost of cleaning up after them.

For example, the Army's Twin Cities Army Ammunition Plant in Minnesota produced 16.5 billion rounds of ammunition between World War II and the Vietnam War. Trichloroethylene (TCE), a common industrial solvent and suspected carcinogen used in the manufacturing process, has leached into the groundwater from waste pits and landfills where it was deposited. A contaminant plume extending more than four miles from the plant has forced the Army and EPA to provide new drinking water systems for thirty-two thousand people in two neighboring towns. PCBs and heavy metals are also found in soils at the base. In what it describes as one of its success stories, the Army has treated more than 3 billion gallons of groundwater to remove 320,000 pounds of volatile organic compounds, and it has excavated 1,400 cubic yards of soil containing PCBs. Yet of 19 contaminated sites at the base, only one has been completely cleaned up, and restoration of the entire facility is not expected before the year 2000, at a cost now estimated at $154 million.

Unfortunately, the Army's record at Twin Cities is not unique or even unusual. By the end of 1993, the Defense Department had identified more than 21,000 sites at active and formerly used military bases around the country that were contaminated with hazardous materials. Cleanup had been completed at 743 sites. Of the total, 12,000 sites were declared by DOD to pose no threat to human health or the environment. Restoration efforts were either underway or planned for the future at the remainder. Ninety-three active DOD installations, plus 15 formerly used military facilities, were listed on the National Priorities List (NPL). Another 14 were proposed for listing. The Defense Department is also a potentially responsible party under CERCLA at dozens of privately owned sites.

The military began cleaning up after itself in earnest in 1975, when the Army created an Installation Restoration Program (IRP), principally to address contamination at two of its facilities, the Rocky Mountain Arsenal in Colorado, where it manufactured chemical weapons, and the Weldon Springs Ordnance Works in Missouri, where explosives were made. The following year, the Pentagon established a DOD-wide IRP. The initial goals of the program were modest: to locate and evaluate contaminants, and confine them before they could migrate off-base. Actual cleanup was not contemplated unless contaminated property was to be conveyed to a nonmilitary proprietor. When CERCLA was enacted in 1980, the Defense Department's IRP was tailored to meet the new statutory requirements.

In the Superfund Amendments and Reauthorization Act of 1986 (SARA), Congress created the Defense Environmental Restoration Program (DERP).[120] DERP expanded the Defense Department's existing duties under CERCLA to include cleanup of active military bases, as well as formerly used defense sites. Unexploded ordnance was added to the list of dangerous materials that must be located and disposed of. DOD was also charged to conduct "research, development, and demonstration" of new technologies for cleaning up hazardous wastes. Each DOD service branch—Army, Navy (including the Marines), and Air Force—as well as the Defense Logistics Agency, administers its own IRP at the facilities it controls.

Like any federal agency conducting a CERCLA response, DOD must sign an enforceable interagency agreement (IAG) with EPA to clean up each NPL site. State and local governments must be consulted and often are made parties to these agreements. Actual cleanup must get underway as soon as possible, but in no case later than fifteen months after completion of the RI/FS. At the end of 1993, there were eighty-seven IAGs for military facilities. SARA also authorizes the military to enter into agreements with state and local governments and other federal agencies to help clean up off-site contamination, and to reimburse the other parties for any "services" rendered.[121] In addition, the Defense Department encourages each state to enter into what it calls a Defense and State Memorandum of Agreement (DSMOA). A state signatory may be reimbursed for technical services for up to one percent of the total cleanup costs at facilities within the state. The DSMOA also provides a process for work at non-NPL sites. Nearly every state and territory has signed such an agreement.

Unfortunately, it almost always takes far longer to clean up a site than to contaminate it. Following an inspection to sample and analyze the existence of contamination, the remedial investigation and feasibility study (RI/FS) required by CERCLA can take years to complete at a complex site. For example, at McClellan Air Force Base, which is listed on the NPL, a base-wide RI/FS begun in 1984 will not be completed until the year 2002. Such extensive preliminary work is designed to make the actual cleanup as efficient as possible, and to avoid creating new environmental hazards in the process. However, this kind of protracted planning at facilities nationwide has brought complaints from base neighbors worried about imminent threats to their health, from community leaders eager to convert base properties to non-military uses, and from congressional critics worried about spiraling costs. Military officials are also concerned about budget priorities that might compromise their defense missions. Through 1993, 43 percent of all expenditures for environmental restoration were for this kind of preliminary work.

Responding to the delays, Congress created the Defense Environmental Response Task Force in 1990 to find ways to improve cooperation between state and federal authorities in response actions, and to streamline cleanups at bases scheduled for closure or realignment.[122] The task force is made up of officials from DOD, EPA, and other federal agencies, a state environmental official, and representatives from the National Association of Attorneys General and an environmental organization. In a 1991 report, the task force recommended new procedures and criteria for the use and transfer of contaminated DOD lands, as well as the integration of overlapping regulatory requirements.[123]

In 1992, the Secretary of Defense established a Pilot Expedited Environmental Cleanup Program. At fifteen different bases, five for each service branch, DOD has sought to demonstrate full compliance with environmental laws, the use of existing authorities such as CERCLA for extensive cleanups, turnkey contracts for more than one phase of a cleanup, and competition in contracting—all with a view to accelerating the restoration process.[124] Later in 1992, Congress authorized an additional pilot program at another eighteen active and closing bases to expedite the performance of on-site environmental restoration.[125] The emphasis in this second program is on innovative contracting techniques and new methods for making decisions and reaching agreements.

Even more broadly, DOD has begun to concentrate its efforts on interim cleanup actions (removals and interim remedial actions, in CERCLA terminology). This approach permits contaminated sites to be stabilized promptly, and actual cleanup to begin without waiting for extensive preliminary study and planning. In 1993, for the first time, more money was spent on actual cleanup activities than on investigation and planning. The Defense Department is also working to develop generic remedies for the most common contaminants, in order to avoid some of the delay associated with site-specific cleanup plans. DOD installations scheduled for closure or realignment have been placed on their own fast track for cleanup, to help speed their conversion to nonmilitary uses and promote the economic revitalization of the communities in which they are located.[126]

The 1986 legislation amending CERCLA also established the Defense Environmental Restoration Account (DERA) to finance military base cleanups.[127] Congress earmarks funds for that account each year to prevent their diversion from cleanup to other defense purposes. DERA also contains amounts recovered under CERCLA from nonmilitary parties responsible for contamination of DOD properties. Congress has created two special Defense Base Closure Accounts to pay for cleanups at closing bases, one for the 1988 round of base closures, and another for all subsequent rounds, leaving funds in DERA for responses at active facilities.[128] By the end of fiscal year 1993, DOD had spent more than $7.8 billion in the cleanup effort—$6.5 billion from DERA, and another $1.3 billion from the Defense Base Closure Accounts.

Even these heroic legislative and executive initiatives have not satisfied some critics, who have invoked various legal authorities in an effort to speed the cleanup even further. The following case study provides one example.

Case Study: The MESS at McClellan

McClellan Air Force Base is a major aircraft maintenance facility covering 2,950 acres in Sacramento, California. It provides logistical support for aircraft, missile, space, and electronics programs. Air Force investigators have discovered 258 sites on the base contaminated with a list of hazardous wastes that includes organic solvents, metal plating residues, caustic cleaners and degreasers, paints, lubricants, photochemicals, phenols, chloroform, acids, and PCBs. By the end of 1993, the Air Force had cleaned up just 2 sites, although it had concluded that another 34 posed no danger to human health or the environment.

A CERCLA response was initiated at McClellan in 1981, and the entire base was placed on the National Priorities List in 1987. In 1990, the Air Force signed an interagency agreement with EPA and the State of California. Several removal actions have begun, including groundwater pumping and treatment, and vapor extraction of contaminants in soils, and an alternative drinking water system has been installed for more than five hundred nearby residents. The Air Force has entered into several partnerships to promote development and testing of innovative cleanup technologies that could be applicable at McClellan and elsewhere. One includes EPA and the California environmental agency. Another involves seven private companies and a nonprofit coordinator called Clean Sites. The Western Governors Association and several national laboratories have been drawn into other collaborative efforts.

The Air Force expected to spend $45 million at the base for environmental restoration during fiscal year 1995, although it was careful to avoid projections of costs for the entire cleanup, which in any event is not scheduled for completion until the year 2034.

Not content to wait for a protracted CERCLA response, a group of concerned citizens living near the base filed a suit in federal district court in 1986 alleging violations of the Clean Water Act, RCRA, and several California environmental laws. Calling themselves McClellan Ecological Seepage Situation (MESS), they asked the court to enjoin the violations and levy civil penalties against the Air Force. In the first of several rulings in this litigation, the court decided that neither the federal facilities provision nor the citizen suit provision of either federal statute waived the government's sovereign immunity from liability for civil penalties.[129]

Two years later, the court ruled that it lacked jurisdiction to hear complaints that the Air Force was violating California hazardous waste and water quality laws.[130] The court based its decision on the fact that California had no EPA-approved RCRA permit program at the time that would have empowered the court to enforce state restrictions under RCRA's citizen suit provision. The court also found that it had no Clean Water Act authority to enforce California statutes and regulations, since they provided neither "precise standards capable of uniform application" nor "effluent standards or limitations" that had been incorporated into McClellen's NPDES permit.

In the same lengthy decision, the court ruled that the Air Force was in violation of permit limitations on discharges of chemical oxygen demand, pH, bioassay, and perhaps other pollutants. However, it refused to rule on other claimed violations, because the plaintiffs had failed to give the Air Force the sixty days' notice required by the statute before filing suit. The notice provisions will be "particularly strictly enforced where the Federal government is the defendant in a citizen suit," said the court. The court went on to rule that it had no authority under the Clean Water Act to punish the Air Force for its earlier unpermitted wastewater discharges to drainage ditches on the base, since there was no evidence that any such violations were continuing. Finally, the court decided that the Air Force was in compliance with EPA regulations for RCRA interim status facilities concerning the storage of hazardous wastes in drums. The drums were held in an area surrounded by concrete berms to contain spills and prevent intrusions of water from the outside.

The following year, in its third ruling in the MESS case, the court decided that the presence of hazardous wastes in unlined burial pits on the base did not violate RCRA's prohibition on the "storage" of such wastes without a permit.[131] No waste had been added to any of the pits at the base since November 19, 1980, when RCRA took effect. Before that, the court found, the Air Force had "permanently disposed" of the wastes, not stored them. The only current activity at the waste pits was the ongoing CERCLA cleanup, which is expressly exempted from any permitting requirements.[132]

In 1990, the court, now with a different judge, heard charges that hazardous waste leaking from the Air Force's burial pits into the underlying aquifer was making its way into the nearby American and Sacramento Rivers.[133] When the Air Force presented evidence that the contaminant plume within the aquifer was being directed away from the rivers by heavy groundwater pumping, the court ruled that no Clean Water Act permit was required for the leakage.

When the plaintiffs appealed that decision, the government for the first time argued that the case should be dismissed because it represented a "challenge" to an ongoing CERCLA response action.[134] If the Air Force had to obtain permits or report discharges under RCRA and the Clean Water Act, it said, that would cost money and could involve restrictions that would impede the federal cleanup. According to the plaintiffs, however, CERCLA's ban on challenges to cleanups is designed to prevent responsible parties from bringing dilatory suits to postpone their liability for cleanup costs, not to bar actions intended to promote compliance with the law. The appeals court decided that under CERCLA it lacked jurisdiction to enforce RCRA or Clean Water Act requirements that might interfere with the CERCLA cleanup of McClellan's inactive pits and sites, but that CERCLA does not bar judicial enforcement at active waste sites.[135]

No doubt some in the environmental community were chagrined when, in 1993, President Clinton cited McClellan Air Force Base for its progress in cleaning up part of the base: "By streamlining government and working together, you have performed a cleanup that, under the old rules, would have taken six years and $10 million. You did it in eight weeks at a fifth of the cost. And we intend to do that all over America, copying [McClellan's] leadership."[136]

At military bases around the country, disposal of unexploded ordnance has presented a special challenge, not only because of the unique danger to workers, but also because of the immense quantities involved. For example, at the former Raritan Arsenal, in a heavily developed industrial and commercial area of Edison, New Jersey, the Army has removed and destroyed more than 113,000 pieces of unexploded ordnance and 12,000 pounds of TNT.[137] Cleanup at the Army's Jefferson Proving Ground in Indiana will be much more difficult. That facility has been used primarily to test artillery ammunition; some 23 million rounds have been fired there since 1941. Unfortunately, about six percent (more than a million rounds) failed to explode, and lie buried at depths of up to 25 feet over a 51,700-acre area. Some 14,000 depleted uranium shells are also scattered around the site. Removal of all the explosive and radioactive materials would require carefully strip-mining the entire base to a depth of about 30 feet, at a cost of as much as $5 billion.[138]

In the Federal Facility Compliance Act of 1992, Congress directed EPA to develop special regulations to determine when munitions held in DOD inventories *become* hazardous waste subject to RCRA regulation.[139] The issue of timing is important, since the Defense Department might otherwise maintain stocks of ordnance for many years after they have ceased to have any military significance, under conditions far less stringent than what RCRA would require. The new regulations, more than a year past due at this writing, are to cover both chemical and conventional munitions. Whether and how the regulations will affect the Army's plans to incinerate all its chemical weapons is unclear.[140] Once they are characterized as hazardous wastes, of course, obsolete munitions may be subjected to state and local controls more stringent than EPA's regulations.

One special DOD cleanup program deserves mention here. The Army is responsible for environmental remediation of more than eight thousand formerly used defense sites (FUDS) that may be contaminated with hazardous or radioactive wastes. FUDS are properties that were previously owned or leased by or were under the control of the Defense Department. For example, the former Walker Air Force Base in Roswell, New Mexico was a flight training school for the Army Air Corps during World War II, and later a Strategic Air Command (SAC) base. Now it is a commercial airfield. The Corps of Engineers is currently working to remove abandoned underground storage tanks and contaminated soil, and to recover TCE and other hazardous chemicals from military operations that have leached into groundwaters around the former base. FUDS cleanups, like those at active military bases, are paid for out of the Defense Environmental Restoration Account.

In addition to the annual reports required of every federal facility by CERCLA, the Pentagon must prepare a special report to Congress each year listing contaminated military sites and describing efforts to clean them up, along with current and anticipated expenses.[141] Another annual report describes DOD's compliance with the various environmental laws,[142] while a third details payments to defense contractors for response actions.[143]

As we saw at the end of the last chapter, DOD has consistently taken the position that United States domestic environmental laws are not enforceable at its nearly four hundred bases overseas. A 1978 executive order calls on federal agencies responsible for construction or operations abroad to comply with pollution controls applicable in each host country.[144] Until 1992, when the Defense Department put out guidelines for more stringent controls in some cases,[145] it was guided by local requirements, if any; by status of forces agreements between the United States and host nations; and by environmental regulations of the individual military service branches. However, without EPA supervision, and with no involvement by either United States or host nation citizens, the result has been a pattern of environmental degradation reportedly exceeding even that at domestic bases.[146]

In 1990, Congress directed the Defense Department to develop a policy for cleaning up contamination at military installations abroad.[147] In response, DOD has adopted two policies. One concerns installations scheduled for closure and turnover to host nations. It directs base commanders to "eliminate known imminent and substantial dangers to human health and safety" and to inform host nations about known contamination that is not cleaned up.[148] A different policy for active bases is still under development at this writing, but it is described as "consistent" with the policy for closing bases. Contamination that does not pose an "imminent and substantial threat to health and safety" is to be left in place. Threats to historic sites and mere eyesores, for example, will not necessarily be disturbed. Base commanders are to set priorities and carry out the cleanups.

Unlike the DOD policy for pollution control abroad, neither cleanup policy for overseas bases contains any numerical performance standards, and neither one provides for public notice or participation. Whether and how well a particular base is cleaned up therefore depends on the good intentions and expertise of the base commander, whose first priority is national defense, and who may not feel that he has adequate resources for environmental remediation. Cleanup funding for overseas facilities comes from DOD general operating and maintenance accounts, not from DERA. Some members of Congress have argued that in the absence of treaty obligations host countries should bear more of the cost of their own defense, including the expense of cleaning up U.S. bases.

Several years ago, Senator Sam Nunn and others called for a "Greening of the Pentagon," in which defense personnel would be retrained and assigned to clean up the environmental wreckage of the Cold War.[149] Some in Washington feared that military readiness might suffer if personnel were diverted to nondefense duties. Others worried that DOD simply could not be trusted to clean up its own mess. Nevertheless, many uniformed and civilian DOD employees are currently responsible for environmental compliance and restoration. And Congress has authorized the Defense Department to institute several programs of environmental education for its staff and for dislocated defense workers and unemployed young adults.[150]

The Pentagon says it is determined to expand the role of communities and members of the public in decision-making about cleanups. Toward this end, it is in the process of setting up Restoration Advisory Boards at most of its closing and realigning bases, and at active bases where there is sustained community interest. These boards include representatives from local communities, the Defense Department, and regulators. Understandably, DOD's motives are related to the performance of its military mission. As one report noted, "The earlier the public is involved in the process, the sooner their concerns will be incorporated, consequently reducing delays that might result from communication barriers and public dissatisfaction with the process."[151] But increased public participation can also be expected to improve the quality of decisions, since local residents are most familiar with local environmental conditions. They also have the strongest interest in truly permanent solutions.

With the cleanup of America's military bases, we have begun to learn a very expensive but vital lesson. The environmental bill for nearly a half-century of Cold War has come due. Payment must be made in spite of shrinking defense budgets and an unprecedented drawdown in military personnel and equipment. At the same time, American armed forces must remain ready to meet a variety of challenges around the globe. The Pentagon has shown that it is serious about environmental restoration. It has created a new Office of Environmental Security to oversee the cleanup effort, and it has named a general or flag level officer to supervise environmental compliance within each service branch. Congress has also demonstrated a growing appreciation of the problem. With continued insistence and support from a better-informed public, military planners will be better able in the future to keep America strong without jeopardizing the health and environmental security of future generations of Americans.

Cleaning Up DOE's Weapons Complex[152]

Throughout the Cold War, the Department of Energy and its predecessor agencies (the Army's Manhattan Project from 1942 to 1946, the Atomic Energy Commission from 1947 to 1974, and the Energy Research and Development Administration from 1975 to 1977) were responsible for designing, manufacturing, and testing nuclear weapons. The work was carried out by one hundred thousand people at fourteen major facilities in thirteen states. In a 1991 study, the congressional Office of Technology Assessment (OTA) found that

> [a]t every facility the groundwater is contaminated with radionuclides or hazardous chemicals. Most sites in nonarid locations also have surface water contamination.

Millions of cubic yards of radioactive and hazardous wastes have been buried throughout the complex, and there are few adequate records of burial site locations and contents.[153]

According to OTA, these conditions can be traced to a

history of emphasizing the urgency of weapons production for national security, to the neglect of health and environmental considerations; ignorance of, and lack of attention to, the consequences of environmental contamination; and decades of self-regulation, without independent oversight or meaningful public scrutiny.[154]

Since 1954, the Atomic Energy Act has subjected civilian nuclear reactors to a well-publicized permitting procedure and ongoing public regulation. But the nuclear weapons complex has always been exempt from that process. Even after Congress began in 1969 to enact the modern federal environmental laws, some DOE facilities systematically ignored those laws for years and took pains to conceal their violations.[155] Workers at the weapons plants were not warned of grave health hazards.[156] Program managers deliberately released radioactive materials into the atmosphere.[157] But because almost all information about the nuclear weapons program was highly classified, states, municipalities, workers, and facility neighbors had few opportunities to discover or respond to the growing environmental threat.[158]

With the specter of nuclear war now beginning to recede, DOE has stopped producing nuclear weapons. No new weapons are currently under development, so far as is publicly known. DOE has now shifted much of the focus of its defense work from weapons production to environmental restoration. According to Energy Secretary Hazel R. O'Leary,

We are redirecting the national commitment that built the most powerful weapons the world has ever known, toward addressing the resulting widespread environmental and safety problems at thousands of contaminated sites across the country. We have a moral obligation to do no less, and we are committed to producing meaningful results.[159]

DOE has pledged to operate "all facilities in full compliance with applicable laws and regulations and [to clean up] inactive sites and facilities so that no unacceptable risk to the public or the environment remains."[160]

Despite DOE's new resolve, however, success is far from assured. No one has ever undertaken environmental restoration on a scale even approaching this one. The technology needed to clean up some of the most dangerous wastes has not even been invented. Critical cleanup standards do not yet exist to measure DOE's progress. While working to develop appropriate treatment and disposal methods, DOE is trying to contain its wastes in situ or retrieve and store them safely for indefinitely long periods of time. By 1994, DOE was spending more than $6 billion a year on environmental programs. The latest estimate of the cost of cleaning up the entire weapons complex is more than $200 billion,[161] approaching the $250 billion already spent to manufacture nuclear warheads. However, with no clear idea of the amount of work required, the technology to do it, or the level of cleanup to be achieved, any such figure is bound to be unreliable.[162]

Consider the scope of the problem. According to DOE, the nuclear weapons complex currently contains 137 installations contaminated with wastes in several categories:

1. "High-level" wastes that were generated when uranium and plutonium were separated from other materials for use in nuclear weapons. There are 397,000 cubic meters of high-level wastes, mostly in liquid or sludge form, containing about 99 percent of all the radioactivity in weapons production wastes. Some will remain dangerously radioactive for millions of years.
2. "Transuranic" wastes, more than 250,000 cubic meters altogether, which contain manufactured elements heavier than uranium. Some, like plutonium, decay so slowly that they must be securely isolated from the environment for hundreds of thousands of years.
3. "Low-level" wastes, which include every type of radioactive waste other than high-level and transuranic wastes and uranium mill tailings. Typically large in volume and sometimes highly radioactive, the 387 million cubic meters of material in this category includes contaminated rags, filters, tools, containers, and protective clothing. Low-level wastes may contain such elements as uranium U-238, which has a half-life of 4.5 billion years.
4. Nonradioactive hazardous waste. The total for all sites in this category has not been estimated.

We have "more than 1.4 million drums of buried or stored waste," according to one DOE official. "If you just take the stored waste and start piling those drums on a football field, it literally would go six miles high. That's just the stored waste we already have."[163] Over the years, many of the older containers have begun to leak into the open environment. Even more troubling, much of the waste resulting from the manufacture of nuclear weapons was not put into containers, but was simply dumped in unlined ditches and pits.

Much of the radioactive waste is "mixed" with nonradioactive hazardous wastes, making storage, treatment, and disposal especially difficult. The total in 1993 was estimated at 600,000 cubic meters.[164] All high-level and most transuranic waste fall into this category. The hazardous components of mixed wastes are so varied and complex that most of them have not been fully identified or measured yet. Spent nuclear fuel from defense reactors is a separate category of extremely radioactive material that presents unique handling problems. Approximately 2,700 metric tons of it are held in DOE storage pools, some of it rusting and in danger of dispersal into the environment.[165] The amount of waste in each category will increase in coming years as weapons are dismantled and facilities are cleaned up. Environmental remediation might generate another 920,000 cubic meters of mixed waste by the end of 1998.

The Atomic Energy Act of 1954 authorized the Atomic Energy Commission to "prescribe such regulations and orders as it may deem necessary . . . to protect health and to minimize danger to life or property."[166] A 1959 amendment directed the establishment of programs to control radioactive hazards. And in 1974 DOE's predecessor was given authority to set health, safety, and environmental protection standards for radioactive waste.[167] Exercise of this authority, however, was made entirely discretionary. States are given no regulatory role, as they are under the various environmental statutes, and there is no citizen suit provision to compel enforcement. DOE and its predecessors responded by adopting broadly drawn internal directives and orders that allowed the most expedient disposal of the most dangerous wastes, with little concern for public safety, and no thought for the long-term consequences. No opportunity was provided for the usual public notice and comment required by the Administrative Procedure Act for agency regulations.

The effectiveness of this self-regulation may be judged by the results. High-level mixed wastes were poured into single-wall steel tanks without always preserving an accurate record of the contents. Transuranic wastes were buried in drums without marking the disposal sites. Low-level wastes were placed in cardboard boxes and buried in unmapped trenches. Liquid mixed wastes were poured into unlined excavations or burned in incinerators without effective controls on stack gases. Workers at the weapons plants were directly exposed to radioactive materials. These practices continued even after EPA began in 1977 to promulgate standards for protecting the environment from radioactive hazards.[168]

In 1988, Congress created an independent Defense Nuclear Facilities Safety Board to provide oversight for the nuclear weapons complex.[169] The Board's safety recommendations need not be followed by DOE. And the public has no formal role in the Board's proceedings, except to comment on its recommendations. It should be noted here that from the beginning DOE's weapons plants and laboratories have been operated almost entirely by commercial contractors (such as DuPont, Westinghouse, and Rockwell International) and universities (such as the University of California) under DOE supervision.

The very process of cleanup has important environmental implications. The disturbance of contaminants in place may result in their spread to other locations. If the contaminants are removed from the open environment and containerized, they must be transported with great care. Eventually they must be safely stored or disposed of. Since we have neither the know-how nor the resources to clean up the entire complex at once, priorities must be set to minimize risks of additional or lasting harm. Establishing such priorities involves both scientific uncertainty and political controversy. Whatever resources we devote to the cleanup will have to be taken from other important public investments. Just drawing up plans for the cleanup has provided an unprecedented challenge, as the following case study indicates.

Case Study: Planning the Weapons Complex Cleanup

In 1988, citizen groups and several members of Congress asked DOE to prepare a programmatic environmental impact statement (PEIS) for its plans to clean up and rebuild the entire weapons complex. They contended that such a statement was required by the National Environmental Policy Act (NEPA).[170] Only such a comprehensive analysis, they argued, could furnish the perspective needed to set sound priorities and make the most efficient allocation of limited resources. A PEIS would reveal common environmental problems at various sites, avoiding the need to develop unique responses on a case-by-case basis, and speeding the overall cleanup. Equally important, they insisted, members of the public would be given an opportunity to participate in the planning process.

When DOE failed to respond, the citizen groups asked a federal court to order the preparation of a PEIS.[171] They maintained that "one of the largest industrial rehabilitation programs ever undertaken" required public comment and scrutiny. Failures in planning such a massive cleanup, they pointed out, could lead to further contamination of the air and water and exposure for workers and the public.

Six months later, before a trial could begin, DOE agreed in a settlement of the litigation to prepare one PEIS for the cleanup process, and another for a "modernization" of the weapons complex.[172] The PEIS for the cleanup is to contain a broad environmental assessment of the program's plans. In 1991, public hearings were held in twenty-three cities to determine the range of issues that will be addressed in the statement. DOE received a number of comments concerning the need for greater public participation and oversight, public and worker health and safety, adequate resources for the cleanup, alternative technologies, and environmental standards.

> The PEIS proved to be so massive and complex that DOE conducted six regional workshops just to develop a detailed strategy for completing it. DOE then proposed to eliminate environmental restoration alternatives from the PEIS, since cleanup decisions at individual sites must reflect local conditions and involve state regulators and the public.[173] After additional public workshops, a draft impact statement will likely be issued in mid-1995, to be followed by public hearings around the country and release of the final PEIS sometime thereafter.[174]

In 1989, even before it agreed to prepare a programmatic impact statement, DOE announced the creation of a new Environmental Restoration and Waste Management (EM) Program to spearhead cleanup of the weapons complex. EM is divided by function into several smaller programs, including:

1. Waste Management, which is concerned with treatment, storage, and disposal of waste, and reduction in the amount of new waste produced by DOE.
2. Environmental Restoration, which carries out assessments and cleanups of inactive sites and facilities.
3. Technology Development, which conducts research into new waste management and cleanup technologies.
4. Facilities Transition, which is responsible for closing production facilities that are no longer needed and preparing them for decontamination and reuse.

DOE announced at the time that all environmental restoration would be completed within thirty years, that is, by the year 2019.

To help meet this goal, DOE issued the first in a series of annual Five-Year Plans spelling out activities in the various program areas and describing objectives for budget management and environmental restoration at every DOE installation.[175] The Five-Year Plan, published in three volumes, became the central program-planning document for the cleanup. It contains a wealth of information about the program, including public comments on the plan itself.[176] The Five-Year Plan is being replaced in 1995 by the Baseline Environmental Management Report, which will place special emphasis on program costs.[177]

However difficult the planning, actually cleaning up the weapons complex has presented a far greater challenge. Eight of DOE's fourteen major facilities have been placed on the CERCLA National Priorities List. Treatment of mixed wastes (radioactive and nonradioactive hazardous wastes mixed together) has proved to be especially demanding. In some instances it is possible to separate the radioactive and nonradioactive components. Some mixed wastes can be treated to reduce their toxicity. Others can only be reduced in volume to facilitate their storage or disposal. Reduction is accomplished by compacting, melting, evaporation, or incineration. Some radioactive liquids can be solidified, to help prevent their migration, by mixing them with cement or glass (vitrification), or by converting them to a dry powder form (calcining), then sealing them in stainless steel containers. Many nonradioactive hazardous wastes are sent for treatment and disposal to commercial facilities holding permits under the Resource Conservation and Recovery Act (RCRA).

RCRA forbids the land disposal of some extremely dangerous wastes, such as solvents and dioxins, until their hazardous constituents are reduced to specified concentration levels or until it is certain that the wastes will not migrate from their disposal sites for as long as they pose a threat to human health and the environment. RCRA also bans the storage of such wastes, except to accumulate sufficient quantities for treatment. EPA regulations extend these restrictions to many Department of Energy mixed wastes.[178]

For the Department of Energy, the land disposal restrictions present an intractable dilemma, since DOE holds a huge inventory of mixed wastes, which it mostly lacks either the technology or the capacity to properly treat, and which it cannot legally store. In an effort to avoid sanctions for violating the restrictions, DOE has entered into agreements with EPA to bring four of its largest facilities—Savannah River, Rocky Flats, Oak Ridge, and Hanford—into compliance. The agreements call for temporary storage of the wastes but contain enforceable milestones for the development and implementation of suitable treatment technologies. Other facilities have obtained permits with compliance deadlines, while still others are operating under judicial or administrative orders to safely treat and dispose of their wastes.

In the 1992 Federal Facility Compliance Act, Congress provided additional breathing room for DOE.[179] The act amends RCRA to enable DOE to store mixed wastes otherwise banned from storage without incurring state fines or penalties while it develops the needed treatment capacities and technologies. For each facility not already subject to a permit, agreement, or order concerning such wastes, DOE must complete a site-specific plan for treating them and get the plan approved by state regulators or EPA before October 1995.[180] In the meantime, DOE has given EPA and state governors a detailed inventory of mixed wastes at each of its facilities, along with an assessment of its ability to treat those wastes.[181]

DOE has redoubled its efforts to identify new methods for separating some hazardous and radioactive wastes and for converting other mixed wastes to a stable form for safe disposal. In particular, a search is underway for practical alternatives to incineration, in light of technological and political objections to burning mixed wastes. Among many alternatives under study is the use of heated gases such as carbon dioxide to dissolve and carry off hazardous organic compounds attached to nonhazardous solids. DOE is also investigating the in situ oxidization or microbial reduction of organic constituents, melting and solidification of mixed wastes using microwaves, and encasing the wastes in plastic to retard their dispersal. Commercial contractors have been invited to join in this search.[182] Nevertheless, according to a recent report from the General Accounting Office, little of the new technology has found its way yet into DOE's cleanup actions. That failure can be traced, the report says, to a breakdown in communications within DOE, as well as to regulatory officials and members of the public who are afraid to employ unproven methods, and contractors with large investments in older technologies.[183]

Finding a safe, permanent repository for radioactive wastes has proved to be especially difficult. Efforts to develop a high-level waste disposal site at Yucca Mountain in Nevada have met with stiff resistance from state officials. Likewise, completion of the Waste Isolation Pilot Plant (WIPP) for permanent storage of transuranic wastes in deep salt formations near Carlsbad, New Mexico has been thwarted by numerous legal challenges. With its huge volume of low-level waste, DOE's recent practice has been to reduce it in size as much as possible, containerize it in barrels, then bury it on-site in trenches, with caps to carry off water that could corrode the containers or leach their contents into the underlying groundwater. A large quantity of low-level waste has been shipped to the Nevada Test Site for disposal. Congressional directives and DOE programs for disposal of each category of radioactive waste are described in greater detail in the last section of this chapter.

Congress has given DOE the responsibility for cleaning up millions of tons of uranium mill tailings at twenty-four sites in ten states. The wastes from mining and milling uranium ore give off radioactive radon gas. They also contain hazardous materials that can erode or leach into groundwater. Under the Uranium Mill Tailings Radiation Control Act,[184] DOE is working to stabilize tailings piles, or in some instances is moving entire piles to unpopulated areas. DOE is also charged with cleaning up another five thousand properties where tailings were used for fill or construction.[185]

While most information about the design and manufacture of nuclear weapons necessarily remains secret, DOE has pledged to be open, responsive, and accountable in cleaning up the weapons complex. The EM program provides a number of opportunities for public involvement. For example, DOE has established a cleanup advisory group at each facility that includes local government representatives, environmental groups, and technical staff from the facility. Local group meetings have been organized to review each Five-Year Plan. At the national level, DOE has created a State and Tribal Government Working Group to provide a representative voice for those constituencies. In 1991, DOE adopted a comprehensive American Indian policy calling for government-to-government relations and coordinated planning that takes into account, among other things, traditional religious practices of the tribes.[186]

In spite of these innovations, the EM program has experienced a number of setbacks. The cleanup at Hanford Reservation in Washington State illustrates some of the problems, as we see in the following case study.

Case Study: DOE's Cleanup at Hanford

One of DOE's biggest cleanup efforts is taking place at the Hanford Reservation, a 560-square-mile nuclear production facility in southeastern Washington State. Hanford was built in the 1940s as part of the Manhattan Project.[187] For nearly 50 years its primary mission was to produce plutonium for the United States nuclear arsenal. When production ended in 1989, hazardous and radioactive wastes were left at some 1,170 locations around the reservation ranging in size from one square foot to 1,800 acres. By 1995, cleanup costs were running more than $1.4 billion per year and rising.

Hanford's main product, plutonium, is a highly radioactive metal. The process for manufacturing plutonium was extremely dangerous. Hanford employees were continually at risk of exposure, and some buildings and equipment there are so contaminated that robots are being used to clean them up. Even though plutonium's hazards were recognized early on, quantities of the material were deliberately released into the air at Hanford and elsewhere to study patterns of dispersal and effects on unsuspecting human populations downwind. For nearly three decades, contaminated cooling water from Hanford's production reactors was discharged directly into the Columbia River.[188] Spills and discharges of liquid wastes into unlined pits over many years have left the soil and groundwater for miles around polluted with plutonium and its manufacturing residuals. Many local residents believe that they and their families have been seriously injured by these releases.[189]

Officials at Hanford are especially concerned about 177 underground tanks used to store 410,000 metric tons of liquid high-level and transuranic wastes. Almost 10 percent of the entire DOE environmental budget for 1994 was spent just on these tanks. Some hold more than a million gallons each. At least 68 of the tanks are believed to be leaking. Workers who filled the tanks failed to sample the wastes or keep proper records, so no one is certain today what is in them, and DOE estimates that it will take about ten years to find out. Collecting and analyzing a single core sample from one of these tanks can cost as much as

$500,000. The technology needed to remove some semi-solid materials from the tanks for treatment does not yet exist. More stable liquids in some single-wall tanks are being transferred to new double-wall tanks until they can be refined and immobilized for permanent disposal. The contents of some tanks are so hot that they have to be refrigerated. Some contain explosive hydrogen gas.[190] One in particular, Tank 101-SY, holds 1.14 million gallons and emits a large "burp" every hundred days or so. Gases form in sludge at the bottom of the tank, then rise beneath a hard crust at the top with such force that officials fear the tank will rupture.[191] The short-term solution has been to stir the mixture in the tank so the gases are released more gradually.

The current plan is to separate as much low-level and nonradioactive waste from the liquids as possible, in order to reduce the volume, then mix what remains with molten glass. This mixture will be poured into stainless steel containers where it will cool to form glass "logs." In this form, DOE hopes, the chances of its physical or chemical breakdown and migration from a permanent disposal site will be small, at least for the thousands of years that the logs will remain dangerously radioactive.

Several obstacles have arisen to slow the cleanup. DOE originally planned to separate and treat the non-radioactive components in an existing plant at Hanford, but that plant was found to be ill-suited to the task. A facility to convert radioactive wastes into glass logs is currently on the drawing board. It will draw on DOE's experience with a new vitrification facility at the Savannah River complex, which is not expected to be fully operational before 1996.[192] However, there is enough waste in the Hanford tanks alone to make about thirty-eight thousand glass logs ten feet long and two feet thick, while the planned high-level waste repository at Yucca Mountain, Nevada will have space for only five thousand to seven thousand logs from all DOE facilities.[193]

In 1989, DOE entered into an agreement with EPA and the Washington State Department of Ecology for cleanup of the Hanford reservation. Each phase of the cleanup process was to begin by 1999 and be completed twenty years after that. Just four years later, in 1993, DOE announced that it could not meet some of the deadlines. A new agreement gives DOE another ten years to get started, and calls for completion of all work by the year 2033. DOE has now promised to encase low-level as well as high-level wastes in glass, to accelerate its cleanup of contaminated groundwaters, and to give state regulators a larger role in the cleanup process.[194]

Almost half of the 560-square-mile reservation not contaminated with radioactivity may be opened for other uses. A noisy debate is underway about how those lands should be used. Farmers claim that the reservation contains some of the best agricultural land in the region. Conservationists want to see a 50-mile stretch of the Columbia River running through Hanford given Wild And Scenic Rivers Act protection, and to set aside other areas untouched by development during the half century of DOE operation. An environmental impact statement will have to be prepared for any disposition, in accordance with the National Environmental Policy Act.

Although DOE maintains that its contaminated sites pose no "imminent threat" to public health, information about the location and characteristics of pollutants is still very limited. In particular, data about exposure of populations off-site and about chronic health effects of both inadvertent and deliberate releases are not available. Without such information, according to the Office of Technology Assessment, DOE cannot set health-based environmental restoration priorities, or avoid public skepticism of its efforts.[195] Equally troubling, OTA reported in early 1993 that "thus far, DOE and its contractors have devoted little attention to cleanup worker health and safety. . . . Policies and programs to protect cleanup workers are not yet in place."[196]

Some in Congress have complained that so far we have little to show for an expenditure of billions of dollars in cleanup funding.[197] The shortage of concrete successes to date may stem in part from the sheer enormity of the task, and from the fact that so many studies and safeguards are required before actual cleanup can begin at a given site. Moreover, much of the technology needed to do the job is only just being developed. DOE complains that it has confronted a serious political and scientific dilemma in setting priorities:

> The trade-off DOE faces is whether to commit funds to clean up sites that do not pose an immediate threat using current technology that may only be moderately effective and costly, or to commit funds to develop technology that may in the future provide a more effective and less costly solution, while stabilizing the contaminants until that technology becomes available.[198]

Choices are limited, says DOE, by current laws that force the immediate restoration of lower-risk sites using antiquated (but available) methods. Frequent changes in those laws also increase costs and complicate planning. These laws are "obstacles to reducing risks in a cost-effective, efficient, and timely fashion."[199] According to DOE, what is needed is an answer to the question "How clean is clean?" and agreement on what constitutes an "acceptable risk," based on projected future uses of each site, so that DOE will know better how to focus its efforts.

DOE thinks it might be better *not* to clean up some sites at all, at least for the time being. For example, a 2,600-acre lake at the Savannah River complex used for production reactor cooling water has become contaminated with highly radioactive cesium 137. Draining the lake would displace animals and migratory birds, and scraping the contaminated sediments from the bottom could endanger workers and nearby residents. The radioactivity will decline by ten to one hundred times, DOE says, if the lake is simply fenced off for a couple of hundred years.[200]

In certain instances, the obstacles to environmental restoration may prove to be insurmountable. At the Nevada Test Site near Las Vegas, for example, and at other locations from Mississippi to the Pacific, DOE set off 1,054 nuclear test explosions during the Cold War before testing was halted in 1991.[201] Prior to the signing of the Limited Test Ban Treaty in 1963, more than 200 of these tests were carried out in the atmosphere, contaminating large areas downwind and injuring government personnel and nearby residents. (Claims against the government for damages are examined in chapter 8.) Radioactivity was also vented from at least 20 underground tests.[202] Work is underway in Nevada to clean up more than 5 square miles of soil contaminated with plutonium. But at underground test sites activity has been confined to the installation of groundwater monitoring wells. At least in Nevada, DOE has relied on the "remoteness of the sites and the rigidly controlled access [to] prevent inadvertent public exposure."[203] In the end, those sites may have to remain off-limits to all human use forever.

In the face of nearly overwhelming difficulty, expense, and danger, it now appears possible that we will be able to clean up the nuclear weapons complex, at least to some degree. The Department of Energy has experienced a genuine change of culture over the last few years. This new way of thinking is illustrated by Energy Secretary O'Leary's dramatic release of millions of previously secret records of Cold War operations.[204] One-third of DOE's fiscal year 1995 budget is devoted to environmental management and restoration. The Department has shown that it is serious about waste minimization and recycling, and it is working hard to develop new technologies that can be used throughout American industry. Regulatory responsibility has begun to shift from DOE to EPA and the states. It remains for the American people to insist that they be fully informed about the cleanup and given an opportunity to participate in decisions affecting public health and the environment. They must also pay the staggering cleanup bill, and learn from this experience to count the environmental costs while planning for the nation's defense in the future.

Disposal of Defense Nuclear Wastes

Getting rid of millions of pounds of radioactive wastes from the production of nuclear weapons has turned out to be incredibly difficult. The technical challenges confronting DOE officials and state environmental regulators are simply unprecedented. These wastes generally cannot be treated to reduce their radioactivity; only the passage of time can accomplish that. Some can be separated from hazardous chemical wastes with which they are mixed in order to simplify their handling. Others can be reduced in volume by compaction or incineration so that they occupy less space in a disposal facility. Still others will be mixed with molten glass or cement, or encased in plastic, to retard their interaction with their surroundings. These processes are not risk-free, and there is no guarantee that the wastes, once treated and placed in a permanent repository, will not eventually migrate into the open environment.

The political challenges are no less daunting. Everyone will be happy to see all the radioactive waste safely disposed of. Yet no one wants to live near a disposal site, and no one wants the waste transported through their town to a disposal site somewhere else. Local opposition to treatment facilities for separation and reduction of these wastes is strident, and nobody is comfortable having radioactive materials stored in the neighborhood while the search continues for a permanent repository.[205] Government officials worry that a disillusioned electorate will be unwilling to continue footing the bill.

With the end of the Cold War, we are also left with hundreds of tons of plutonium and highly enriched uranium that were to have been fashioned into nuclear weapons or that remain from the dismantling of such weapons. Not everyone agrees that these materials should be irretrievably disposed of, since they might be wanted to fabricate new weapons in the future. However, their retention poses the danger not only of dispersal into the environment, but also of diversion or theft by terrorists who could manufacture their own weapons of mass destruction. Because we currently lack the means to safely and permanently dispose of such materials, we have been able to avoid confronting this policy dilemma.

The Search for a Final Resting Place

Throughout the Cold War, DOE and its predecessors stored some of their radioactive wastes "temporarily" in barrels or tanks on the sites where they were produced, hoping that someday effective methods for treatment and disposal would be found. Many of these containers have been leaking for years into the surrounding soil and groundwater. Other radioactive wastes were deliberately released into the environment by placing them in unlined pits, injecting them into deep holes, or burning them in incinerators.

As early as the mid-1950s, however, the need for a more nearly permanent solution had become apparent. The search for that solution has tested the ingenuity and persistence of a generation of federal and state officials and concerned citizens. In 1957, for example, the National Academy of Sciences identified salt beds beneath a number of states as promising sites for radioactive waste disposal. But a $100 million program to store radioactive wastes in an abandoned salt mine near Lyons, Kansas had to be dropped in 1971 when it was discovered that the site was riddled with unmapped oil and gas wells, through which water might migrate, dissolving the salt and dispersing the radioactivity. Another study to bury the wastes in the deep ocean seabed was halted in 1986.

In 1979, Congress approved construction of the Waste Isolation Pilot Plant (WIPP) deep underground in a salt bed just east of Carlsbad, New Mexico.[206] WIPP's mission is "to demonstrate the safe disposal of radioactive wastes resulting from the defense activities and programs of the United States exempted from regulation by the Nuclear Regulatory Commission." Now earmarked for permanent disposal of defense-related transuranic wastes (mainly wastes contaminated with plutonium), the WIPP facility was completed in 1988. But it has yet to open, for technological and political reasons described in the case study below.

In 1982, Congress passed the Nuclear Waste Policy Act, establishing a separate program to identify and develop two permanent repositories for high-level radioactive wastes and spent fuel from both civilian and defense-related activities.[207] The act originally directed the DOE Secretary to identify states with potentially suitable sites, and to issue "guidelines" for use in recommending three possible sites to the President for "characterization."[208] In this context, characterization means further study, including test borings, surface excavations, and the digging of shafts and tunnels. The guidelines are supposed to take into account the population density around a candidate site, hydrology, geophysics, seismic activity, proximity to valuable natural resources and water supplies, and the location of National Parks, Wildlife Refuges, Wild and Scenic Rivers, Wilderness Areas, and Forests.[209]

A political firestorm erupted when Secretary Herrington recommended, and President Reagan approved, characterization of sites at Yucca Mountain, Nevada, Deaf Smith County, Texas, and Hanford, Washington. Congress immediately amended the Nuclear Waste Policy Act to direct characterization of the Yucca Mountain site alone, and to halt the search for other possible sites.[210] Once the Secretary has characterized the site, conducted public hearings, and completed an environmental impact statement, she may recommend its approval to the President. If the President finds the site qualified, he may recommend it to Congress. The State of Nevada then has sixty days to indicate its disapproval to Congress. If that happens, the project must be abandoned unless within ninety days thereafter Congress passes a joint resolution approving the President's recommendation.[211] Site characterization at Yucca Mountain is currently underway, but the controversy over its selection as the nation's high-level waste repository continues unabated, as we see in the case study below.

Until 1974, facilities that generated and disposed of defense-related radioactive wastes were exempt from licensing under the Atomic Energy Act. In that year, Congress prescribed licensing by the Nuclear Regulatory Commission (NRC) for long-term storage of high-level defense wastes.[212] The licensing requirement was expanded in the Nuclear Waste Policy Act of 1982 to cover permanent disposal of such wastes.[213]

The 1982 Act ordered EPA to set "standards for protection of the general environment from off-site releases from radioactive material in repositories." It also directed the NRC to promulgate technical requirements and criteria for the construction, operation, and closure of repositories, utilizing multiple barriers and restricting the retrievability of the materials disposed of.[214] Otherwise, the statute furnishes no guidance for either EPA or NRC.

NRC's regulations for high-level waste repositories say that packages containing the waste must remain "substantially complete" for at least three hundred years, and must release no more than 0.001 percent of their radioactive contents in any of the first thousand years they are held in the repository.[215] Since some of the materials placed in the repositories will remain intensely radioactive for tens of thousands of years, these regulations offer scant protection for future generations of Americans.

EPA published its standards for the Nuclear Waste Policy Act in 1985.[216] Those standards call for the establishment of a "controlled area" around each repository site within which no limits would exist on contamination. Such a sacrifice zone is intended to provide a buffer between the disposed waste and the surrounding environment. The EPA standards set maximum cumulative totals of various radionuclides, such as uranium, that can be released from a controlled area over a ten-thousand-year period.

Two aspects of the EPA standards were particularly controversial. One provided that for at least one thousand years, radionuclide contamination of underground sources of drinking water should not exceed limits set in EPA regulations for the Safe Drinking Water Act. The other prescribed limits on individual human exposure to radiation during the same period. In a suit brought by four states and several public interest organizations, both provisions were struck down by a federal court as arbitrary and capricious because they were inconsistent with standards contained in EPA's regulations for the Safe Drinking Water Act.[217] New standards were adopted in 1993 that extend from one thousand to ten thousand years the limits on human exposure and pollution of drinking water sources.[218] However, regulations for the proposed Yucca Mountain facility will be promulgated separately as a result of legislation passed in 1992.[219]

Both the WIPP and Yucca Mountain facilities have proven to be extremely controversial. The history of their development is traced briefly in the following case studies.

Case Study: Whither WIPP?

The WIPP facility is located on 10,240 acres of federal land 26 miles east of Carlsbad, New Mexico. WIPP's current mission is to provide a repository for the safe disposal of plutonium-contaminated wastes from weapons production that are held in temporary storage at other DOE facilities in 7 states. The plan is to permanently seal the waste in a vast salt deposit 2,150 feet underground.[220] WIPP has been developed independently of the strictures of the Nuclear Waste Policy Act, although it must conform to EPA regulations developed under that act and the Atomic Energy Act.[221]

Construction of WIPP was completed in 1988. Yet not a single drum of radioactive waste has been placed in it so far.

> Over the years, the project has faced a variety of challenges. In 1980, President Carter sought to cancel WIPP, in part because Congress exempted it from Nuclear Regulatory Commission licensing and oversight. Twice during the Reagan administration, the White House Office of Management and Budget tried to eliminate WIPP because of its near billion-dollar cost. Then, in 1986, the U.S. General Accounting Office revealed that WIPP will hold only a small fraction of DOE defense wastes and that the facility will not be available for the wastes most in need of cleanup—solid wastes buried in shallow pits or liquid wastes drained into the ground.[222]

Soon after the facility was completed, water was found seeping into what had been thought to be absolutely dry salt chambers. This discovery suggested that radioactive wastes might be transported into nearby water supplies, including the Pecos River, when drums in which the waste would be stored eventually corroded and decomposed.[223] To make matters worse, scientists discovered a large body of pressurized water beneath the site that could flood the disposal area and disperse the wastes. There is also concern about the possible escape of gases generated by the decomposing waste.

Despite these revelations, DOE pressed forward with a plan to begin storing radioactive wastes at the site experimentally in 1988. However, a large fraction of the radioactive wastes proposed for storage at WIPP are mixed with non-radioactive hazardous wastes subject to regulation under the Resource Conservation and Recovery Act (RCRA). Although DOE had by this time conceded the applicability of RCRA to such mixed wastes, it had not obtained a RCRA disposal permit for the site. This failure, together with doubts about the quality of the facility's construction and its ability to withstand earthquake damage, caused DOE officials to place the demonstration plan on hold. There were also fears that DOE might be unable to retrieve the wastes if the experiment failed.[224]

Efforts to open WIPP took on new urgency in October 1988 when Idaho Governor Cecil D. Andrus barred new shipments of plutonium wastes into his state for interim storage at the Idaho National Engineering Laboratory. Three years later, in 1991, DOE announced new plans for a test involving up to nine thousand barrels of radioactive waste. The states of New Mexico and Texas, joined by several environmental groups, then obtained a court order to halt the test.[225] The Court of Appeals for the D.C. Circuit upheld the lower court's injunction because Department of the Interior land on which WIPP is situated had not been properly "withdrawn" from public access to enable its use as a repository.[226] The court also found that WIPP might be eligible for interim status under RCRA permitting requirements, but stopped short of concluding that WIPP had achieved interim status. RCRA interim status would mean that disposal could proceed without first obtaining a permit from the State of New Mexico.

In 1992, Congress enacted legislation effecting the necessary land withdrawal.[227] Under the new act, as well, EPA must approve a revised DOE plan for experimental emplacement of radioactive waste, along with a plan for retrieval of the waste if that becomes necessary. Those plans must demonstrate the stability of rooms in the salt cavern, and must show that any migration of radioactivity from the site over thousands of years would be limited in accordance with EPA regulations under the Nuclear Waste Policy Act.[228] The 1992 act specifically calls on DOE to comply with applicable provisions of the Clean Air Act, the Safe Drinking Water Act, the Toxic Substances Control Act, RCRA and CERCLA.

The State of New Mexico has not yet recognized WIPP's interim status under RCRA. Unless the federal district court changes its earlier ruling denying interim status, DOE must obtain a RCRA permit from the state before proceeding with any disposal. The permitting process may be complicated by the fact that no records were kept of what is in most of the hundreds of thousands of barrels containing plutonium wastes. If an accurate assay of their contents is required, it could take decades to complete and cost billions of dollars, because of the danger that in reopening the containers plutonium will be released into the environment.

DOE officials have now dropped plans for a test involving actual radioactive wastes, hoping instead to use laboratory experiments to show that the facility can comply with EPA and New Mexico environmental requirements. DOE is also engaged in a vigorous public education and community outreach program in an effort to allay public fears about the project.[229] While it works to meet environmental requirements for opening WIPP, DOE is busy retrieving, analyzing, and repackaging its transuranic wastes, in order to store them safely for the time being.

Case Study: Politics and Groundwater at Yucca Mountain

Yucca Mountain is a barren, flat-topped ridge in the Nevada desert just outside the boundary of the Nevada Test Site, about one hundred miles northwest of Las Vegas. When Congress decided in 1987 to designate Yucca Mountain as the sole site to be characterized for use as the nation's first high-level radioactive waste repository, Nevadans were shocked and angry. There was talk of a conspiracy among more populous and politically powerful states that desperately wanted to avoid playing host to the repository.[230] Not so, said sponsors of the measure, who claimed that Yucca Mountain's remoteness and desert conditions made it a likely first choice anyway. Furthermore, they claimed, the selection process urgently needed to be streamlined.

A series of lawsuits was aimed at derailing the selection of Yucca Mountain. Two asked for federal funding of state efforts to resist placement of the repository in Nevada.[231] In another, the court found that Congress had ample authority under the Property Clause of the U.S. Constitution to designate federal lands in Nevada alone for study, and that Congress' disparate treatment of Nevada did not offend the Tenth Amendment or other constitutional guarantees.[232] The same court decided that a Nevada legislative ban on the storage of high-level nuclear waste in that state was preempted by the federal Nuclear Waste Policy Act. Nevada then unsuccessfully challenged the Bureau of Land Management's grant of a right-of-way to DOE to perform characterization studies at the site.[233] Two county governments gained designation as units of local government affected by the project.[234] An attack on DOE's "guidelines" for site selection was then dismissed as premature.[235] Finally (as of this writing), Nevada sought to enforce a Nuclear Waste Policy Act requirement that DOE prepare an environmental assessment before characterizing any site; the court said that requirement was rendered moot by Congress' 1987 mandate to characterize only the Yucca Mountain site.[236]

Meanwhile, as studies of the site began to pile up, concern also mounted that it might not after all be safe as a permanent repository. A report by a DOE scientific panel in 1991 revealed a sharp division of opinion as to the suitability of the site.[237] Some experts argued that groundwater beneath the mountain might be driven upward by an earthquake and into the disposal area, dispersing the radioactive waste. Their fears were heightened by a strong earthquake close by the site in 1992, although a DOE official maintained that Yucca Mountain itself had been "relatively stable" for two million years.[238]

A lively debate is underway about the applicability of the Safe Drinking Water Act at the site. Nevada operates an EPA-approved Underground Injection Control (UIC) program. As part of that program, Nevada has enacted a statute forbidding any "injection of fluids through a well into any waters of the state . . . [o]f any radiological, chemical or biological warfare agent or high-level radioactive waste . . . [w]hich would result in the degradation of existing or potential underground sources of drinking water."[239] EPA regulations under the Safe Drinking Water Act's UIC program also prohibit the injection of hazardous or radioactive wastes within one-quarter mile of an underground drinking water source.[240] Because the proposed facility overlies a large aquifer that might someday serve as an underground source of drinking water, the regulations would prevent construction of the repository. However, EPA insists that the Safe Drinking Water Act has no application to activities regulated under the Nuclear Waste Policy Act, since the high-level waste and spent fuel are solids, not liquids. In litigation challenging EPA regulations under the latter act, the First Circuit Court of Appeals declared otherwise.[241]

Late in 1992, Congress sought to ease the regulatory burden by directing EPA to adopt regulations for Yucca Mountain that specify maximum permissible doses of radiation for any individual off-site.[242] However, the regulations, based on a National Academy of Sciences study, are to contain no limits on the total dose for the population as a whole. This measure represents a sharp break from established procedures for calculating risks to public health.

In the same legislation, Congress ordered DOE to remain in custody of the repository for the indefinite future, rather than abandoning it after it is filled, so that distant generations would more likely be warned of the hazard.[243] The National Academy of Sciences was given the task of determining:

> . . . whether it is reasonable to assume that a system for post-closure oversight of the repository can be developed, based upon active institutional controls, that will prevent an unreasonable risk of breaching the repository's engineered or geologic barriers or increasing the exposure of individual members of the public to radiation beyond allowable limits; and
>
> . . . whether it is possible to make scientifically supportable predictions of the probability that the repository's engineered or geologic barriers will be breached as a result of human intrusion over a period of 10,000 years.[244]

It should be remembered that the Republic itself has only been in existence for a little over two hundred years, and that all of recorded history spans no more than five thousand years. The notion that we might devise some "active institutional control" to prevent human encroachment during the millennia that the high-level wastes remain extremely dangerous can only be described as bizarre—proof of the growing desperation to locate a final resting place for these wastes. Long-term predictions about the physical integrity of the repository site, climatic changes, and other natural events are currently based on observations carried out over only a few months or years; such predictions seem scientifically indefensible.[245] If, as seems likely, the National Academy of Sciences is unable to come up with satisfactory answers, we will be well advised to delay the opening of Yucca Mountain or any other geologic repository while we mount a search for safer alternatives.

Even if the Yucca Mountain site is finally approved for use as a high-level waste repository, only 10 percent of the available space has been allocated to defense wastes; the balance will be devoted to wastes from civilian nuclear power plants. That 10 percent will hold only about one-fifth of all DOE high-level wastes, with no allowance for the large volume of spent nuclear fuel that must be disposed of. The inescapable conclusion is that a *second* high-level waste repository will have to be constructed, with political, economic, and environmental consequences that can hardly be imagined.

DOE currently possesses a large quantity of highly radioactive spent nuclear reactor fuel. Some of it comes from research reactors at universities and DOE research laboratories, some from the propulsion systems of Navy warships. A great deal more comes from DOE production reactors. When DOE sought to collect this material and store it at its Idaho National Engineering Laboratory (INEL) until a permanent repository could be found, the Governor of Idaho in 1991 banned further shipments of spent nuclear fuel into the state. A court ordered the ban lifted on grounds that it improperly interfered with interstate commerce and was preempted by provisions of the federal Atomic Energy Act and Hazardous Materials Transportation Act.[246] The Governor was then rebuffed in his effort to have the shipments halted as violations of Idaho's air quality standards.[247]

The State of Idaho then persuaded the court in the first action that DOE had violated NEPA by failing to prepare a comprehensive EIS for the shipment, processing, and storage of spent nuclear fuel at INEL.[248] The court found that DOE had exercised bad faith in misrepresenting the availability of alternative storage sites for the fuel. It enjoined further shipments into the state (except for limited amounts from Navy propulsion reactors) until an EIS was prepared, setting a deadline of April 30, 1995. In its opinion, the court scolded the Department of Energy:

The court has been continually surprised and dismayed by DOE's reluctance to perform full NEPA analyses of the actions questioned by Idaho in this litigation. DOE's strenuous opposition, and the tremendous effort and taxpayer expense associated with such opposition, does not seem an appropriate course for an agency charged with overseeing such important, yet hazardous activities. DOE simply does not seem to understand that this nation is depending on it to protect the health and safety of all Americans from the dangers associated with its activities. . . . DOE must realize that its own failure to comply with the law, as well as its unwillingness to plan for the possibility of an injunction being entered in this action, are responsible for the difficulties it now faces.[249]

A draft EIS was published on June 24, 1994, covering the management of spent fuel both at INEL and nationwide.[250]

Much of DOE's immense inventory of low-level radioactive waste is buried at the facilities where it was produced, although a great deal of it has been shipped to the Nevada Test Site for storage, apparently in the hope that it can be left there permanently. New waste will be added to the inventory as cleanup of the weapons complex progresses. Disposal of low-level waste is exempt from Atomic Energy Act licensing requirements, as well as from regulation under the Low-Level Radioactive Waste Policy Act, which covers only civilian wastes.[251] While DOE is working to reduce the volume of these wastes to facilitate their disposal, the Department continues to rely on shallow burial in "vaults" engineered to resist encroachment by the elements, at least for a time. However, storage and disposal of low-level wastes, like other radioactive wastes, remains subject to regulation under the Clean Air Act, the Safe Drinking Water Act (if it is injected into the ground), and the Resource Conservation and Recovery Act (if it is mixed with hazardous chemical waste).

The Special Problem of Plutonium

Not all of the radioactive material left over from the nuclear arms race is waste, strictly speaking. That is because some of the fifty tons of plutonium and hundreds of tons of highly enriched uranium in United States inventories might someday be wanted to make new nuclear weapons.[252] At least part of the highly enriched uranium will eventually be used as naval or civilian reactor fuel.[253] But as the manager of one DOE facility put it, "I don't think people have a really good answer for what is going to happen to the plutonium."[254]

The plutonium is held in liquid, solid, and powder form at thirteen sites around the country. Some is stored in metal or plastic containers that have begun to break down from the intense radiation and leak into the environment. Some is trapped in air ducts and piping, where it poses a risk of setting off a lethal chain reaction. A great deal more is in the form of plutonium "pits" from the disassembly of nuclear weapons. A 1994 DOE study found that "the Department's inventory of plutonium presents significant hazards to its workers, the public, and the environment; that the hazards have not been aggressively addressed; and that the Department needs a strong, centrally coordinated program to achieve safe interim storage of plutonium."[255]

Measured by the cost of its production, plutonium is enormously valuable, especially when one counts the as yet undetermined expense of cleaning up after the manufacturing process. It could also be of incalculable value to terrorists, who could use it to assemble a nuclear bomb. In the words of a recent National Academy of Sciences report, it represents a "clear and present danger to national and international security."[256] As long as the plutonium is maintained in its current form, therefore, it must be kept absolutely secure from theft. It also must be protected from inadvertent release into the environment; exposed to the atmosphere, plutonium can spontaneously ignite and disperse into the air, as happened several times at Rocky Flats. Moreover, retaining the plutonium indefinitely in storage would send a negative political signal for nonproliferation and arms-reduction efforts.

DOE's current plan is to collect all the United States plutonium from stockpiles and place it in interim storage at its Pantex facility near Amarillo, Texas. "Interim" for this purpose is supposed to mean six or seven years. The Pantex plant will also dismantle up to two thousand nuclear warheads each year and store the plutonium "triggers" there. However, local residents worry that the plant could become a de facto permanent repository, since DOE has no fixed plan yet for dealing with the plutonium after the six or seven years have passed, much less for the hundreds of thousands of years that it will remain dangerously radioactive.[257] DOE insists that the storage is not subject to state environmental regulation, although it has agreed to let Texas officials inspect the storage areas.[258] DOE has recently agreed to prepare an EIS for the Pantex storage program.

The Department of Energy began in late 1994 to conduct public scoping hearings on a programmatic environmental impact statement for the storage and disposition of weapons-usable fissile materials.[259] One option under consideration is to dilute the plutonium so that it can be used in civilian nuclear reactors. Unfortunately, the resulting spent fuel would be even more radioactive than the ordinary kind, for which we still have not found a safe repository. Utility executives also fear that members of the public might come to associate nuclear power with the weapons program. Besides, they currently have a plentiful supply of cheap uranium for fuel. Utilities in Europe and Japan could provide a market for the plutonium, but the risk of diversion or accidental dispersal would probably make transportation overseas extremely controversial.

A second option is vitrification—mixing the plutonium with high-level wastes and molten glass to produce glass "logs" for eventual disposal in geologic repositories. Plutonium in this form would be more difficult to divert for weapons use. But some DOE scientists fear that thousands of years after its burial, the plutonium could interact with the surrounding rock and produce an explosion equal to that of a large hydrogen bomb.[260]

Finally, authorities are considering the possibility of disposing of the plutonium without treatment in extremely deep bored holes in the ground. The plutonium could then be retrieved only by the country in control of the borehole. However, questions about possible migration into the wider environment have not been answered.

Compounding the problems associated with disposal, private experts believe that as much as 1.5 metric tons of plutonium manufactured over the last forty years cannot be accounted for and may have been lost. That is enough to make three hundred nuclear weapons. While DOE disputes the extent of the shortage, there is fear that some of the missing material may have been diverted to another country.[261]

The United States bears a special burden of leadership in developing an effective means of disposal for its weapons-grade materials, not only for political reasons related to arms control and nonproliferation, but also to help provide practical solutions for Russia to dispose of its huge inventories of these very dangerous substances.

Chapter 5

Military Base Closures and Realignments

Long before the Cold War ended, the Pentagon had begun to close obsolete and redundant military bases around the country. Now that the United States remains as the sole military superpower, that process has accelerated. A 1991 government report explains why:

> The end of the Cold War, evidenced by the fall of the Berlin Wall in 1989 and the formal dissolution of the Warsaw Pact in 1991, fundamentally altered the military threat posed by the Soviet Union and its allies. These events had dramatic impacts on U.S. military requirements. In addition, the growing U.S. budget deficit provided an impetus to cut U.S. military spending. Therefore, DoD is planning to decrease the U.S. military by approximately 25 percent over the next five years.
>
> Clearly, fewer forces require fewer bases. By eliminating unnecessary facilities, limited dollars can go to vital military needs. Balancing the base structure with the new force-structure plan will make DoD more efficient, streamline the defense infrastructure, and enhance national security.[1]

We might suppose that these base closings would automatically produce net benefits for the environment. But it is not that simple. The Defense Department controls 25 million acres of land in this country alone. Some of it includes critical wildlife habitat, aquifer recharge zones, and areas of unique beauty. When the military departs, these properties may be converted to other government uses, or they may be given over to commercial or industrial development, with varying environmental consequences.[2] Most bases hold accumulations of toxic or radioactive wastes that must be cleaned up to protect future users and nearby residents.[3] Moreover, each base closing results in the loss of both military and civilian jobs, with impacts on the local economy ranging from insignificant to catastrophic. A current example illustrates the political, socioeconomic, and environmental complexity of the problem.

Case Study: Nuclear Disarmament in Small Town America[4]

On September 27, 1991, when President Bush made his historic announcement that the United States would "eliminate its entire worldwide inventory of ground-launched . . . nuclear weapons,"[5] Americans everywhere breathed a sigh of relief. But the citizens of Seneca County, in rural western New York, also felt a deep sense of foreboding. For many years the Seneca Army Depot there had as its principal mission the storage of "special weapons," including ground-launched nuclear missiles and nuclear artillery shells. The President's message came on the heels of an earlier decision to transfer other functions of the Seneca Army Depot to bases elsewhere.

Together, these developments meant that 70 percent of Seneca Army Depot's civilian employees—562 of 847—would lose their jobs, while all but three of 487 uniformed military personnel would be transferred to new assignments. Civilian workers would have to find new jobs or retire early, and many would be forced to leave the community, taking their children out of the schools and selling their homes. Student enrollment in the local school district would drop by 30 percent, and Seneca County would lose 15 percent of its population of 32,000. Property values could be expected to drop sharply, along with the ad valorem tax base. Also, two local communities would lose water and sewer services, as well as fire protection, long provided by the Army.

Two years earlier, the Army had begun investigating reports that toxic wastes—chlorinated organic solvents—were leaching into groundwaters from incinerator ash dumped into a landfill at the Depot, and that some of the polluted groundwater had migrated off the post. Areas used for open burning and detonation of ordnance were also found to be contaminated with heavy metals.[6] Altogether, hazardous wastes were identified at 57 sites. In 1990, Seneca Army Depot was placed on the Superfund National Priorities List, and by late 1992 more than $4 million had been spent just on investigation and planning for the cleanup.

The Army insisted that the nearly complete shutdown of the Depot would produce no significant environmental impact, and that the environmental review procedure spelled out in the National Environmental Policy Act could be avoided. Neither, it maintained, would these reductions require compliance with special legislation concerning base closings and realignments. Thus, the Army has made no systematic assessment of environmental costs or risks. No inventory has been made of plants or wildlife on the 10,600-acre facility, nor any analysis of the effects that new activities there might have on them. There has been no study of the impacts on air and water quality or other environmental amenities. And no effort has been made to estimate the consequences for the local economy.

Several business and citizen groups, a labor union, and the county government filed suit to stop the layoffs and transfer of functions. The federal district court ruled that the 1990 Base Closure and Realignment Act (described below) provided the "*exclusive* means for closing and realigning bases," and it ordered the Army not to carry out the planned reductions except in compliance with the act.[7] Its preliminary injunction was dissolved on appeal, however. The appellate court decided that the total elimination, rather than relocation, of jobs related to storage of "special weapons" did not fit the definition of "realignment" for purposes of the Base Closure and Realignment Act. It also decided that while the Army's activities at the Depot had a significant impact on the environment in the past, and despite the effect of the reductions on the economic health of the region, the proposal reflected no new risk to the physical environment that would require preparation of an EIS. Moreover, the court observed, the reduction in civilian staffing levels is one of those actions categorically excluded from NEPA analysis by the Army's NEPA regulations.[8]

When functions and personnel are shifted from one base to another, and not simply eliminated, there will be impacts at both ends of the transfer process. Part or all of each base's operations may fall under a new command structure, different equipment and procedures can be installed, and the types of waste generated may change, for better or for worse. The environmental implications of all these changes must be carefully considered.

Applying NEPA to Base Closures and Realignments

Throughout the 1960s and 1970s, the Secretary of Defense sought to trim the military budget by closing or realigning a number of bases. Following passage of the National Environmental Policy Act (NEPA) in 1969, the Secretary's efforts provoked extensive NEPA litigation. (NEPA's operation is described in chapter 2.) Many of the plaintiffs in those suits seemed to be concerned as much with the threatened loss of jobs as with the impact on wildlife habitat and water quality. One court responded brusquely that "NEPA is not a national employment act."[9] That sentiment is echoed in regulations of the Council on Environmental Quality (CEQ), which provide that "economic and social effects are not intended by themselves to require preparation of an environmental impact statement."[10] Other courts were more sympathetic, ruling in one case that the movement of 2,000 military and 850 civilian workers from one base to another required the preparation of an EIS.[11]

The CEQ regulations also state that "[w]hen an environmental impact statement is prepared and economic or social and natural or physical environmental effects are interrelated, then the environmental impact statement will discuss all of these effects on the human environment."[12] Accordingly, a number of courts ruled that where there was a significant impact on the natural environment, socioeconomic effects should be included in the environmental analysis.[13] The outcomes in individual cases tended to turn on their particular facts, either as to the need for an EIS or as to the adequacy of one that had already been prepared.[14]

Members of Congress, fearing the loss of jobs and businesses within their districts, resisted the efforts of the Secretary of Defense to close individual bases. They also suspected that the closure process was being used by the administration in power to discipline uncooperative members.[15] In 1977, Congress passed legislation forbidding the Secretary to close or realign a base without first notifying the Armed Services Committees of both houses and furnishing an evaluation of the economic, environmental, budgetary, and strategic consequences, so that Congress could halt the Secretary's action if it disapproved.[16] Compliance with NEPA's environmental review procedure was expressly required.[17]

However, not a single major base was closed pursuant to the 1977 legislation, implementation of which continued to be thwarted by political wrangling. Military leaders also complained that the required environmental review unduly hampered their efforts. Defense Secretary Dick Cheney remarked, "I can deactivate an entire Army division overseas, but before I can close a domestic military base, the issue must be studied every which way. The effect of all this red tape is simply to make it hard for us ever to close a base."[18]

In 1988 Congress responded to these concerns by enacting a one-time exception to the 1977 act, establishing an independent commission to recommend the closure or realignment of bases to the Secretary.[19] The Secretary was confined to approving or disapproving the entire list of commission recommendations without change. Following the Secretary's approval, Congress also reserved for itself only the ability to disapprove the entire list, thus saving itself from the politically painful task of singling out individual bases for closure. Congress specifically exempted the *selection* of bases for closure from the NEPA review process, though not other activities related to their closing. Under the 1988 law, eighty-six bases either have closed or are scheduled for closing, and another fifty-nine were ordered realigned.

As the Cold War drew to a close, it became apparent that U.S. military forces would have to be dramatically restructured to reflect the diminished threat of a Soviet attack. Congress acted again in 1990,[20] this time authorizing the Secretary of Defense to recommend the closure or realignment of additional bases, taking into account current force-structure plans, to a reconstituted Defense Base Closure and Realignment Commission. The Commission's job is to consider the Secretary's proposals, then pass its own recommendations along to the President. If the President approves the Commission's recommendations, he must then notify Congress, which has 45 days to reject the recommendations intact in order to avoid their implementation. The 1990 act makes NEPA applicable *only* to the *disposal of base properties* and the *transfer of functions* to new installations. The first recommendations under the 1990 act, involving the closure of 43 bases and the realignment of 28 others, were announced in 1991.[21] A second round of 130 closings and 45 realignments was made public in 1993,[22] with a third and final round to be announced in 1995.

Thus, there are currently three different levels of NEPA review, depending upon which statute is invoked in closing a particular base:

1. The 1988 act made NEPA inapplicable to the selection of any particular base for closing or realignment under that statute. But the Secretary's actions in carrying out the Base Closure Commission's recommendations are subject to NEPA review.
2. For closures and realignments under the 1990 act (that is, until completion of the third round of closures to be announced in 1995), NEPA is applicable only to the disposal of property and the transfer of functions to new installations.
3. Closures and realignments not covered by either of these statutes are subject to NEPA just like any other federal action.

The 1988 and 1990 acts also provide that a suit to enforce NEPA's remaining requirements must be brought within sixty days after a claimed violation, not later.[23] This very short timetable means that concerned citizens and local governments have to keep abreast of current developments and be prepared to respond quickly.[24]

Environmental Regulation of Base Closures

Our focus up to this point has been on the environmental analysis required by NEPA. But the other federal environmental laws are fully applicable to base closures and realignments as well.[25] In closing a base, accumulated hazardous wastes must be cleaned up in accordance with the requirements of RCRA and

CERCLA. EPA also must be satisfied that a closure or realignment will not result in a violation of any of the various environmental laws.[26] In particular, a base closure EIS must address the application of these laws to post-closure uses after the military departs. Moreover, activities transferred to other bases may require the issuance of new environmental permits. The following case study illustrates both the direct application of the other statutes and their complex interaction with NEPA.

Case Study: New Management at Pease Air Force Base

In early 1991, Pease Air Force Base became the first installation to close under the 1988 Base Realignment and Closure Act. Located on the New Hampshire seacoast, the base covered 4,356 acres, more than half of it forested, with some 800 acres of wetlands. Following the statutorily mandated consultation with state and local officials,[27] it was decided that 1,100 acres will be preserved as a National Wildlife Refuge under the control of the Department of the Interior. Other base lands will be used to build a public golf course. The Air Force will also retain a portion for the Air National Guard. But the bulk of the property will be turned into an international airport and industrial park by the state-chartered Pease Development Authority, utilizing existing runways and 1.2 million square feet of empty buildings.

Long before it closed, Pease had a history of environmental problems. In 1977, a well supplying drinking water to 8,700 people on the base was found to be contaminated with trichloroethylene (TCE), although an aeration system to clean it up was not installed until seven years later. Further investigation revealed the presence of the pesticides heptachlor and lindane in surface waters, lead and zinc in drainage ditches, and paint strippers, waste oil, fuel, and various organic solvents at some forty-two sites around the base. Pease was placed on the CERCLA National Priorities List in 1990. To date, the Air Force has spent more than $53 million to clean up the base, but the job will not be completed for a number of years, perhaps decades.

The Air Force prepared one environmental impact statement for the closure of the base, and another for its development and reuse, although EPA protested that a single EIS should have addressed both actions, since the two were so closely related. According to the neighboring town of Newington, New Hampshire and the Conservation Law Foundation, the EIS for development and reuse was deficient in several other critical ways. They filed suit to stop the transfer of base lands.[28]

First, the plaintiffs asserted that the transfer would violate section 176 of the

Clean Air Act,[29] the so-called "conformity" provision. That section imposes on each federal agency an affirmative duty to ensure that its activities will conform to an approved state implementation plan for control of air pollutants. Specifically, an agency must not cause or contribute to any violation of the national ambient air quality standards, increase the frequency or severity of any violation, or delay the attainment of required emission reductions. Nevertheless, the Air Force's EIS for development and reuse states that new aircraft, stationary sources, and traffic will cause levels of carbon monoxide and ozone-forming hydrocarbons to increase dramatically. That, according to the plaintiffs, would make it impossible to achieve federally mandated reductions in ozone and would exacerbate recurring violations of carbon monoxide standards. The Air Force sought to avoid this nonconformity by entering into a memorandum of understanding with EPA, which provides that the State of New Hampshire will take unspecified measures later on to avoid violations. In addition to this Clean Air Act violation, say the plaintiffs, the memorandum of understanding was signed months after completion of the NEPA review, so there was no opportunity for public notice or comment. Therefore, they argue, a supplemental EIS should be prepared before the base property is turned over.

Section 120(h) of CERCLA provides that any "deed" for the transfer of federal property contaminated by hazardous substances must contain a covenant that "all remedial action necessary to protect human health and the environment . . . has been taken before the date of such transfer."[30] But cleanup of all the areas scheduled for transfer at Pease may not be finished for many years. To make base properties available for new community uses as soon as possible, the Air Force decided instead to *lease* them, rather than deeding them outright, to the Pease Development Authority for fifty-five years, a term that gives the new user something very close to outright ownership. This, say the plaintiffs, effectively violated the ban then in existence on conveyance of contaminated property.[31] Moreover, they insist, the EIS for the transfer failed to address the lease arrangement or give the public an opportunity to comment on it, another reason to require the preparation of a supplemental EIS. Six months after the Air Force executed its lease to Pease Development Authority, Congress amended CERCLA to provide that contaminated lands may be transferred if "the construction and installation of an approved remedial design has been completed, and the remedy has been demonstrated to the [EPA] Administrator to be operating properly and successfully."[32]

In August 1994, the trial court ruled that the Air Force had satisfied the Clean Air Act conformity requirements. However, it found that the process of determining conformity—without public review or comment—violated NEPA. It also found that the final EIS was deficient in failing to address air quality impacts in the neighboring state of Maine, in failing to fully describe possible mitigation measures, and in failing to discuss the effects of redevelopment on adjacent wetlands. Finally, the court decided that the Air Force had violated even the recent amendments to CERCLA by transferring contaminated property without an approved plan for remediation of that property. The court then ordered the Air Force to prepare a supplemental EIS but refused to enjoin the transfer while the several violations continued.

The litigation described here, still pending in the federal district court at this writing, is being watched closely by military planners and environmentalists around the country. It is expected to set the pattern for application of the environmental laws when hundreds of other bases close over the next few years.

CERCLA contains two additional provisions intended to assuage the fears of nongovernment transferees that they might someday be found responsible for cleaning up contamination remaining when the military departs.[33] A contract or deed for the transfer of base property must disclose details of any hazardous waste activities at the site that are known to the transferring agency.[34] Whether or not hazardous wastes were ever present at the site, such a deed must also include a covenant that the government will perform any cleanup that may be required in the future.[35] Separate legislation authorizes the government to indemnify a transferee of closed base properties for any future cleanup liability.[36]

Other Windows on the Base Closure Process

Limitations on systematic NEPA review under both the 1988 and 1990 legislation sharply reduced opportunities for citizens and state and local governments to become involved in the planning for base closures. However, a variety of other measures direct the Secretary of Defense to keep Congress and the public informed about the environmental implications. Under the 1977 act, no closure or realignment can take place until the Secretary provides the Senate and House Armed Services Committees with an evaluation of the environmental consequences.[37] Progress toward closing bases selected under the 1988 act must be reported in each year's DOD budget request, along with an assessment of the environmental effects.[38] Annual environmental assessments are required under the 1990 act, as well, beginning with the budget request for fiscal year 1993.[39]

Under the 1990 act, the Secretary was required to formulate and then apply criteria for identifying bases for closure or realignment.[40] Despite the specific statutory exemption of the base selection process from NEPA review, the Department of Defense included as one of eight such criteria the "environmental impact."[41] This presumably means that a NEPA-style analysis must be conducted for each base selected, but without the extended public notice and comment procedure required by NEPA.

The 1990 act also calls on the Secretary to make available to Congress, the Commission, and the Comptroller General "all information used by the Secretary" to prepare his recommendations for base closings and realignments.[42] This information would of course include an explanation of his compliance with his own eight criteria for selection, among them the environmental impact. The Comptroller General is charged to prepare a detailed, independent analysis of the Secretary's selection process and recommendations, and send it to the Commission and Congress.

Failure by the Secretary of Defense to transmit all this information provoked two lawsuits. One involved the closure of Loring Air Force Base in Maine. The plaintiff charged that the Air Force had "deviated substantially" from the published base closure criteria and failed to deliver all the required information to Congress and the Comptroller General. The First Circuit Court of Appeals held that neither the Base Closing Commission's recommendations to the President nor the process leading up to their formulation constituted a "final agency action" subject to judicial review under the Administrative Procedure Act,[43] and so dismissed the suit.[44]

By contrast, the Third Circuit Court of Appeals, in a case growing out the proposed closure of the Philadelphia Naval Shipyard,[45] enjoined the Secretary to deliver such information to Congress. The court distinguished between the *substantive* decision to close a particular base, which the court found was left by Congress to the President's discretion, and the nondiscretionary *procedure* used in reaching his decision, which was subject to judicial review. Furthermore, said the court, regardless of the "finality" of the President's actions for purposes of the Administrative Procedure Act, and whether or not the President is regarded as an "agency" under that statute, compliance with an act of Congress is subject to a "common law" right of judicial review. That review is also constitutionally grounded in the separation of powers doctrine: "The President's power, if any . . . must stem from an act of Congress or from the Constitution itself," and the court is bound to say whether the President has remained within statutorily mandated limits.

The Supreme Court overturned that decision in 1994.[46] Chief Justice Rehnquist, writing for the Court, declared that the recommendations of the Secretary of Defense to the Base Closure Commission did not constitute "final agency action" subject to judicial review under the Administrative Procedure Act. Moreover, said the Court, the final action here was that of the President, whose action is not reviewable because he is not an "agency" for Administrative Procedure Act purposes. The Court ignored the distinction between substantial and procedural actions, and ignored the strong presumption of judicial review of executive actions. It held simply that because Congress had granted the President discretion to act, judicial interference was not warranted: "How the President chooses to exercise the discretion Congress has granted him is not a matter for our review." Four other justices concurred on grounds that the 1990 act as a whole implicitly precluded judicial review of any aspect of the selection process. Thus, the Court slammed shut one window on the base closure process that might have informed Congress and the public about environmental implications.

Several other sources of information deserve mention here. The 1988 Base Closing Act, but not the 1990 act, directed the Secretary to prepare a comprehensive five-year plan for environmental restoration at installations scheduled for closing or realignment.[47] The 1990 act sets up an Environmental Response Task Force, whose job it is to make recommendations on ways to expedite cleanups at closing bases, then to publish annual reports on progress in implementing its recommendations.[48] The 1991 DOD Appropriation Act establishes a Legacy Resource Management Program, calling on the Secretary to identify and manage "all significant biological, geophysical, cultural and historical resources" on DOD lands.[49] The Army has developed an Environmental Early Warning System, a computer planning program that measures the "carrying capacity" of some bases in terms of housing, schools, utilities, endangered species, noise, landfills, and other environmental amenities.[50] Also, each military installation routinely prepares base planning and management documents, and many conduct periodic environmental audits. Finally, on July 2, 1993, President Clinton announced a five-point plan for revitalizing base closure communities.[51] The plan calls for rapid redevelopment of base properties with a view to the creation of new jobs. It also orders base cleanups to be placed on a "fast track," utilizing a local environmental coordinator at each base to keep the public informed and involved.[52] Information from all these sources, available for the asking or through Freedom of Information Act requests, may provide baseline data for an EIS or reveal conditions at individual bases that need cleanup or perhaps continuing management. The following case study shows another way such information might be used.

Case Study: Showdown at Saylor Creek

When the Air Force announced in 1987 that it would close George Air Force Base in California, it said it planned to move ninety-four F-4 fighter aircraft and associated personnel and equipment to Mountain Home Air Force Base in Idaho. It also proposed to add 1.5 million acres to the Saylor Creek bombing range near the Idaho base, to accommodate expanded training there for the F-4s.[53] In addition, it wanted to increase the air space devoted exclusively to pilot training. To do this it proposed lowering minimum altitudes and raising ceilings for training flights over public lands in three states and over a nearby Indian reservation.

In an effort to halt the proposed expansion of the bombing range, the Natural Resources Defense Council and other public interest groups carefully reviewed congressional committee testimony and other public documents, and discovered that the F-4 was scheduled for retirement within a short time. They were able to show that existing practice ranges nearby were adequate for projected needs, and that the end of the Cold War made the increased training unnecessary in any case. That information was presented in public testimony both locally and before Congress and in written comments on a draft EIS. It was also widely reported in the news media. The proposed expansion of the Saylor Creek Range is currently "on hold," pursuant to a DOD moratorium on all large land purchases.[54]

Paying for Environmental Restoration

The expense of cleaning up a closing military base, no less than cleaning up an operating one, can be enormous. The Pentagon has estimated the cost of cleanup just for bases closed under the 1988 Act at $900 million, and the total could run much higher.[55] In these times of fiscal stringency, it has not been easy to find the funds needed to leave base properties safe and environmentally sound. In the 1988 and 1990 base closing legislation Congress provided that cleanup for closing bases can only be funded from two special Defense Base Closure Accounts.[56] One purpose of this provision was to spare military leaders from having to decide whether to spend limited resources on defense missions or on environmental restoration. Another was to eliminate competition for funds in the Defense Environmental Restoration Account, described in chapter 4, which pays for cleanups at operating bases. The Defense Base Closure Accounts contain monies specifically authorized and appropriated to them by Congress, funds appropriated for other purposes but transferred to the accounts by the Defense Secretary after notice to Congress, and proceeds from the disposal of base properties. There is considerable political pressure to fund these special accounts adequately, since closing base properties cannot be given over to redevelopment until they are cleaned up or until restoration efforts are well underway.

Chapter 6

Environmental Protection During Wartime

When war begins, the environment, like truth, is usually one of the first casualties. Each belligerent uses bullets, rockets, bombs, and perhaps chemical weapons in an effort to alter the environment of its enemy. Its purpose is to kill, disable, and terrorize the enemy's troops and to deprive them of hospitable places to hide, rest, eat, move, or launch an attack. Military assets are also targeted.

Nations sometimes use more force than needed to achieve the political goals for which they go to war, or they direct their military might against civilian populations and other targets that have no significant involvement in the armed struggle. These excesses are sometimes inadvertent, sometimes deliberate. Occasionally, a nation's political goals include revenge or even genocide. What is true for nations as belligerents may also be said of parties to civil wars. The conflicts in Bosnia, Rwanda, and Chechnya provide horrifying recent examples.

Even when warfare is confined to the achievement of legitimate ends, however, the environmental costs can be staggering. With advances in technology, modern weaponry has become vastly more destructive and more widely distributed. The potential for lasting harm to the environment has increased accordingly.[1]

While the threat of a nuclear holocaust has begun to recede somewhat with the end of the Cold War, that threat is still very acute. United States and former Soviet arsenals continue to hold more than twenty-thousand nuclear weapons. The danger is increased by the proliferation of such weapons among developing nations. Nor can we ignore the possibility that a nuclear device will fall into the hands of terrorists or common criminals. We know from direct experience that just one nuclear weapon can produce an unprecedented environmental calamity. Detonation of a few thousand of them (only a fraction of the total remaining) could precipitate a "nuclear winter," dramatically changing the environment worldwide and killing billions of people.[2] Yet as we saw in chapter 2, we have no assurance that this terrifying prospect has affected the United States development of new nuclear weapons, or influenced strategic planning for their use.

A new generation of "conventional" weapons also poses unprecedented environmental threats.[3] Fuel/air explosives, first employed in Vietnam and used more recently in the Persian Gulf War, kill humans, animals, and plants over a wide area by igniting a cloud of flammable material, such as propane, creating a huge overpressure and using up all the oxygen.[4] Another "weapon of mass destruction" employed in the Gulf War is a fifteen-thousand-pound high-explosive bomb known as the Daisy Cutter, which simply demolishes everything within hundreds of yards. Also developed for use in Vietnam, the Daisy Cutter is so large that it can only be launched by rolling it out the back door of a C-130 cargo plane.[5]

Over the years, the amount of ordnance expended for each enemy casualty has steadily increased from one conflict to the next. The intensity of bombardment has also escalated. United States forces dropped about 22,000 tons of bombs per month in Korea, 34,000 tons in Southeast Asia, and 59,000 tons per month in the Persian Gulf War.[6]

Unexploded ordnance is a special environmental threat in the wake of every armed conflict.[7] A recent State Department report says that an estimated 85 to 90 million uncleared landmines worldwide may be "the most toxic and widespread pollution facing mankind."[8] In Cambodia, for example, after more than two decades of war, landmine accidents rank with malaria and tuberculosis as one of that country's greatest public health hazards.[9] Bombs, rockets, and grenades continue to kill villagers elsewhere in Southeast Asia, where it is estimated that about 10 percent of all high-explosive munitions expended by the United States during the Vietnam War failed to explode.[10] In Kuwait and southern Iraq, the desert is littered with thousands of unexploded "bomblets" from United States cluster bombs, and with radioactive "depleted uranium" artillery shells.[11]

There is also concern that accidental or even deliberate attacks on industrial targets, such as oil tankers and atomic reactors, could release dangerous materials into the environment, causing widespread, indiscriminate harm, perhaps far beyond the borders of warring nations. The destruction of high dams poses a similar risk. The peacetime environmental catastrophes at Bhopal, India and Chernobyl, in Ukraine, as well as Saddam Hussein's environmental depredations during the Persian Gulf War, illustrate the peril.[12]

It might seem terribly naive to suggest that in the midst of battle military leaders should have to worry about protecting the natural environment, or be distracted in any way from the immediate task of winning. But because, as we have noted, the environment itself is worth fighting to protect, environmental consequences must be considered in making tactical decisions. Fortunately, much that happens during a war is determined far in advance, from planning and training for combat, to the design of weapons. There is ordinarily plenty of time for reflection and debate about the wartime environmental implications of these preparations. One former infantry officer summed up the responsibility of military leaders this way:

> [C]ommanders must take strong positive steps to limit environmental damage. They must plan campaigns with the avoidance of damage in mind. For example, they should avoid, if at all possible, especially fragile areas. They should prohibit mass destruction of the land (such as the use of Agent Orange in Vietnam) as a method of warfare. They must make their subordinates aware of the environment, and they must issue orders prohibiting damage. They must continually assess the effects of their campaigns on the environment. Finally, they must insure that positive steps are taken to heal environmental damage in areas that they conquer and occupy.[13]

Both the advance planning for combat and the moment-to-moment conduct of a war are subject to a variety of legal constraints that may help to prevent environmental excesses. In this chapter we examine those laws, as well as political and strategic considerations in enforcing them, and assess the need for additional protections. We begin with a review of environmental impacts of the recent Persian Gulf War, then discuss international laws and practices that make up the law of war. Next we look at United States rules and regulations for combat, and at prospects for applying United States domestic environmental laws to war. Finally, we consider mechanisms for possible waiver of those laws in great national emergencies.

Environmental Spoils of the Persian Gulf War

The war in the Persian Gulf provides a sobering illustration of the environmental impacts of modern "conventional" warfare.[14] Whether Iraqi President Saddam Hussein was acting out of desperation or simple malevolence, he used the environment itself as a weapon. Saddam had threatened to set fire to Kuwaiti oil fields if attacked, and he did so after coalition air strikes began on January 16, 1991, apparently hoping that the resulting smoke would thwart air navigation. United States military leaders knew from a variety of sources, including a Pentagon-funded study, that if Saddam carried out his threat the result could be "a massive and unprecedented pollution event," affecting "the ecology of the Persian Gulf and [causing] fallout on a wide swath across Southern Iran, Pakistan, and Northern India."[15] How this knowledge affected coalition planning, if at all, has not been revealed.

In an apparent effort to prevent United States Marines from landing on Kuwaiti beaches (or perhaps simply as an unprecedented act of vandalism), Saddam also created one of the largest oil spills in history, killing wildlife and destroying fisheries in the Persian Gulf. In addition, Iraqi troops sowed half a million mines in the Kuwaiti desert, rendering large areas off limits to human use and dangerous to wildlife for decades to come.[16]

United States and coalition forces, by contrast, took great pains to avoid damage to nonmilitary targets in Iraq and Kuwait.[17] Even when they were placed in harm's way, such targets were spared, as when Iraqi fighter aircraft were parked near an ancient temple at Ur and when anti-aircraft artillery was mounted atop schools and hospitals.[18] United States pilots were instructed to pull up from assigned targets without releasing their bomb loads when they determined on their own that collateral (that is, unintended) damage would be unacceptable.[19]

Collateral damage was almost certainly limited, as well, by the unprecedented use of high-tech, laser-guided "smart" bombs and missiles, which reportedly hit their targets about 80 percent of the time.[20] The United States decided early on that it would not use nuclear weapons in the conflict.[21] For all the publicity given precision guided weapons, however, 93 percent of the ordnance delivered by coalition air forces during the Persian Gulf War consisted of unguided "dumb" bombs,[22] and many of those missed their marks. A United Nations inspection team found that 9,000 Iraqi homes were destroyed, leaving 72,000 civilians homeless.[23] Altogether, it is estimated that between 5,000 and 15,000 civilians died in coalition attacks.[24] Damage to the infrastructure of Iraqi society—drinking water systems, electrical power supplies, factories, and bridges—was also extensive. Given the coalition's limited political objective of evicting Iraqi forces from Kuwait, serious questions remain about the military necessity for deliberately destroying such civilian targets.[25]

Information about long-term damage to the natural environment from coalition operations remains sketchy. Both during and after the war, the United States government obstructed efforts to publicize environmental and other damages.[26] Coalition bombing of Iraqi oil facilities apparently contributed to the pall of black smoke rising from the region.[27] It still is not known for certain whether attacks on two Iraqi nuclear research reactors near Baghdad released any radioactive materials into the environment.[28] It is known that aircraft used in each of 110,000 coalition air sorties purged their fuel tanks with halon, a fire retardant gas that destroys stratospheric ozone,[29] and that the desert in Kuwait and Iraq remains strewn with tons of unexploded coalition ordnance. Heavy traffic in armored and other military vehicles on both sides also severely damaged vegetation and the surface of the desert itself. Behind battle lines, military forces made large withdrawals from groundwater sources that some say are near exhaustion, and they left behind tons of solid waste—everything from used motor oil to packaging for ready-to-eat meals.

Would adherence to existing international or domestic laws have helped to protect the natural environment during this conflict? Did the military forces on either side violate any such laws? If so, are there any effective sanctions for violations? Are new laws needed to discourage gratuitous injuries to the environment in the future?[30] The following sections suggest answers to these questions.

International Law Limitations on Wartime Destruction of the Environment

It might seem that the only important consideration in waging war is winning. Once the fighting begins, support for the troops always becomes the first article of political faith. We might expect that rules to protect the environment would simply be ignored. Some have even argued that instead of trying to limit wartime environmental impacts, we ought to make war as terrifyingly unpredictable and destructive as possible, to encourage the enemy to surrender sooner.[31]

However, limits on wartime destruction have been recognized at least since biblical times. For example, the Old Testament provides this instruction: "If you besiege a town for a long time, making war against it in order to take it, you must not destroy its trees by wielding an ax against them. . . . You may destroy only the trees that you know do not produce food."[32] Such limits were well understood by early scholars of international relations, such as Hugo Grotius, here paraphrased by a modern writer:

> First, do not destroy anything in areas you occupy and the enemy does not. Second, do not destroy anything when it appears that victory is likely and imminent. Third, do not destroy anything the enemy can obtain from somewhere else. Fourth, do not destroy anything the enemy cannot use to wage war. Finally, man-made objects—what he calls "sacred things" or "consecrated things"—are to be treated in accordance with the first four principles.[33]

If "trees could speak," Grotius noted, "they would cry out that since they are not the cause of war it is wrong for them to bear its penalties."[34] These ideas inform the modern law of war, a collection of customary practices and formal agreements among nations concerning limitations on the initiation and conduct of armed conflict.[35]

Several international agreements are deliberately aimed at protecting the natural environment from the effects of war.[36] The best known is the 1977

Environmental Modification Convention (ENMOD), which was prompted by United States use of chemical defoliants to destroy forests and croplands in Vietnam, and by its efforts to manipulate weather patterns over that country.[37] ENMOD provides in part:

> Each State Party to this Convention undertakes not to engage in military or any other hostile use of environmental techniques having widespread, long-lasting or severe effects as the means of destruction, damage, or injury to any other State Party. . . .
>
> . . . [T]he term "environmental modification techniques" refers to any technique for changing—through the deliberate manipulation of natural processes—the dynamics, composition or structure of the Earth, including its biota, lithosphere, hydrosphere and atmosphere, or of outer space.[38]

Despite the convention's broad language, its framers understood it to apply only to the creation of phenomena such as earthquakes, tsunami, cyclones, and changes in weather patterns, climate, and ocean currents—all tactics that are probably beyond the capabilities of modern military science.[39]

Another 1977 agreement, Protocol I to the 1949 Geneva Convention, calls for wide-ranging protections for noncombatants and their property, as well as for the natural environment. It prohibits using "methods or means of warfare which are *intended*, or may be *expected*, to cause widespread, long-term and severe damage to the natural environment." It also admonishes belligerents not to so damage the environment as to "prejudice the health or survival of the population," and it prohibits attacks "against the natural environment by way of reprisals."[40]

Protocol I places the burden on warring nations to analyze the environmental implications of their military operations and to avoid serious environmental harms, just as those nations are bound to avoid injuries to noncombatants. Yet one critic dismisses the protocol as a "vague, impractical, and unworkable . . . effort to prevent all collateral ecological damage." He complains that by prohibiting reprisals against noncombatant and environmental targets, it removes the "best deterrent to illegal conduct in war." Worse, in this critic's view, it might forbid the first use of nuclear weapons.[41]

Protocol I has not yet been formally ratified by the United States, although it may reflect a "developing customary international law" which is binding on all nations, regardless of whether they are signatories to any treaties.[42] Indeed, the Army's field manual, "The Law of Land Warfare," recognizes as customary international law some of the very principles set out in Protocol I.[43] For example, the manual defines "permissible objects of attack" as including only combatants and "objects which by their nature, location, purpose, or use make an effective contribution to military action and whose total or partial destruction, capture or neutralization, in the circumstances ruling at the time, offers a definite military advantage."[44] Identical language appears in Protocol I.[45]

Other international agreements are meant to protect specific geographical areas from the effects of particular weapons or from armed conflict generally. Nuclear weapons are prohibited altogether on the floor of the sea,[46] in outer space,[47] and in Latin America.[48] All weapons of war are banned from Antarctica.[49]

Still other law of war principles are concerned mainly with protection of noncombatants and their property, but their application may also help to protect the environment from needless destruction.[50] A 1907 convention declares that the "right of belligerents to adopt means of injuring the enemy is not unlimited."[51] An even earlier agreement concludes that the only permissible objective of a nation at war is to weaken the military forces of its enemy.[52] The selection of targets is strictly limited by military necessity. In other words, any destruction must be indispensable for securing the prompt submission of the enemy with the least possible expenditure of resources. However, military necessity is no defense for acts that violate the customary or conventional law of war, since that law incorporates the concept of military necessity.[53] In one of the Nuremberg war crimes trials, the court declared that the "rules of international law must be followed even if it results in the loss of a battle or even a war."[54]

Attacks on noncombatants and their property are almost always forbidden, whether deliberate or inadvertent. Injury to noncombatants is permitted only when enemy forces or other military targets are located nearby, and then only when the injury is proportionate to the military advantage gained by the attack. The United States interprets customary international law as providing that civilian property may not be attacked unless it "effectively contributes to the enemy's war-fighting or war sustaining capability."[55]

Just as attacks on noncombatants are usually forbidden, destruction of the environment that supports those noncombatants is also outlawed. Thus, civilian foodstuffs, agriculture areas, livestock, and drinking water supplies are protected from attack, unless they directly support enemy military operations. This principle is accepted by the United States as reflecting customary international law.[56] According to the Army's field manual, the poisoning of water wells and streams, as well as pillage and purposeless destruction, are punishable as war crimes.[57] Facilities dedicated to art, religion, science, or charity, must also be spared, along with hospitals and historic monuments, since their destruction would serve no military purpose.[58]

Installations such as dams, nuclear power plants, and loaded oil tankers may not be attacked, because they contain dangerous forces that could seriously injure the civilian population and the environment if released.[59] The prohibition is waived, however, when such a facility provides "regular, significant and direct support of military operations," and there is no other feasible way to terminate such support.[60]

Weapons that cause superfluous injury or unnecessary suffering, such as biological and chemical weapons, are prohibited. For example, the 1925 Geneva Gas Protocol prohibits the use in war, but not the possession, of asphyxiating, poisonous, and other gases, all analogous liquids, materials, and devices, and bacteriological methods of warfare.[61]

There is disagreement about whether the Geneva Gas Protocol covers irritants, such as tear gas, and herbicides like those used extensively by United States forces for defoliation in Vietnam.[62] In 1975, President Ford renounced the first use of riot control agents and chemical herbicides in war, "except in defensive modes to save lives."[63] Moreover, such materials may not be used except upon the express order of the President. The use of tear gas and nausea-inducing agents was authorized by President Bush during the Persian Gulf War.[64] But they were never actually directed against Iraqi forces, apparently out of concern that Saddam would retaliate with deadly chemical weapons.

The Biological Weapons Convention of 1972 prohibits the development, production, or stockpiling of biological or chemical toxin weapons.[65] While the United States insists that it holds no offensive biological weapons, it has long conducted experiments with such weapons for defensive purposes. In 1984, for example, the Army asked Congress to fund construction of new facilities for testing biological and toxin agents at its Dugway Proving Ground southwest of Salt Lake City. Fears that the expanded facility would be used for recombinant DNA research and the development of a new type of genetically engineered biological weapon prompted NEPA litigation to force an assessment of the environment risks.[66]

More than 150 nations have now signed the 1993 Chemical Weapons Convention.[67] It expressly prohibits the development, production, possession, or use of chemical weapons, including their retaliatory use, and requires the destruction of existing stockpiles and production facilities. It also forbids the use of riot control agents in warfare. The United States has signed, but not yet ratified, the convention.

Even weapons that are not illegal per se must be used in a way that does not cause unnecessary harm. For example, the 1981 Inhumane Weapons Convention imposes restrictions on the use of certain conventional weapons, such as mines, booby traps, and incendiary devices, that could cause excessive or indiscriminate injuries. That convention also forbids incendiary attacks on forests and other plant cover, unless such vegetation is used to conceal military targets.[68] The United States has signed, but not ratified, the Inhumane Weapons Convention, and does not consider itself necessarily bound by the terms of the convention.

There is a long-running debate about the legality of nuclear weapons.[69] On one side, it is argued that the use of such weapons would inevitably cause many unnecessary injuries, in violation of the law of war principles of discrimination, proportionality, and humanity. Some say that such use would cause wide-spread, long-term, and severe damage to the environment as well, which is forbidden by Protocol I. The United States and other nuclear powers insist that since there is no explicit prohibition of nuclear weapons, their use is permitted. It is Army policy that, absent contrary instructions from national command authority, field commanders who utilize nuclear weapons must restrict noncombatant casualties to 5 percent at the margins.[70] As with other weapons, however, even that amount of damage must not be disproportionate to the military advantage gained from their use. In May 1993, the governing body of the World Health Organization asked for an advisory opinion from the International Court of Justice (World Court) about whether the use of nuclear weapons would violate international law.[71] The matter is still pending before the court at this writing.

The law of war thus furnishes an incomplete and unpredictable bulwark against excessive environmental damages in armed conflict. Some of the law's most important principles have yet to be codified in formal agreements among nations, and there is disagreement about the meaning of some provisions. Many nations are not yet parties to all of the formal agreements. Only about one-third of all countries have ratified the Environmental Modification Treaty, while fewer than 60 percent have approved Protocol I. Some nations have qualified their ratification with weakening conditions. For example, before signing the Chemical Weapons Convention in 1993, the United States had not forsworn the use of chemical weapons in war, as required by the 1925 Geneva Gas Protocol, but only promised that it would not use them first.[72] The agreements that do exist fail to address a number of critical issues, such as the legality of nuclear weapons. And, except for some outright prohibitions, they contain few clear standards to guide military planners or commanders in the field. One commentator sums up the existing authorities this way:

> The legal norms that exist are scattered, and are either very general and vague, as well as subject to "military necessity" exceptions, or more specific and relevant, but not directed at prevailing belligerent practices of the sort most likely to generate environmental harm. Furthermore, the status and relevance of principles of customary international law are quite indefinite, as is the related matter of whether treaty norms reflect and embody customary norms.[73]

These difficulties have led to calls for a "Fifth Geneva" Convention on protection of the environment in wartime.[74]

Even where the law is clear, it is not always followed. Military commanders understandably tend to accumulate and use all the firepower they can, both to accomplish their assigned missions and to protect their troops from harm. Sometimes they use more force than necessary, or miss their targets because of poor planning, flawed intelligence, enemy deception, or equipment failure.

Criminal sanctions for law of war violations (war crimes) may be imposed by special courts convened by the victorious nations to punish serious infractions. The Nuremberg Tribunal at the end of World War II is an example.[75] War crimes trials may also be conducted by representatives of other nations. The International Tribunal for the Former Yugoslavia and the newly created Tribunal for Rwanda are current examples.[76] In addition, war crimes may be tried in United States military courts.[77] There have been numerous calls for war crimes trials of Saddam Hussein and members of his military for their gratuitous pollution of air and waters, for atrocities committed in Kuwait, and for missile attacks on Israeli cities.[78] Thus far, however, such calls have gone unheeded.

In theory, at least, compensation may be exacted from the guilty party. In 1991, the United Nations Security Council adopted a resolution declaring that Iraq was "liable under international law for any direct loss, damage, including environmental damage and the depletion of natural resources . . . as a result of Iraq's unlawful invasion and occupation of Kuwait."[79] The International Court of Justice, or World Court, is authorized to hear such claims for damages, but only when the nations involved submit themselves to the jurisdiction of the court.[80]

Rather than relying on threats of punishment for violations, the law of war depends mostly for its influence on the self-interest of nations. Each nation can be expected to refrain from using weapons or tactics that it would consider unjust if used against it, hoping that other nations will follow suit. To encourage uniform respect for the law, however, each nation must discipline members of its own forces who fail to comply.

United States Military Rules for Armed Conflict

Compliance with the law of war is a central tenet in United States military doctrine. The Department of Defense has directed United States armed forces to "comply with the law of war in the conduct of military operations and related activities in armed conflict, however such conflicts are characterized."[81]

While the United States has not given unqualified approval to every international agreement that might protect the environment in wartime, it has pledged to suppress war crimes generally, and it has adopted rules to ensure that persons within its jurisdiction comply with the law of war. Military or political figures of any nation who violate the law may be punished under the Uniform Code of Military Justice.[82]

Every United States soldier, sailor, airman, and marine receives extensive training in the elements of the law of war. In addition, each Defense Department service branch publishes a manual that reproduces, paraphrases, and explains various international agreements and common law principles by which the United States regards itself as bound.[83] These manuals provide important guidance for military planning, training, and actual combat operations.

As we have seen, the law of war applies not only to the actions of military personnel, but also to the weapons they employ in combat. The Defense Department reviews the legality of each new weapon system to ensure that it complies.[84] Environmental impact statements or environmental assessments are also prepared for some weapons in accordance with NEPA. However, as noted in chapter 2, such studies generally do not assess the environmental consequences of using the weapons in wartime.

Each year the National Security Council publishes the President's "National Security Strategy of the United States," containing an assessment of threats to the nation and a broad prescription for addressing those threats. The National Security Strategy is implemented by various orders and directives, including "operation plans" and "operation orders" for the organization, deployment, and management of military forces during hostilities. These operation plans and orders give rise in turn to peacetime and wartime "rules of engagement" describing circumstances and limitations under which United States forces may initiate or continue in combat with enemy forces. Rules of engagement might, for example, include prescriptions for the use of particular weapons and tactics upon specified targets. During actual hostilities, rules of engagement may be promulgated at several levels of command and amended as developments require, but all must conform to rules of engagement issued by the highest civilian authority—the President and Secretary of Defense—to help ensure that they are consistent with national policy.[85]

According to a 1983 directive from the Joint Chiefs of Staff, all operation plans and orders and rules of engagement must be reviewed by military legal advisors to ensure that they comply with domestic and international law, including the law of war.[86] During the Persian Gulf War, military lawyers even sat on committees to approve targets for each day's bombing raids.[87]

The formal law of war review procedure should be expanded to include a systematic environmental analysis of proposed plans and alternatives. Such plans are typically drawn up well in advance of the need to execute them, leaving ample time to assess local conditions and avoid unnecessary environmental destruction. For example, the general operation plan for Operation Desert Shield was based on a draft prepared a month before Iraq invaded Kuwait.[88]

Application of Domestic Environmental Laws to Armed Conflict

The United States environmental laws provide a possible third source of limits on the use of force in armed conflicts, complementing the law of war and military rules and regulations. While there is growing sensitivity in the United States military to the environmental impacts of warfare, there has been no systematic effort to apply these laws in combat situations. The Pentagon has directed its forces to comply with "domestic" law in drawing up operation plans and rules of engagement, but its focus has been on domestic laws relating to logistics, arms exports, and authority for the use of force.

There is serious, continuing debate about whether the domestic environmental laws apply abroad.[89] The Defense Department expressly disavows the applicability of NEPA to military actions outside the nation's borders, especially to armed conflict.[90] The Pentagon's "Overseas Environmental Baseline Guidance Document" claims only to have "considered" United States environmental laws and regulations, not to be governed by them, and it does not apply to "deployments for operations," that is, to warfare.[91] The environmental laws themselves are silent on the question, and the legislative histories are almost as enigmatic. Aside from the usual presumption that domestic laws are not meant to apply abroad unless Congress expressly states otherwise, there is no compelling evidence that Congress intended to exclude their application to armed conflict. Indeed, these laws should be applied to warfare. As a practical matter, they provide convenient, familiar mechanisms for evaluating and minimizing risks to the environment in time of war just as they do in peacetime. Applied with a practical flexibility, they need not interfere with military operations.

No one has suggested that the Defense Department ought to have prepared the kind of formal environmental impact statement required by NEPA before deploying troops and equipment in the Persian Gulf, even though Operations Desert Shield and Desert Storm were undeniably "major federal actions affecting the human environment." The political objectives of freeing Kuwait and protecting Saudi Arabia from further Iraqi advances might well have been frustrated by delays inherent in the usual public notice and interagency review process. Very much to its credit, the Pentagon did not ignore the environmental risks altogether. But it failed to undertake, even internally, the kind of systematic, coordinated environmental evaluation that NEPA requires.

We cannot expect the environmental laws to apply the same way on the battlefield that they do in planning a highway or operating a sewage disposal plant. A field commander whose forces come under attack cannot stop to prepare an environmental assessment or apply for a Clean Water Act permit before mounting a counteroffensive. Because of the need for speed and secrecy, members of the public cannot expect to receive advance notice or have an opportunity to comment on proposed tactics. Citizen enforcement will be nearly impossible; we will have to rely on the military branches to police their own operations and personnel, aided by oversight from their inspectors general. It may not even be practical for our field commander to fully document his consideration of environmental effects, making accountability more problematic. Yet even in a combat setting, our commander can apply performance standards and follow procedures set out in the domestic environmental laws as closely as circumstances permit.

Much that takes place on the battlefield is planned far in advance. Operation plans, rules of engagement, and standardized tactics should be routinely vetted for compliance with domestic environmental law standards, just as they are now reviewed for conformity with the law of war, even though for security reasons neither the planning process nor the plans themselves can be made public. The designs of weapons and other equipment, and protocols for their use on the battlefield, should also conform to requirements of the environmental laws. Just as the law of war proscribes weapons that cause unnecessary suffering, application of the environmental laws ought to prevent the deployment of weapons that cause unnecessary injury to the environment. Thus, the Navy should only deploy ships that have the capacity to treat or store their solid wastes while at sea, instead of dumping them overboard in violation of the Ocean Dumping Act or the Clean Water Act.[92] The Army has decided that if chemical herbicides are used in combat, they "must be employed in accordance with federal laws and regulations which would govern their use within the United States. . . . Environmental Protection Agency regulations pertaining to dilution, droplet size, protective clothing, etc. are binding on U.S. forces."[93] The EPA regulations are promulgated under the Federal Insecticide, Fungicide, and Rodenticide Act (FIFRA).[94]

The military services are beginning to incorporate environmental compliance into combat training, just as they now train every soldier, sailor, and airman to be familiar with the law of war. As one high-ranking Air Force officer put it, "we fight the way we are trained." Long before reaching the battlefield, for example, a tank commander needs to learn not to drive through the middle of a wetland if a path across high ground offers the same tactical advantage. The same commander should be instructed to carry along not only a change of oil for his tank's engine, but also a safe receptacle for the old oil, so it will not have to be drained onto the ground, as was done in the Persian Gulf War.

Finally, environmental compliance on the battlefield itself will not necessarily make combat units less effective in carrying out their military missions. A recent Army-financed study concluded that successful introduction of pollution prevention initiatives into combat doctrine and planning would actually enhance fighting strength by increasing each unit's self-sufficiency, reducing disease and nonbattle injury, and reducing the unit's visibility to the enemy.[95]

Environmental Law Waivers in Wartime

Crises may arise in which strict conformity with environmental performance standards or procedural requirements would truly jeopardize the nation's security. Congress has inserted provisions into each of the major environmental laws (except NEPA) permitting its waiver in exigent circumstances. These waiver provisions enable military leaders to avoid or, even better, adjust their compliance in order to protect the environment to the extent possible, while at the same time carrying out their defense missions. While we hope that the need for such a waiver will never arise, we want to be assured that if it does, environmental values will not simply be ignored, and that military leaders will be accountable for the environmental consequences of their decisions.

None of the waiver provisions of the regulatory environmental statutes has yet been invoked in wartime. But the planning provisions of NEPA were formally waived during the Persian Gulf War. A few weeks after Saddam Hussein's Iraqi army invaded Kuwait in August, 1990, the Defense Department wrote to the Council on Environmental Quality (CEQ)[96] asking for agreement that an "emergency" existed within the meaning of the following CEQ regulation:

> Where emergency circumstances make it necessary to take an action with significant environmental impact without observing the provisions of these regulations, the Federal agency taking the action should consult with the Council about alternative arrangements. Agencies and the Council will limit such arrangements to actions necessary to control the immediate impacts of the emergency. Other actions remain subject to NEPA review.[97]

In light of the urgent deployment of United States forces in the Gulf, the CEQ agreed.[98]

The agreement was invoked on two occasions to avoid the preparation of an EIS. The first time, the Department of Energy wanted to test the ability of fuel-air explosives to clear mine fields. The experiment, using five hundred gallons of propane, was conducted on a dry lake bed at the Tonopah Test Range in Nevada after it was determined that the test site was not inhabited by either the desert tortoise or the kit fox, two protected species in the area, and after review of an earlier environmental assessment for detonation of a much larger volume of liquefied natural gas at a similar facility nearby.[99] The second Gulf War waiver of NEPA is described in the following case study.

Case Study: Sleeping Through the Persian Gulf War

In 1987, residents of the area surrounding Westover Air Force Base in Massachusetts filed suit to challenge an Air Force EIS for relocation of sixteen C-5A cargo aircraft to the base. They complained that noise from takeoffs and landings of the huge planes would severely interfere with schools, churches, and other aspects of normal life on the ground, especially sleep. The court approved an EIS that contemplated training operations limited to twenty takeoffs and landings per week between 7:00 A.M. and 10:00 P.M.[100]

When Saddam Hussein's Iraqi army invaded Kuwait in the summer of 1990, hundreds of thousands of U.S. troops and their equipment had to be sent quickly to the Persian Gulf. Westover was chosen as a major shipment point, using C-5A aircraft. Without preparing a new EIS, the Air Force increased flights at the base to between ten and nineteen per day around the clock. Cargo planes departing for the Middle East were also noisier than training flights because they were fully loaded and climbed more slowly, sometimes just clearing rooftops. Eager to support our troops in the Gulf, local citizens were slow to complain. But after weeks of sleepless nights, they began to attribute a variety of health problems to the incessant noise, including high blood pressure, ulcers, and hearing loss. Their protests to the Air Force were turned back with a claim of military necessity and invocation of the earlier NEPA waiver agreement between CEQ and DOD. "A supplemental EIS is not required in this situation," came the reply, "and base personnel must keep their attention focused on accomplishing their operational missions. That by itself is keeping them busy enough, and if they need not prepare an EIS, I will not ask it of them."[101]

When the Persian Gulf war came to an end the following February, the Air Force informed the neighbors that round-the-clock flights would continue for at least six more months as troops and equipment were brought home, and that a continuing "emergency" prevented the completion of an environmental evaluation.

After the neighbors gave notice of intent to sue both CEQ and the Air Force, the CEQ's General Counsel pointed out to Air Force officials that the "regulation for emergency circumstances is not a waiver of all compliance with NEPA; rather, it provides for alternative arrangements for procedural compliance with NEPA in emergency circumstances."[102] She asked the Air Force to prepare immediately an environmental assessment of increased flights at Westover, including noise impacts, alternate east coast landing sites, alternate flight patterns, and other mitigation measures. Meanwhile, the flights continued day and night, unabated.

Suit was then filed to halt the nighttime flights until a supplemental EIS could be prepared. The court accepted the Air Force's argument that an emergency still existed and dismissed the suit, finding that strict compliance with NEPA was waived under the CEQ regulation and was unnecessary under the terms of the statute itself.[103]

Despite the fact that CEQ regulations are presumed to be authoritative interpretations of NEPA's procedural requirements,[104] there is serious disagreement about whether those requirements can be waived, even in an emergency.[105] Prior to the Persian Gulf War, the CEQ emergency regulation had been invoked only nineteen times since its promulgation in 1978, none of them having anything to do with national security.[106]

Most courts have interpreted NEPA's mandate to assess the environmental consequences of agency actions "to the fullest extent possible" as permitting exceptions only when preparation of an EIS would create an "irreconcilable and fundamental conflict" with agency duties under another statute.[107] Such a conflict might arise, for example, when other statutory duties expressly or implicitly leave too little time for formal, public NEPA review. In the litigation described in the case study above, the Air Force claimed that compliance was impliedly excused by the January 12, 1991, Joint Resolution of Congress authorizing the use of United States armed forces in the Persian Gulf,[108] although that measure postdated by months the decision to waive preparation of a supplemental EIS. The Air Force also cited a series of executive orders cutting off trade with Iraq, deploying defensive forces, and ordering reserve units to active duty. Each order, it said, was based on statutory authority that precluded compliance with NEPA.[109]

It is worth noting that two other CEQ regulations might have been used to put the analysis and documentation required by NEPA on a fast track. One provides that an agency can set time limits for review reflecting the "degree of public need for the proposed action, including the consequences of delay." Those limits must be "consistent with the purposes of NEPA and other essential considerations of national policy," including, presumably, national security. The other enables EPA to shorten the time period for review "upon a showing by the lead agency of compelling reasons of national policy."[110] Each of these regulations assumes that some significant environmental analysis will be made. However, neither was invoked during the Persian Gulf War.

It might be supposed that the President has inherent constitutional authority as Commander in Chief to ignore the requirements of NEPA or the other environmental laws when the national security is threatened. But the President almost certainly lacks any such authority, unless she is forced to act quickly to repel an enemy attack. While there is general agreement that Congress may not properly involve itself in day-to-day and minute-by-minute tactical decisions on the battlefield, the President is bound to "faithfully execute" the laws

Congress makes, including the environmental laws, even when the nation is at war.[111]

Obviously, the nation's very existence should not be imperiled by a wooden adherence to formal rules and procedures for environmental protection. Yet any variance should be carefully tailored to the emergency, promptly publicized (unless it qualifies for withholding under the Freedom of Information Act), and documented to ensure agency accountability and enable judicial review.

Chapter 7

Environmental Protection in the Courts

Efforts to apply the environmental laws to national defense activities have often resulted in disagreement. Sometimes the dispute is about what Congress meant when it wrote particular language into a statute. Since the President and his executive agencies are bound by Article II of the Constitution to "take Care that the Laws be faithfully executed," the search for legislative intent is critically important. At other times the question is whether an agency has fairly complied with its congressional mandate. Occasionally, agency officials insist that they are not bound by a statute, because its enforcement would interfere with the President's independent prerogatives as Commander in Chief of the armed forces.

When these controversies cannot be resolved by discussion or negotiation, or in administrative proceedings before EPA or state regulatory agencies,[1] it falls to the judicial system to settle them. The courts play a key role in saying what the law requires and in fashioning remedies for its enforcement. That role is the subject of this chapter.

In the first part of the chapter, we see that judicial enforcement of the environmental laws may be extremely controversial when the national security is implicated. Sometimes courts are accused of meddling in military matters that are beyond their practical competence or constitutional authority. At other times, they are perceived as unduly deferential to the judgment of military officials. Occasionally a court is accused of usurping Congress' authority to establish national policy, as when the court refuses to exercise its equitable powers to halt an agency violation, citing the importance of the agency's actions to the national defense.

Later in the chapter, we review efforts by individual citizens, public interest organizations, and state governments to ensure that national defense activities conform to the environmental laws. Because the Justice Department has long maintained that the Environmental Protection Agency cannot properly sue another federal agency for environmental violations, nonfederal plaintiffs may provide the only means for judicial enforcement. Before one of these plaintiffs can maintain a suit in court, however, the court must be satisfied that the plaintiff has a judicially cognizable stake in the outcome and that the lawsuit is not essentially political, making it nonjusticiable. Once in court, the plaintiff may find itself stymied by a lack of access to critical classified evidence. States as plaintiffs may be hampered in their efforts to impose civil penalties for agency violations by the federal government's assertion of sovereign immunity. At the end of the chapter, we consider prospects for holding government officials and defense contractors personally accountable for environmental law violations through the imposition of criminal sanctions.

Judicial Deference in National Security Disputes

Courts have always displayed a sensitivity to both constitutional and practical limitations on their involvement in national security controversies. In one early NEPA case involving the Navy, the court declared that "substantive decisions relating to the national defense and national security lie within that narrow band of matters wholly committed to official discretion both because of the delicate security issues they raise and the constitutional delegation of those concerns to the political departments of our government."[2] The judicial constraint described here is both constitutional and prudential.

Of course, judges ordinarily lack the training and experience in national security matters that military and intelligence professionals bring to their work. Nor are courts equipped to gather critical information needed to establish policy or make day-to-day decisions about defense issues. Still, as the Supreme Court pointed out in an internal security case that the Justice Department called "too subtle and complex for judicial evaluation," the courts "regularly deal with the most difficult issues of our society. There is no reason to believe that federal judges will be insensitive to or uncomprehending of the issues involved."[3] The same can be said for environmental cases implicating the national security.

Nevertheless, courts have often deferred broadly to defense agencies in environmental matters. Thus, in litigation described in the next case study, involving a Navy facility for communicating with submerged submarines, the court accepted without further inquiry the Navy's argument that the facility was "essential to the national defense and that any delay in its construction is contrary to national defense interests. We have no basis," the court said, "to ignore those executive representations."[4] Another case, decided soon after Iraq invaded Kuwait, challenged the relocation of an Army National Guard helicopter unit to land adjoining a state park. Refusing to issue a preliminary injunction for an alleged NEPA violation, the court said it took "judicial notice that there is a public interest in [allowing the operation to continue] in light of the crisis in the Mid East. . . . [T]he world is living in a very threatening situation."[5] It has been suggested that "the judicial deference paid to national security arguments has consistently been conditioned on the DOD's good faith efforts to comply with the relevant statutes,"[6] though many in the environmental community would disagree.

As a matter of practical necessity, Congress must describe the task of defending the country in fairly broad terms. It cannot anticipate every threat to the nation's security or prescribe every aspect of each agency's work. An agency fills in some of the missing details when it adopts administrative regulations to implement a statutory program. Other details are supplied in the agency's daily interpretation of its duties. The Supreme Court recognized the practical value of this arrangement when it declared that "considerable weight should be accorded to an executive department's construction of a statutory scheme it is entrusted to administer." When an agency's actions are called into question, the court's role is limited to deciding whether an agency has adopted a "permissible" construction of the statute.[7] On the other hand, the Court has declared itself ready to "reject administrative constructions of a statute . . . that are inconsistent with the statutory language or that frustrate the policy that Congress sought to implement."[8] The problem for the court comes in determining whether a particular construction is "permissible."

What of cases in which the executive and legislative branches come into clear conflict? It is one thing to say that the applicable laws must be complied with. It is quite another to order the President to carry out some congressional mandate when he claims to be exercising his independent, inherent powers as Commander in Chief. To be sure, the conflict is rarely couched in such stark terms. The President is more likely to argue that national security considerations are of transcendent importance. A court may then exercise its equitable discretion to allow a violation of the law to continue, as we see in the next section.

Equitable Remedies for Statutory Violations

We begin with the premise that in our government of separated functions the legislature is entrusted with the job of making laws. The executive must implement those laws. The courts are responsible for saying what the law is and for providing remedies for violations. Yet we know that in practice each of these functions is necessarily shared to some degree by all three branches. Thus, executive agencies make some of the law they implement when they interpret enabling legislation and promulgate administrative regulations. Courts make law when they decide cases for which the legislature has provided no clear guidance.

What about the case where Congress has articulated its policy choice in apparently clear terms? Is a court nevertheless free to adopt its own view of what the nation's best interests require when it fashions a remedy for a violation?[9] Consider the court's answer in the following case study.

Case Study: Run Silent, Run Deep in Wisconsin

In 1977, the Navy prepared an environmental impact statement (EIS) for its upgraded and expanded extremely low frequency (ELF) submarine communication facility. ELF uses radio antennas spread over a large area of northern Wisconsin and the Upper Peninsula of Michigan to transmit instructions to submerged submarines around the world. When new information subsequently came to light about possible adverse biological effects of such low frequency radiation, the State of Wisconsin brought suit to compel the Navy to prepare a supplemental EIS before the facility was completed. The trial court found that a supplemental EIS was indeed required. It also found that the Navy had failed to overcome a presumption that violations of NEPA should be enjoined, and that no remedy short of an injunction would serve the purposes of the act.[10]

The court of appeals decided that a supplemental EIS was not needed.[11] It found that the new information failed to raise concerns of such gravity that another formal, in-depth look at the environmental consequences was required. In dictum, the court went on to declare that even if there had been a NEPA violation,

> NEPA cannot be construed to elevate automatically its procedural requirements above all other national considerations. Although there is no national defense exception to NEPA, and the Navy does not claim one, the national well-being and security as determined by the Congress and the President demand consideration before an injunction should issue for a NEPA violation.[12]

The court insisted that its job was to "tailor its relief to fit each particular case, balancing the environmental concerns of NEPA against the larger interests of society that might be adversely affected by an overly broad injunction."[13]

Other courts have concluded as well that they could avoid enjoining environmental law violations in national security cases. For example, in litigation involving the Department of Energy's Oak Ridge nuclear weapons facility, noted in a case study in chapter 3, the court found that unpermitted discharges of heavy metals and other hazardous substances violated the Resource Conservation and Recovery Act (RCRA) and the Clean Water Act. But it refused to enjoin the discharges while DOE applied for permits, in part, the court said, because the facility "is a unique and essential element of this nation's system of nuclear defense."[14] In another case, noted earlier in chapter 3, the Supreme Court found that while the Navy's discharge of ordnance into the water at a practice bombing range in Puerto Rico violated the Clean Water Act, "the injunctive relief sought would cause grievous, and perhaps irreparable harm, not only to Defendant Navy, but to the general welfare of this nation."[15]

Some courts have decided to enjoin statutory violations after balancing environmental and national security values. One such case involved the Army's construction of new aerosol and toxin weapons test facilities at its Dugway Proving Ground in Utah. The court declared that in fashioning a remedy for a NEPA violation it was "particularly sensitive that national defense concerns be carefully weighed in the calculus. . . . [A]n undue delay in the construction of the proposed facility may well affect DOD's ability to meet legitimate defense and national security needs, and could, indeed, be devastating." But after balancing environmental and defense concerns, the court ordered the construction stopped until an adequate environmental assessment was prepared.[16]

In at least some of the environmental statutes Congress appears to have already balanced all the relevant interests, leaving the courts with no equitable discretion to allow violations to continue unenjoined.[17] Thus, in one recent case the court ordered the Navy not to conduct dredging for a new homeport facility for the aircraft carrier *Nimitz* in Everett, Washington without a Clean Water Act permit. The act requires an injunction, the court said, for a violation that presents a real danger of pollution.[18] By contrast, the Supreme Court refused to halt the Navy's discharges of ordnance into the water at its Vieques Island bombing range, saying the purpose of the Clean Water Act could be served without granting an injunction.[19]

Other courts have granted injunctive relief for defense facility violations without addressing broader national implications. For example, discharges in violation of Clean Water Act permit limits at McGuire Air Force Base in New Jersey were halted, just as if they had come from an industrial discharger.[20]

An injunction may be denied because a court concludes that halting a statutory violation would pose a greater threat to the environment than allowing it to continue. One such case involved the failure of the Energy Research and Development Administration (DOE's immediate predecessor) to prepare an EIS for the construction of tanks to hold high-level radioactive wastes at the Hanford, Washington reservation. The tanks were intended to replace older ones that were leaking. The court ordered the preparation of an EIS but refused to enjoin the construction, because it felt that the hazard posed by further leakage probably exceeded any injury from the temporary violation of NEPA.[21]

Enforcement by EPA: Congressional Authority and the Unitary Executive

The major regulatory statutes provide that federal agencies must comply with "any process or sanction, whether enforced in Federal, State, or local courts or in any other manner."[22] Despite this language, however, the Justice Department took the position early in the Reagan administration that what it calls the "unitary executive" prevents EPA from issuing administrative compliance orders or filing suit against other federal agencies for violations.[23] The President cannot and should not sue himself, the argument goes, but may, as Chief Executive, establish mechanisms within the executive branch to solve interagency environmental enforcement disputes. The unitary executive theory has prudential and constitutional dimensions. The Justice Department argues that it would be a waste of time, judicial resources, and taxpayer money for one executive branch agency to sue another, especially since the President is already bound by Article II of the Constitution to "faithfully execute" the law. Furthermore, according to the Justice Department, such a conflict within a single branch of government would not present a "case" or "controversy" appropriate for judicial resolution under Article III.[24]

In practice, EPA enforcement efforts against federal agency violators have been confined to persuasion and negotiation, and implemented through compliance agreements and consent orders. A 1978 executive order directs each agency to comply with the environmental protection laws and to cooperate with EPA, as well as state and local officials, in their enforcement. It also directs the Administrator of the EPA to resolve conflicts between agencies whenever possible, otherwise to refer matters for resolution to the Director of the Office of Management and Budget.[25] But while the executive order provides that "[t]hese conflict resolution procedures are in addition to, not in lieu of, other procedures, including sanctions, for the enforcement of applicable pollution control standards," it has been the Justice Department's position that the President has a right to resolve such problems internally without court interference. EPA has developed an elaborate procedure for internal resolution of such disputes.[26]

The 1992 Federal Facility Compliance Act now explicitly authorizes "administrative enforcement actions" by EPA against other federal agencies for RCRA violations.[27] This means that EPA may issue administrative orders against other agencies. Since 1983, such orders have not been imposed under any of the environmental statutes because of concern that they would invite judicial intrusion into what amounted to an interagency dispute. That could happen if a citizen or state brought suit under a citizen suit provision authorizing judicial enforcement of such an "order." The 1992 act also adds the United States to the list of "persons" against whom EPA may institute judicial proceedings for RCRA violations.

Can EPA levy a penalty against another agency? Constitutional and congressional bans on expenditures of unappropriated funds,[28] as well as limits on executive branch impoundments of appropriated funds,[29] present questions about the propriety of such a levy. EPA policy is not to do so "under most environmental statutes."[30] This limitation does not apply to violations of interagency cleanup agreements under the Comprehensive Environmental Response, Compensation, and Liability Act (CERCLA),[31] although EPA does not explain why. Thus, when the Department of Energy repeatedly missed deadlines in an agreement for cleanup at its Fernald Feed Materials plant in Ohio, it agreed to pay $100,000 in penalties to EPA, plus $150,000 for environmental restoration, all contingent on the authorization and appropriation of funds for these purposes.[32] Although agency managers are not personally responsible for the penalties, their distaste for explaining mistakes to Congress may provide a powerful incentive for compliance.

Congress has yet to address the unitary executive question for environmental statutes other than RCRA. In other settings, however, one federal agency has often been represented by the Justice Department in litigation against another.[33] EPA has sought to avoid further speculation about this point by proceeding directly against defense contractors that operate agency facilities whenever possible.[34] Like other nonfederal entities, these contractors must obey administrative compliance orders and respond in court to suits initiated by EPA or by states with EPA-approved programs.[35] In some instances, however, fines imposed on contractors end up being paid by the agencies that hired them.[36]

Enforcement by the States: Waiver of Sovereign Immunity

As long as the unitary executive policy is followed, direct enforcement of the regulatory statutes other than RCRA will be left to the states or to citizens proceeding under the "citizen suit" provisions of those statutes. EPA has a clear policy not to interfere with state enforcement actions against federal facilities.[37] However, while states have been able to obtain declaratory and injunctive relief for federal agency violations, they have had only limited success in collecting fines for violations or imposing administrative orders against federal agencies. The difficulty lies in the principle of "sovereign immunity."[38]

It is well settled that unless Congress unequivocally waives the federal government's sovereign immunity, the government will not be exposed to lawsuits or obligated to pay penalties levied by states for statutory violations. The Supreme Court addressed this issue in a 1992 case construing the language of the Clean Water Act and RCRA. That case involved efforts by the State of Ohio to impose civil penalties against the Department of Energy for years of violations at its Fernald Feed Materials plant. The Court decided that neither statute showed Congress' clear intent to allow the imposition of punitive state penalties.[39]

Later that same year Congress, provoked in part by the Supreme Court's decision in the Fernald case, passed the Federal Facility Compliance Act. The act expressly waives sovereign immunity for state penalties, but only for RCRA violations.[40] The Act also expressly authorizes states to impose administrative orders or seek judicial relief against federal agencies to enforce compliance with federal, state, and local solid or hazardous waste programs.

Whether and to what extent state penalties may be levied against federal facilities under the other statutes has yet to be addressed by the nation's highest court. Several lower courts have upheld such penalties for Clean Air Act violations.[41] However, one court has relied on the Fernald case to hold that a state could not impose a civil penalty for a CERCLA violation by the Navy.[42] Further litigation of this question seems likely.

Another court has held that the Navy was not responsible for cleanup costs or natural resource damages resulting from a spill of hazardous wastes it sent to a commercial disposal facility. The court decided that RCRA's provision making federal agencies subject to state "requirements" includes only "regulatory" requirements, that is, "specific, precise standards capable of uniform application," and not such open-ended claims.[43]

The principle of sovereign immunity offers no protection for defense contractors that break the law. One court has decided that a private corporation operating an Army ammunition plant was not shielded from state penalties for violating state water pollution control laws.[44]

Enforcement by Citizens

According to the Department of Defense,

> Our installations must comply with a mix of Federal, state and local laws. Because we are public servants, we must be forthright with communities and regulatory authorities in addressing our compliance objectives. Everywhere open and constructive dialogue has been established, [and] problem solving has been undertaken in a cooperative rather than an adversarial atmosphere.[45]

Despite such commitments by the Pentagon and DOE, however, implementation of national policy has always depended to some degree on judicial enforcement at the behest of interested citizens, public interest organizations, and state and local governments. So-called "citizen suits" may be instituted to resolve disagreements about agency responsibilities or to remedy deliberate misbehavior by federal officials.

Many of the regulatory environmental laws contain express provisions authorizing suits by citizens to enforce their terms. For example, the Clean Water Act states that, with some qualifications,

> any citizen may commence a civil action on his own behalf—
>
> 1. against any person (including (i) the United States, and (ii) any other governmental instrumentality or agency to the extent permitted by the eleventh amendment to the Constitution) who is alleged to be in violation of (A) an effluent standard or limitation under this Chapter, or (B) an order issued by the [EPA] Administrator or a State with respect to such a standard or limitation, or
> 2. against the Administrator where there is an alleged failure of the Administrator to perform any act or duty under this Chapter which is not discretionary with the Administrator.[46]

Suit may not be instituted if EPA or a state is already "diligently prosecuting" an enforcement action in the courts, or if prior notice has not been given to EPA, the state, and the alleged violator.[47]

In the absence of an explicit citizen suit provision, the Administrative Procedure Act (APA) authorizes judicial challenges to final agency actions not "committed to agency discretion by law," provided the plaintiff has exhausted its administrative remedies before resorting to litigation.[48] For example, the APA furnishes the jurisdictional grounds for suits to enforce compliance with NEPA, which has no citizen suit provision.

Having a Proper Interest in a Proper Case: Standing to Sue

A plaintiff in any sort of litigation is said to have standing to sue only when, according to the Supreme Court, it can meet several criteria: (1) it must show "injury in fact," that is, some "actual or threatened" personal injury resulting from the action it seeks to have the court adjudicate; (2) the injury "fairly can be traced to the challenged action"; and (3) a favorable decision by the court will likely redress the injury.[49] Thus, as a prerequisite to litigation (and as a constitutional matter), the standing doctrine seeks to ensure that only those persons with a direct stake in the outcome of a dispute can invoke a court's remedial jurisdiction. The doctrine is grounded in Article III of the Constitution, which extends the federal judicial power only to the resolution of "cases" and "controversies."

Until recently, courts have been extremely lenient in determining standing in environmental enforcement cases. Injury to environmental interests has been sufficient to satisfy the standing requirement. In 1972, the Supreme Court noted that

> Aesthetic and environmental well-being, like economic well-being, are important ingredients of the quality of life in our society, and the fact that particular environmental interests are shared by the many rather than the few does not make them less deserving of legal protection through the judicial process.[50]

It has been enough to show that the plaintiff would suffer injury to those interests in its use of a particular land area affected by the challenged agency action. A public interest group has been able to establish organizational standing by showing an injury in fact either to its members or to the organization itself.[51] Thus, in an action to prevent the restart of the K Reactor at DOE's Savannah River nuclear weapons plant in violation of the Clean Water Act, the court held that the Natural Resources Defense Council could maintain its suit by demonstrating that the use of the river by one or more of its members would be harmed by discharges of heated water from the reactor.[52]

Two recent Supreme Court decisions may have made it more difficult to demonstrate standing to sue. In 1990, the Court held that an environmental organization lacked standing because it failed to state with sufficient specificity facts to support general allegations of injury to its members.[53] The Court also held that instead of seeking "wholesale corrections" of agency actions, the organization should have sought redress from the agency itself or Congress, rather than the courts.[54] Two years later, another public interest organization brought suit to challenge a federal agency's rescission of certain regulations. Without reaching the merits of the case, the Supreme Court decided that none of the organization's members had suffered injury in fact, because none had shown that they had definite plans to visit the specific geographical areas affected by the regulatory change. The Court also ruled that there was no necessary connection between the injury complained of and the change, and that it was not likely that a favorable decision would redress the alleged injury.[55]

Justice Scalia, writing for the majority in the 1992 case, also suggested that the "cases" and "controversies" requirement of Article III of the Constitution is jurisdictional, leaving no authority for a court to decide a matter in which the plaintiff cannot satisfy the traditional criteria for standing, even when Congress has sought to confer standing on "any citizen" to institute enforcement litigation. Congress may not create

> an abstract, self-contained, non-instrumental "right" to have the Executive observe the procedures required by law. . . . [A] plaintiff raising only a generally available grievance about government . . . and seeking relief that no more directly and tangibly

benefits him than it does the public at large—does not state an Article III case or controversy. . . . To permit Congress to convert the undifferentiated public interest in executive officers' compliance with the law into an "individual right" vindicable in the courts is to permit Congress to transfer from the President to the courts the Chief Executive's most important constitutional duty, to "take Care that the Laws be faithfully executed."[56]

Two members of the seven-member Supreme Court majority expressed reservations on this point. A third felt that the plaintiff had standing, and would have resolved the case on other grounds.[57]

These developments may simply mean that citizen plaintiffs will have to be more specific in describing their stake in disputes. A clear nexus must be demonstrated between a plaintiff's personal health or welfare, or that of an organizational plaintiff's members, and the adverse effects of some agency action. Careful attention must also be paid to the requirement of present injury. The injury must be described as concrete, particularized, and imminent. For example, a public interest organization with members living immediately downwind from DOE's Hanford nuclear weapons plant was found to have standing to complain about DOE's failure to provide notice of its releases of hazardous substances, as it was required to do under RCRA and CERCLA. The organization included specific allegations in its complaint and submitted detailed affidavits, naming its affected members and describing their injuries in considerable detail.[58] A citizen plaintiff must also be able to show that it is within the zone of interests Congress intended to protect when it enacted the statute.[59]

States as plaintiffs must also establish standing to sue. A recent base closure case illustrates the difficulty a state may encounter. The State of Illinois filed suit to halt the scheduled closing of Chanute Air Force Base and the Army's Fort Sheridan, alleging that violations of the 1988 base closure law and the Administrative Procedure Act would result in serious economic injuries to the state and its citizens.[60] In dismissing the suit, the court ruled that the state had no standing as *parens patriae* to represent the interests of its citizens. The court also found that the economic burden of reduced state taxes, diminished federal funding, and increases in social welfare programs, even if fairly traceable to the closures, would fall not on the state as state, but on its citizens. Finally, the court characterized the state's claim as a "generalized grievance shared in substantially equal measure by a large class of citizens, that does not warrant exercise of jurisdiction."[61] By contrast, another court recently held that the State of Idaho had standing to seek judicial review under NEPA and the Administrative Procedure Act when DOE failed to prepare an EIS for its handling of spent nuclear fuel at the Idaho National Engineering Laboratory.[62] According to the court, standing was provided by threatened injuries to the state's economy and natural resources and health of its citizens.

Environmental Enforcement as a Political Question

It should come as no surprise that private citizens and public interest groups who seek to apply the environmental laws to national security activities sometimes have mixed motives. They may seek to use the environmental laws to help advance broader political agendas. The Air Force characterized the plaintiffs in one NEPA suit this way:

> At first glance, [the] Complaints appear to be less "political," and similar to the hundreds of lawsuits brought annually against agencies of the Federal Government by individuals and groups pursuant to the National Environmental Policy Act of 1969, except that their challenge is to a missile project which the President and the Congress have determined is needed to help achieve defense and foreign policy goals. . . . [But] it is precisely because of the nature of the governmental action challenged in the three Complaints—deployment of the MX/Peacekeeper system—and the judicial relief sought—declaratory judgments and injunctions barring the scheduled deployment—that this case is not and cannot be treated as just another NEPA case. . . . The plaintiffs and intervenors want the court to alter foreign and defense policies formulated by Congress and the President.[63]

But in an earlier suit involving the MX missile, the public interest plaintiffs characterized their efforts to force compliance with the procedural requirements of NEPA as follows:

> Procedures may be said to be what make a system of ordered liberty work. Plaintiffs here seek not to overturn decisions best left to coordinate branches of the government, but rather seek to insure that the Executive Branch follows the procedures the Legislative Branch has prescribed. In that way the public will be involved in the decisionmaking, the relevant information will be widely available, and consideration of the environmental factors—including the most potentially severe ones—will assume their proper place in Executive, Legislative, and public debate and decision.[64]

In that case the plaintiffs included not only the environmental groups Friends of the Earth, Greenpeace, Environmental Action, and the Sierra Club, but also a farmers' union, the Committee for a Sane Nuclear Policy, the Council for a Livable World, Nebraskans for Peace, and the United Church of Christ.

Neither the Administrative Procedure Act nor any of the citizen suit provisions disqualifies a citizen plaintiff who has a broader agenda encompassing, say, fiscal restraint or matters of foreign policy. As a practical matter, of course, it would be impossible to determine the true motivations of every person who accuses the Navy of threatening endangered species habitat, or who files suit to force the Energy Department to comply with the Clean Water Act. Yet even if such psychological screening were possible, the environmental laws could hardly be misused by insisting on their enforcement, since they are authoritative expressions of national policy. Indeed, Congress recognized that the effective implementation of that policy depends in substantial part on the enforcement efforts of citizen plaintiffs, whatever their motives.

Nevertheless, suits to compel compliance with the environmental laws have occasionally been challenged on grounds that they raised nonjusticiable "political questions."[65] A federal district court explained its reluctance to enjoin construction of a homeport for the Trident submarine this way:

> There is a "textually demonstrable constitutional commitment" of the conduct of national defense to the Congress in Article I, §8, and the President in Article II, §2. Secondly, the courts are not the proper forum for debate on national security and defense issues. Third, the policy determination to proceed with a particular approach is not within the ambit of the court's expertise or discretion, and, if so undertaken, would be a usurpation of the powers of the Congress and the President who have the duty under the Constitution to develop such policies. . . . [T]he substantive decision to choose one alternative of national defense over another lies with the political branches of government and not with the courts.[66]

Political question challenges were raised and rejected in two NEPA cases involving national security activities. The first, described in a case study in chapter 2, was brought by the State of Colorado and several environmental groups against the Air Force to test the adequacy of environmental impact statements for deployment of the MX missile in Minuteman silos. The Air Force contended that the court was without power to decide the case because it was "interwoven with political issues going to the heart of foreign policy and national defense which have already been resolved by the President." But the court found that the relief sought by the plaintiffs was merely to have the executive branch follow clear congressional directives. The "issues presented," said the court, "are purely legal ones of statutory interpretation, the resolution of which are a part of our constitutional duty."[67]

The second suit challenged the sufficiency of an EIS for the Air Force's Ground Wave Emergency Network (GWEN), described in another case study in chapter 2. The Air Force sought to halt the litigation by claiming that the court was without any "judicially discoverable and manageable standards to review the merits of national defense policy." The court held that the basis of the suit was to ensure compliance with NEPA rather than to review the merits of national defence policy. Ruling against the Air Force on this point, the court declared that "a lawsuit challenging development of a defense installation on the grounds that the responsible agency did not discuss environmental impacts causally related to the installation raises justiciable questions."[68]

Protecting Military Secrets

A number of regulatory environmental statutes provide for monitoring of compliance and publication of the results. For example, the Resource Conservation and Recovery Act calls on the EPA or a state with an EPA-approved program to inspect permitted federal facilities that store, treat, or dispose of hazardous wastes at least once each year and make its findings available to the public.[69] Several of the statutes require regulated agencies to monitor their own performance, as well, and to file regular public reports of activities that contaminate the environment.[70] Information in these reports can provide the basis for a state enforcement action or citizen suit against the permitted facility.

What if disclosure of the information might be helpful to the nation's enemies?[71] Among the environmental statutes, only CERCLA squarely addresses this question:

> Notwithstanding any other provision of law, all requirements of the Atomic Energy Act and all Executive orders concerning the handling of restricted data and national security information, including "need to know" requirements, shall be applicable to any grant of access to classified information under the provisions of [CERCLA or SARA Title III].[72]

Each of the major regulatory environmental statutes provides that when it is in the "paramount interest of the United States" (or words to that effect), the President or a high government official may exempt a federal facility or activity from its requirements, including, presumably, the public reporting mandate.[73] Apparently, however, this exemption authority has never been invoked to withhold environmental information from public view.

Without invoking any express statutory authority, EPA has adopted a policy for each of the regulatory statutes that requires EPA or state inspectors to obtain appropriate security clearances where necessary, and to protect properly classified information from disclosure.[74] In at least one instance, a regulated defense agency has sought to prevent the release of a permit application. That happened when DOE applied for an RCRA permit at its Rocky Flats facility. DOE insisted that the application contained "unclassified controlled nuclear information," which is protected by the Atomic Energy Act.[75]

Independently of the publication requirements in the environmental statutes, the Freedom of Information Act (FOIA)[76] provides access to information about the environmental effects of agency operations. Each federal agency must disclose any record in its files, with certain important exceptions, to anyone who requests it, without regard for the requester's purpose. FOIA exempts several classes of documents from disclosure. Perhaps most important here are records properly classified under the current executive order on national security information.[77] For example, the Army might withhold data concerning toxic air pollutants from a new anti-tank weapon, on grounds that an enemy could learn something of value about its performance or the composition of propellants. Also subject to withholding are records specifically exempted from disclosure by another statute.[78] The Atomic Energy Act, for one, requires protection of "restricted data" concerning the design, manufacture, or use of atomic weapons or the production of materials for their manufacture.[79] Another FOIA exemption permits withholding of interagency or intra-agency communications that would not ordinarily be available through discovery in litigation.[80]

Classified information need not be included in an environmental impact statement prepared under NEPA, but may be set out instead in a classified annex.[81] The very existence of the EIS itself may be concealed if its disclosure could jeopardize national security. The leading case, described in a case study in chapter 2, involved the Navy's construction of ammunition magazines near Honolulu that were admittedly "capable" of storing nuclear weapons. The Supreme Court required the Navy to disclose neither the existence of actual plans to store nuclear weapons at the facility, nor the fact that it had or had not prepared an EIS for such storage.[82]

Information about agency activities affecting the environment may also be available through discovery once litigation is instituted to test compliance with the environmental laws. However, the government may try to withhold data from production or from introduction into evidence by asserting its "state secret" privilege, on grounds that disclosure could injure the national security.

Federal courts now routinely look at sensitive documents in camera (in chambers) when they rule on the propriety of government secrecy claims in FOIA and state secrets cases.[83] Also in the context of FOIA litigation, they consider agency claims that the government can "neither confirm nor deny" the existence of a certain document or activity.[84] Early in NEPA's history, in a case involving underground testing of nuclear weapons, a lower court rejected the President's effort to withhold information about agency compliance with NEPA by invoking his "executive privilege."[85] The court described its role this way:

> An essential ingredient of our rule of law is the authority of the courts to determine whether an executive official or agency has complied with the Constitution and with the mandates of Congress which define and limit the authority of the executive. Any claim to executive absolutism cannot override the duty of the court to assure that an official has not exceeded his charter or flouted the legislative will.[86]

The court then avoided public disclosure by reviewing the relevant documents in camera, out of the presence of the plaintiff or its lawyers.[87] In the same manner, a court should be willing to examine classified agency records in camera to ensure compliance with environmental requirements.[88] But in a NEPA case involving the Navy's proposal to build a homeport facility for the battleship *Iowa* on Staten Island, the court refused to review classified documents in camera that might have revealed plans to bring nuclear weapons into New York Harbor, saying that a ruling in favor of the environmental plaintiffs would effectively disclose the existence of such weapons aboard Navy ships.[89]

One answer to the reluctance of federal courts to decide such cases on their own (since public interest plaintiffs, usually lacking security clearances, could not be involved in a true, adversarial presentation of the case) would be to create a special court with expertise in environmental affairs to rule on compliance with environmental laws whenever agency records are withheld from public view.[90] In the alternative, a special environmental counsel, empowered to operate independently of the executive branch, but with access to classified materials, might be appointed to inquire into such cases and bring enforcement actions where appropriate.[91]

One other type of case deserves mention here. It is exemplified by recent litigation growing out of the Iraqi Exocet missile attack on the United States frigate *Stark* in the Persian Gulf in 1988, which left thirty-seven American sailors dead. When surviving family members filed suits against the manufacturer of the Aegis missile defense system that was supposed to protect the ship from such attacks, the courts dismissed the suits, declaring that the trials would very likely result in disclosure of sensitive information about the design of the system.[92] The Supreme Court expressed the same concern when it refused to look into the Navy's compliance with NEPA in the construction of a nuclear weapons-capable munitions magazine near Honolulu.[93] So in a future case involving environmental degradation from some defense-related activity, a court might refuse even to hear the case on grounds that the litigation itself would expose important national secrets. If that happened, there would be no way to enforce compliance with environmental laws, and no accountability for violations, since the public would not have access to the information it needed to respond politically.

Funding for Environmental Litigation

No citizen plaintiff, public interest organization, or even state government can ever hope to marshal the financing, legal expertise, or information base that the federal government brings to litigation involving the environmental laws. However, because Congress recognized the need to enlist nonfederal parties in the effort to enforce these laws, the major regulatory statutes contain provisions that may enable successful citizen suit plaintiffs to recover their litigation expenses. The Clean Water Act is typical: "The court, in issuing any final order in any [citizen suit], may award costs of litigation (including reasonable attorney and expert witness fees) to any prevailing or substantially prevailing party, whenever the court determines such award is appropriate."[94] For nonfederal plaintiffs seeking enforcement against federal agencies of statutes without such provisions, such as NEPA, the Equal Access to Justice Act may provide similar reimbursement.[95]

Criminal Sanctions for Statutory Violations

Several of the regulatory environmental statutes provide criminal sanctions for violations. Under the Clean Water Act, for example, anyone who either negligently or knowingly violates provisions of the act or any permit issued under it can be fined up to $100,000 per day of violation, depending on the circumstances, and imprisoned for as long as six years.[96]

In 1989, defense officials were shocked to learn that three civilian employees at the Army's Aberdeen Proving Ground in Maryland had been convicted of illegally storing and dumping hazardous wastes from research on chemical weapons, in violation of the Resource Conservation and Recovery Act.[97] The court held that the defendants were not immune from the criminal provisions of the statute simply because of their status as federal employees working at federal facilities. A different court had earlier dismissed charges against the administrator of a federal hospital when there was no evidence of his direct involvement in RCRA violations at the facility.[98] However, another court had upheld a federal employee's criminal conviction under CERCLA without addressing the immunity question.[99]

These cases provoked a spate of scholarly speculation about the potential criminal liability of federal officials for both state and federal environmental law violations.[100] At least for RCRA violations, their questions were laid to rest by passage of the Federal Facility Compliance Act in 1992. That statute amends RCRA to provide that an "agent, employee, or officer of the United States shall be subject to any criminal sanction (including, but not limited to, any fine or imprisonment) under any Federal or State solid or hazardous waste law. . . ."[101] One federal appeals court has recently upheld a federal employee's criminal conviction under the Clean Water Act.[102]

In criminal prosecutions, the state or federal government is always the plaintiff. The Department of Justice, representing the federal government, may sometimes find itself in the awkward position of pursuing criminal sanctions against high government officials in other agencies whose actions it would ordinarily defend. One notorious case involved criminal conduct by the Department of Energy's contract operator of the Rocky Flats nuclear weapons plant. The corporate defendant, Rockwell International, insisted that it acted in good faith to comply with DOE procedures and directives when it burned mixed wastes in an incinerator without a RCRA permit and dumped untreated toxic wastes on plant grounds. While a criminal investigation was underway,

Rockwell sought a court order that would have protected it from criminal or civil liability for future performance of its contract with DOE; the order was denied.[103] A settlement was finally reached that allowed Rockwell to plead guilty to five felony and five misdemeanor charges, and to pay a fine of $18.5 million.[104] But the settlement halted ongoing deliberations of a federal grand jury and prevented indictments of responsible corporate or government officials. Secrecy surrounding the proceeding prompted charges that the government had not been sufficiently aggressive, and even that the settlement was a "sweetheart deal." One former Justice Department official has suggested that "there is no effective and workable remedy for inadequate prosecutorial zeal."[105] Frustrated members of the Rocky Flats grand jury have tried to publicize their findings in hopes of provoking further action by prosecutors, but they have been threatened with criminal prosecution themselves for disclosing secret information.[106]

Chapter 8

Liability for Environmental Damages

As we saw in earlier chapters, the environmental laws may not always be effective in protecting human health and the environment from defense-related activities. That can happen when the environmental laws do not (or did not) apply to a particular action, when applicable laws provide inadequate protection, or when the laws are violated. Acts of needless environmental destruction may nevertheless be discouraged if government agencies and officials are forced to pay for the losses they cause. The idea of fiscal accountability, with the attendant hope that it will influence behavior, is behind the provisions in the Comprehensive Environmental Response, Compensation and Liability Act (CERCLA) for natural resource damages and liability for cleanup costs.[1] It is also behind the common law tort principle that persons who unreasonably injure others should ordinarily provide compensation.

The specter of potential liability may not always deter misbehavior. A Department of Energy official whose act wrongfully destroys a neighbor's drinking water supply may suffer some embarrassment or even lose her job, but compensation for the loss will usually be covered, if at all, by the taxpayers, for whom the pain of payment will hardly provide a disincentive for further wrongdoing. The corporate contract operator of a DOE facility may perhaps feel greater economic sensitivity and exercise more effective discipline over its staff if it does not expect to be reimbursed by the government for its liability.

Even when fear of liability fails to influence the tortfeasor, however, the tort laws may make the injured party whole to some degree. Compensation is based on our sense of fairness—on the notion that no single individual should have to bear more than a fair share of the costs of providing for the common defense. Nevertheless, if a federal official is exercising reasonable judgement and acting within the scope of her authority when the injury occurs, the injured party may just have to grin and bear it. Likewise, if a defense contractor is scrupulously adhering to instructions from the government agency that hired it, the contractor may be excused from liability for any resulting harm.

This chapter is about efforts to provide fair compensation for injuries growing out of national defense activities that harm the environment. Such injuries have been much in the news in recent years. They range from deaths and disease linked to the Vietnam War herbicide Agent Orange and fallout from atmospheric testing of atomic weapons to environmental contamination from neighboring military bases. We look first at legal principles for recovery of damages from the federal government, then address the liability of defense contractors. In each case, compliance with the environmental laws may be a major determinant in fixing responsibility.

Government Liability for Damages

Sovereign Immunity and the Federal Tort Claims Act

The doctrine of sovereign immunity has long frustrated efforts by plaintiffs to obtain compensation for injuries suffered at the hands of the federal government. As the Supreme Court put it not long ago, "It is axiomatic that the United States may not be sued without its consent."[2] For more than a century, claims for damages against the federal government were addressed almost exclusively by private bills in Congress, pursuant to its power "to pay the Debts . . . of the United States."[3] These bills, still used occasionally today, had to be introduced, reported out of committee, voted on, and signed by the President just like any other appropriation measure.[4] With the arrival of the Industrial Revolution and increased governmental regulation, the number of claims against the government rose dramatically. By the middle of the nineteenth century, one senator complained that two days of each congressional week were spent on private bills.[5] A more expedient method was needed for resolving claims against the government. In 1887, Congress passed the Tucker Act, waiving sovereign immunity for claims against the United States "founded either upon the Constitution [or] in cases not sounding in tort."[6] The Tucker Act may be used, for example, in actions based on contract or to collect just compensation for takings of property rights.

It was not until 1946 that Congress passed the Federal Tort Claims Act (FTCA).[7] The act provides that a plaintiff may bring suit in federal court against the United States

> for money damages . . . for injury or loss of property, or personal injury or death caused by the negligent or wrongful act or omission of any employee of the Government while acting within the scope of his office or employment, under circumstances where the United States, if a private person, would be liable to the claimant in accordance with the law of the place where the act or omission occurred.[8]

The act thus waives sovereign immunity for the claims described. It is meant to expose the government to the same liability it would experience as a private party. The existence and extent of liability are determined by reference not to federal law but to the law of the state in which the claim arises. Congress has also provided for nonjudicial settlements of claims totalling up to $100,000 against the military services or the Coast Guard for property losses, death, and personal injury arising from noncombat activities.[9]

In the FTCA, Congress provided a number of exceptions to its waiver of sovereign immunity. For example, claims for negligent transmission of postal material, admiralty claims, claims for military activities in time of war, and claims arising in foreign countries are not compensable under the act.[10]

Suits for Injuries to Military Personnel: The Feres *Doctrine*

Soon after passage of the Federal Tort Claims Act, it was settled that military personnel injured while on active duty may not recover from the government for their injuries. In a 1950 case entitled *Feres v. United States*,[11] the United States Supreme Court rejected damages claims from three different servicemen. One had died when a fire consumed his barracks; suit was filed by his executrix alleging negligence on the government's part. Another sought compensation based upon an operation in which an Army surgeon sewed "a towel 30 inches long by 18 inches wide" into his stomach. The third serviceman died from alleged medical malpractice by Army surgeons. While acknowledging that there is little guidance in the Federal Tort Claims Act itself, the Court held that the government is not liable under that act for "injuries to servicemen where the injuries arise out of or are in the course of activity incident to service."[12] Thus, the so-called *Feres* doctrine was born.

Since the Federal Tort Claims Act permits recoveries based on state, not federal, law, the *Feres* Court seemed to be influenced by the fact that it could find no law in any state permitting "a soldier to recover for negligence, against either his superior officers or the Government he is serving." The Court also noted that the military constantly relocates soldiers and that it would be unfair to force them to rely upon the tort law of any given state—more unfair, presumably, than denying them recovery altogether. Finally, the Court observed that the military already furnishes compensation for injured soldiers that compares favorably with that provided by most workman's compensation statutes.[13]

The *Feres* doctrine remains good law today.[14] In one case, for example, the court relied on *Feres* to reject a plaintiff's claim for damages suffered when "he and other soldiers at Camp Desert Rock in the State of Nevada were ordered by their commanding officers to stand in a field without benefit of any protection against radiation while a nuclear device was exploded a short distance away."[15]

The Discretionary Function Exception

The most important exception to the Federal Tort Claims Act's waiver of sovereign immunity for our purposes, one that threatens to swallow the entire rule, is known as the "discretionary function" exception.[16] Recovery will not be available from the federal government for any claim based upon "the exercise or performance or the failure to exercise or perform a discretionary function or duty on the part of a federal agency or employee of the Government, whether or not the discretion involved be abused."[17] In the Supreme Court's words, the discretionary function exception "marks the boundary between Congress' willingness to impose tort liability upon the United States and its desire to protect certain governmental activities from exposure to suit by private individuals."[18] Unfortunately, the complexity and ambiguity of this provision have led to nearly five decades of judicial struggle and scholarly dispute.

The Supreme Court first grappled with the discretionary function exception in a 1953 case.[19] In the aftermath of World War II, the federal government shipped fertilizer to Germany, Japan, and Korea to help those countries reestablish their agricultural economies. When two ships loaded with ammonium nitrate fertilizer exploded in Texas City, Texas in 1947, 560 persons were killed, and another 3,000 were injured. The Court found that decisions of lower level government officials concerning the bagging, labeling, and temperature for storage of the fertilizer fit within the discretionary function exception. "Where there is room for policy judgment and decision (no matter which official makes the decision)," the Court said, "there is discretion." The Court distinguished between decisions made at the "planning" level and those made at the "operational" level, declaring that the latter would more likely be actionable.[20]

Thirty-one years later, in 1984, the Supreme Court revisited the issue in a case involving a government agency's negligence in enforcing safety regulations for civilian airliners. Though proclaiming it "unnecessary—and indeed impossible—to define with precision every contour of the discretionary function exception," the Court found that "it is the nature of the conduct, rather than the status of the actor, that governs whether the discretionary function exception applies in a given case." In particular, it declared that the exception applies to the government when it acts "as regulator of the conduct of private individuals."[21]

In 1988, the discretionary function exception was invoked in a suit that alleged government negligence in licensing a polio vaccine when a child inoculated with the vaccine actually contracted the disease. The Supreme Court rejected the notion that all government regulatory acts are protected. The exception applies, the Court found, only when it can be shown that the acting government employee had a choice in taking the injurious action. "Conduct cannot be discretionary unless it involves an element of judgment or choice." Thus, the exception does not protect the government when an employee acts contrary to specifically prescribed conduct. However, even when the actor has a choice, the discretionary function exception may still be overcome if the actor's judgment was not the kind the exception was designed to shield. The exception "protects only governmental actions and decisions based upon considerations of public policy."[22]

Recovery of Damages in Defense-Related Cases

Until recently, the discretionary function exception has presented a nearly insurmountable obstacle for plaintiffs suing the United States for injuries growing out of defense-related activities. Two widely publicized sets of cases put the practical and conceptual problems in perspective: the "Agent Orange" litigation and litigation surrounding the atmospheric testing of nuclear weapons.

Case Study: The Agent Orange Litigation

Few cases have engendered the emotional response of those claiming injury from the United States military's use of the herbicide Agent Orange during the Vietnam War.[23] Some 19 million gallons of Agent Orange and other defoliants were sprayed over Vietnam from 1962 to 1971 to destroy enemy food supplies and remove vegetation hiding enemy troops and supplies.[24] Civilians and military personnel exposed to the herbicide claim that it caused heart disease, cancer, birth defects, fatigue, anxiety, and other health problems.[25] A number of lawsuits for damages were brought against the United States government and the manufacturers of the chemicals, and were eventually consolidated in one federal court in New York. The plaintiffs included soldiers in the United States, Australian and New Zealand armed forces, their families, and others who came into contact with the defoliants. Despite serious doubts about whether a causal link could be established between the herbicide and the reported injuries, the court certified the class action.

In 1984, the manufacturers of Agent Orange and the vast majority of the plaintiffs reached a settlement that established a $180 million trust fund to provide compensation. By mid-1990, 48,000 claims had been filed for payment from this fund.[26] Plaintiffs who opted out of membership in the class that settled got nothing from the manufacturers. The court in the consolidated litigation found that the manufacturers breached no duty to warn the government of hazards associated with the use of the herbicides, since the government was fully aware of the dangers, and were therefore protected under the "military contractor defense" (described more fully in the next section of this chapter). It also decided that the weight of scientific evidence failed to establish a link between Agent Orange and the plaintiffs' injuries.[27]

All of the suits against the federal government met the same fate, but for different reasons. Among the civilian plaintiffs were three employees of the University of Hawaii who allegedly sustained injuries after being exposed to Agent Orange when it was tested on the university's fields. The Court of Appeals declared that "it cannot be seriously contended that the decision to use Agent Orange as a defoliant was anything but a discretionary act." The court pointed out that the "ultimate policy decision to use Agent Orange was made by President Kennedy," although that decision was implemented by military officials down through the ranks. The plaintiffs' allegations of government negligence in labeling, shipping, and handling the herbicide were dismissed with the observation that "the fact that discretion is exercised in a negligent manner does not make the discretionary function exception inapplicable." The court upheld the dismissal of similar claims by Vietnam veterans, declaring that the discretionary function exception precludes "judicial 'second-guessing' in FTCA litigation of discretionary legislative and executive decisions such as those that were made concerning Agent Orange."[28] The court also decided that the military plaintiffs and their families were precluded from recovery by the *Feres* doctrine.[29]

Case Study: Downwinders: The Legacy of Atmospheric Testing

In the 1946 Atomic Energy Act, the Atomic Energy Commission (AEC) was given broad discretionary power to "conduct experiments . . . in the military application of atomic energy."[30] The AEC conducted 235 atmospheric nuclear weapons tests before testing was driven underground by the signing of the Limited Test Ban Treaty in 1963.[31] Each test was approved by a variety of managers within the AEC and the Defense Department and, finally, by the President. The tests were conducted, in part, by private contractors.

Congress also directed the AEC to "protect health [and] to minimize danger from explosion and other hazards to life or property . . . as [the AEC] may determine."[32] This directive was the focus of claims for injuries said to have resulted from the tests.

More than 100 explosions in the atmosphere at the Nevada Test Site are estimated to have exposed 100,000 civilians to radioactive fallout. Other tests in the Pacific rendered dozens of islands uninhabitable and injured hundreds of local people. The Department of Defense estimates that as many as 210,000 soldiers participated in tests at both locations and may have been exposed to radiation.[33]

Two massive lawsuits were brought claiming damages for test-related injuries. In one, nearly 1,200 named plaintiffs sued the United States, alleging some 500 deaths and injuries from exposure to radioactive fallout produced by open-air tests in Nevada. They argued that the government was liable for the negligent failure of officials at the Test Site "to fully monitor off-site fallout exposure and to fully provide needed public information on radioactive fallout." In 1987, a federal appeals court found that while government officials had a general statutory duty to promote safety, that duty was broad and discretionary. It was left to the AEC, the court said, "to decide exactly *how* to protect public safety." The court reiterated the long-standing rule that even negligent discretionary functions are protected. It concluded that "the government is immune from liability for the failure of the AEC administrators and employees to monitor radioactivity more extensively or to warn the public more fully than they did."[34] The court expressed a genuine reluctance to "second-guess" the executive branch:

> The bomb-testing decisions made by the President, the AEC, and all those to whom they were authorized to delegate authority in the 1950s and 1960s, were among the most significant and controversial choices made during that period. The government deliberations prior to these decisions expressly balanced public safety against what was felt to be a national necessity, in light of national and international security. However erroneous and misguided these deliberations may seem today, it is not the place of the judicial branch to now question them.[35]

The other major litigation involved a consolidation of suits brought by more than 3,000 civilians and military personnel against the United States and various defense contractors.[36] They sought more than $4.9 billion in damages for radiation injuries related to the bombing of Hiroshima and atmospheric testing. The government and its contractors were negligent, they said, in failing to take adequate safety precautions. The court found that government officials in charge of the tests had discretion rooted in policy, as evidenced by this 1953 internal memorandum:

> Due to the special nature of field tests it is considered that a policy of strict adherence to the radiological standards for routine work is not

> realistic. The regulations set forth herein have been designed as a reasonable and safe compromise considering conservation of personnel exposures, the international import of the test and the cost aspects of operational delays chargeable to excessive radiological precautions.[37]

> The court noted that one of the objectives of the tests was to study the psychological reactions of troops to nuclear weapons; for this purpose, the government "needed complete control over information supplied to the troops." The fact that safety decisions, including the decision not to provide a wider warning of the hazards, may have been made by scientists or other operational personnel, the court decided, "does not affect the application of the discretionary function exception."[38]
>
> In the same litigation, the court upheld the constitutionality of the 1984 Atomic Testing Liability Act, or Warner Amendment, so named for its Senate sponsor.[39] That measure gives immunity to private contractors by providing that suits against them for injuries resulting from atomic tests are converted to suits against the United States. The court, of course, had already found that suits against the government were barred by the discretionary function exception. That left the claimants empty-handed.[40]
>
> Congress came partly to the rescue in 1990 with passage of the Radiation Exposure Compensation Act.[41] That measure provides $50,000 for each civilian victim of childhood leukemia or other specified diseases living in an area affected by the tests, and $75,000 for military personnel involved in the tests who contracted any of the same diseases. It also compensates uranium miners injured when the government failed to warn them of the danger of radiation.[42]

The discretionary function exception has barred recovery for environmental injuries in a number of other defense-related cases. One grew out of the Army's conduct of a simulated biological warfare attack on the City of San Francisco in 1950. The Bay area was blanketed with a supposedly harmless airborne bacterium, which allegedly caused at least one death. The court noted that the Army general in charge had to weigh numerous factors in selecting the bacterium and carrying out the test, including "concerns for national security, a need for secrecy, the possible risks of urban testing, and applicable medical concerns." (The same general had rejected a plan for a test using the same bacterium in the New York subways.) The court felt that it was not "equipped to weigh the type of factors involved in such a basic policy determination," and that its "review would likely impair the effective administration of government programs believed to be vital to the defense of the United States at the time they are conducted."[43]

Another claim involved the Air Force's construction of a missile site near

Mead, Nebraska.[44] The contractor hired to build the facility disposed of trichloroethylene (TCE) waste from the construction by pouring it on the ground and into a sewer system. When the TCE contaminated a neighboring farmer's groundwater supply, the farmer brought suit, alleging that the Air Force negligently failed to supervise its contractor. Because the construction had been done thirty-four years earlier, the court found that no specific statute or regulation had been violated. Moreover, said the court, the Air Force had "discretion" about how to supervise the construction. This was the kind of discretion addressed by the FTCA exception, since it was "grounded in policy considerations related to the national defense and the personal and economic resources of the nation."[45]

Another recent case grew out of ongoing efforts to clean up Basin F at the Army's Rocky Mountain Arsenal. See chapter 4, "Struggle for Control at Basin F." Basin F was a ninety-three-acre artificial lagoon in which the Army and Shell Oil Company dumped toxic wastes for decades. When the wastes were removed from the lagoon and placed in piles for storage, some of them escaped into the air, injuring the health and property of the neighbors downwind. The plaintiffs challenged "both the execution of the Basin F cleanup and the failure of Army officials adequately to warn of impending toxic air emissions." The court denied relief for the injuries, finding no violation of any "specific and mandatory objectives" and ruling that the Army's actions "involved policy choices of the most basic kind."[46]

The discretionary function exception has not always prevented recovery for environmental injuries. One case in which the exception was found inapplicable involved the owners of property adjacent to the Navy's Trident submarine base at Bangor, Washington.[47] The Navy had pumped water contaminated with chemicals from the disassembly of missiles into open trenches on the base, from whence it migrated into the neighbors' groundwater source. The court found that the Navy had violated a specific, mandatory requirement in an executive order that prohibited any discharge of wastes into waters if they contain "any substances in concentrations which are hazardous to health."[48] The discretionary function exception does not apply, the court said, when the agency actor has no choice—no discretion—to ignore the requirement. Another court refused to apply the discretionary function exception in a case involving the sale of hazardous chemicals when the government violated regulations for packaging and handling such materials.[49]

Still another case grew out of the Defense Department's conveyance of waste oil contaminated with PCBs to an independent contractor. The contractor disposed of the oil in violation of CERCLA and EPA regulations. Holding the government liable for damages resulting from its contractor's misdeeds, the court observed that the discretionary function exception does not apply to "choices which are either outside the policy making context or in an area in which federal law directs a particular course of action."[50] Courts have employed similar reasoning in cases involving radioactive fallout from venting of an underground nuclear weapon test[51] and radiation injuries to workers at the Nevada Test Site.[52]

The discretionary function exception has no application to Tucker Act claims based on takings of property for which just compensation is required by the Fifth Amendment to the United States Constitution. The classic case arose during World War II, when the Army flew bombers, transports, and fighter planes so low over the plaintiff's chicken farm that they sometimes blew the leaves off the trees. Many of the chickens died, and others quit laying. The farmer's family became nervous and frightened and could not sleep. The Supreme Court found that, while the land might still be used for some purposes, the farmer's loss was so great that a compensable taking had occured.[53]

More recently, the Tucker Act was invoked by a homeowner who claimed that toxic chemicals from McChord Air Force Base had polluted her domestic water well. The court found that just compensation would be due under the Fifth Amendment exactly as if the Air Force had "condemned" the water for public use.[54] But when the homeowner was awarded damages on the same facts in a parallel suit brought under the Federal Tort Claims Act,[55] the court dismissed the takings claim, holding that plaintiff must choose which remedy to pursue, and may not recover under both.[56] The court noted that in a takings claim, but not under the Federal Tort Claims Act, the plaintiff would be entitled to interest and attorneys' fees. Conversely, the Federal Tort Claims Act, but not the Tucker Act, would allow recovery for consequential damages, such as damages for emotional distress.

Claims Against Defense Contractors

Private corporations that furnish products and services to the Defense Department and Department of Energy generally do not enjoy the sovereign immunity available to the federal government. In principle, they are financially responsible for the consequences of their negligent or deliberately destructive acts. In practice, however, they may be insulated from liability just as surely as the government if they can avail themselves of what has come to be called the "government contractor" defense.

The government contractor defense was first articulated in a 1940 Supreme Court case. There, a private contractor employed by the Army Corps of Engineers and working under Corps supervision built dikes on the Missouri River that caused the plaintiff's riparian land to erode. The Court said there was no liability where the contractor was merely executing Congress' will.

According to the Court, "The act of the agent is 'the act of the government.'"[57]

The principle was not invoked in tort claims against defense contractors until more than forty years later. A 1983 case involved wrongful death claims against the manufacturer of an ejection system for Navy jet aircraft. The court framed the issue this way: "[G]iven the immunities of the United States in cases such as these, the question arises whether a supplier of military equipment should be required to shoulder directly and immediately the entire burden of the liability to an injured serviceman."[58] The court worried that if contractors were held accountable in such cases they would simply pass the costs on to the government. It also expressed concern that where the government had set or approved the defective design specifications, the judiciary could become involved in second-guessing such military decisions. The court decided that a supplier of military equipment should not be liable for a design defect when:

> 1) the United States is immune from liability, 2) the supplier proves that the United States established, or approved, reasonably precise specifications for the allegedly defective military equipment, 3) the equipment conformed to those specifications, and 4) the supplier warned the United States about patent errors in the government's specifications or about dangers involved in the use of the equipment that were known to the supplier but not the United States.[59]

The court in the Agent Orange litigation, described earlier, expressed the same concerns:

> The chemical companies sold Agent Orange to the United States government, which used it in waging war against enemy forces seeking control of South Vietnam. It would be anomalous for a company to be held liable by a state or federal court for selling a product ordered by the federal government, particularly when the company could not control the use of that product. Moreover, military activities involve high stakes, and common concepts of risk averseness are of no relevance. To expose private companies generally to lawsuits for injuries arising out of the deliberately risky activities of the military would greatly impair the procurement process and perhaps national security itself.[60]

In another case involving the death of a military pilot in a defective aircraft, the court simply refused to get involved, asserting that "in our constitutional system, when a knowing and purposeful decision to employ such products is made by the military, the judiciary may not question it."[61]

The government contractor defense was fleshed out more extensively in a 1988 Supreme Court case that began with the crash of a Marine helicopter off the Virginia coast.[62] One of the crew members drowned when he was unable to exit from the helicopter; the escape hatch opened outward, rather than inward, making it impossible to open under water. The Marine's father brought suit against the builder of the helicopter, alleging a defective design. The Court found that the state tort law under which liability for damages ordinarily would be fixed was preempted by federal law, even though no federal statute addressed the subject. Preemption occured, said the court, because a "uniquely federal interest" was involved in the claim and because there was a "significant conflict" between federal policy and the application of state tort law.

According to the Court, the work of an independent contractor performing its obligation under a procurement contract is "so committed by the Constitution and laws of the United States to federal control that state law is preempted and replaced, where necessary, by federal law of a content prescribed (absent explicit statutory directive) by the courts—so-called 'federal common law.'" Moreover, said the court, "the imposition of liability on Government contractors will directly affect the terms of Government contracts: either the contractor will decline to manufacture the design specified by the government, or it will raise the price. Either way, the interests of the United States will be directly affected."[63]

The Court recognized that in some instances a contractor could comply both with federal contract requirements and state law. In such cases, contractors should not be immunized. In this case, however, the Court found that the state-imposed duty of care (that is, the duty to fit the helicopter with a particular kind of escape mechanism) was "precisely contrary to the duty imposed by the Government contract (the duty to manufacture and deliver helicopters with the sort of escape-hatch mechanism shown by the specifications)."[64] According to the Court, significant conflict between state law and federal policy can be said to exist when state law would condemn actions that, if committed by the federal government, would be protected by the discretionary function exception of the Federal Tort Claims Act. The government's discretion to make policy choices is at stake, the Court indicated, in suits against suppliers when the United States approved reasonably precise specifications and when the equipment conformed to those specifications. However, the supplier must have warned the government of any dangers in the use of the equipment that were known to it but not to the government.

None of the cases described so far has anything to do with environmental protection, as such. However, the government contractor defense may be used by independent suppliers of both goods and services to avoid liability for environmental injuries.[65] A case involving the Department of Energy's Fernald Feed Materials Plant illustrates the potential application of the defense, as well as its limitations.

Case Study: Flawed Fiat at Fernald

In 1951, the Atomic Energy Commission (AEC), a predecessor to the Department of Energy (DOE), contracted with the National Lead Company to operate its 1,050-acre Feed Materials Production Center, located eighteen miles northwest of Cincinnati, Ohio. For the next thirty-four years, National Lead produced weapons-grade uranium at the plant for shipment to other weapons facilities. During this period, the plant discharged as much as 550,000 pounds of uranium into the air.[66] Another 12.7 million pounds were disposed of in pits on plant grounds, while some 167,000 pounds were discharged into the nearby Great Miami River, a tributary of the Ohio.[67]

When word of these practices finally got out in 1984, fourteen thousand of the plant's nearest neighbors filed a class action lawsuit alleging that "defendants failed to prevent the emission of uranium and other harmful materials from the [plant] and that such failure caused emotional distress and diminished property values."[68] The lead plaintiff lived across the road from the plant. DOE admitted that it had discovered uranium in her family's domestic well as early as 1981 but had continued to certify water from the well as safe for drinking.[69]

National Lead did not dispute its contamination of surrounding areas with radioactive wastes. Instead, it argued that the government contractor defense shielded it from any liability. When the case was finally resolved four years later, in 1989, the court decided that National Lead had failed to demonstrate a "significant conflict" between federal policy and the operation of state law, citing the Supreme Court decision described above. When the defendant discharged uranium into the Great Miami River, it violated federal environmental law, specifically, the Refuse Act of 1899, which forbids the unauthorized deposit of refuse in navigable waterways.[70] The court also found that emissions of uranium into the air violated dose limits set by Atomic Energy Commission regulations.[71] These federal environmental protection rules were not in conflict with but rather perfectly congruent with Ohio tort law providing relief for abnormally dangerous conditions or activities. Moreover, the court noted, the government contractor defense would only be available if the court could find that the discretionary function exception to the Federal Tort Claims Act would bar any liability on the part of the United States. While operation of the Fernald plant required the exercise of a measure of discretion, there was no discretion to violate specific environmental standards.[72] This finding ultimately led to a $73 million settlement.[73] The amount will be paid not by National Lead, however, but by the Department of Energy pursuant to a contract provision holding its contractor harmless from any such liability. The settlement left the neighbors

> free to pursue further claims for injuries to health, in addition to compensation for emotional distress and cleanup costs.

In an earlier case, neighbors of DOE's Rocky Flats nuclear weapons plant in Colorado had brought suit against the government and the contractor operators of the facility when their land was contaminated with plutonium.[74] The plutonium was apparently released as a result of normal plant operations and two accidental fires. The court found that there was no conflict between the Atomic Energy Commission's regulation of radiation hazards at the plant and state laws providing compensation for injuries. Despite the importance of national defense, the court concluded, neither the federal government nor the contractors would be shielded from liability if their acts were found to be negligent under Colorado law. The case was finally settled out of court.

Another suit was filed by Rocky Flats neighbors on similar facts in 1990, advancing common law and CERCLA claims, and alleging violations of RCRA and the Clean Water Act.[75] That suit is still pending at this writing.

With the Department of Energy's recent release of millions of documents describing the operation of its nuclear weapons complex over the last half-century, and improved access to military environmental records, we may see many more lawsuits of this sort in the future, naming both defense contractors and the government itself. The outcome in each case will likely depend to some degree on compliance with the environmental laws.

Chapter 9

National Defense vs. Environmental Protection: We Can Have It Both Ways

Responding to critics of his arrogation of power during the Civil War, Abraham Lincoln asked, "Was it possible to lose the nation, and yet preserve the Constitution?"[1] Today we might ask whether it is possible to lose the nation, and yet preserve the environment. Clearly, we would be willing to make some environmental sacrifices if they were necessary for the nation's very survival. Even if the threat were less grave or more ambiguous, we might decide that some extraordinary environmental damage would be justified to keep us strong militarily. Otherwise, the nation's defense establishment generally must follow the same rules for environmental protection that govern everybody else.

It was not always so. Until recently, it was widely assumed that we could not defend ourselves against external threats if we had to worry about the environmental consequences. This attitude was grounded largely in ignorance of the immediate risks to health and long-term costs to future generations of preparations for war. Like most everything else about national security, these environmental impacts were shrouded in secrecy. The secrets were guarded by a huge and seemingly impenetrable expertocracy. Efforts to ferret out information were often met with aspersions about the inquirer's common sense or even patriotism. The mind-numbing complexity of the subject added another barrier to access. Individual citizens, along with state and local governments, were accordingly excluded from decisions about the nation's defense that sometimes affected them directly and intimately.

Now we recognize that, with rare exceptions, we can maintain a strong, effective defense without endangering the public health or destroying our natural resources. Congress has decided as a matter of national policy that the environmental laws apply to defense-related activities pretty much the same way they do to others. Defense officials have developed sweeping new programs to apply those laws, vowing to keep the public better informed and often drawing their neighbors into the decision-making process. Armed with better information, members of the public have begun to take a greater interest and to demand compliance with the laws.

However, just bringing the defense community into full compliance with existing environmental laws will not be easy. For one thing, it will require changes in patterns of thinking and behavior established over forty-five years of Cold War. Environmental protection must be embraced by all segments of American society as a top national defense priority. Continued strong leadership will be required to bring about a genuine change of culture throughout the defense establishment. The change will have to be supported by an unprecedented financial commitment from the American people.

At the same time, existing laws need clarification to remove any lingering doubt that they provide baseline environmental standards for defense activities. Amendments are also needed to make the laws clearly apply to military actions abroad both in peacetime and in war. Finally, we need to develop a procedure for exigent circumstances to determine whether particular environmental injuries are indispensable to the survival of the nation or reasonably justified by national security threats, instead of leaving the decision entirely to the discretion of the President, as we do under existing law.

Prospects for extending and strengthening the environmental laws do not, for the moment, look favorable. The Republican majority in Congress elected in 1994 has promised to try to weaken those laws, even though most Americans still say protection of the environment is important to them. It also seems likely that the President and Congress will agree to reduce funding for environmental compliance by the Pentagon and DOE, and to drastically cut appropriations for defense facility cleanups.

Yet despite daunting political, financial, and technological challenges, there is plenty of reason for optimism that we can maintain a strong national defense that is also environmentally sound. It seems likely, too, that we will find a way to restore lands and waters already contaminated by our preparations for war.

Reason for Optimism

The American people have more information than ever before about this nation's defense efforts and about how those efforts affect the environment. They are certainly no less patriotic than in the past, nor less willing to pay what it costs to keep America strong and free. They are, however, less tolerant of needless threats to public health and the natural environment. They have begun to demand compliance with the environmental laws whenever possible, and explanations when compliance is impossible.

Government officials responsible for the nation's defense have also become far more sensitive in recent years to the importance of faithful compliance with the environmental laws. They have learned, for example, that military units may be more effective in combat if they pollute less, and that in general pollution prevention costs much less than cleaning up later on. These officials are determined to protect the environment for future generations of Americans whenever possible.

This confluence of interests is reason for optimism that we can avoid most unnecessary defense-related environmental injuries in the future and repair at least some of the damage from earlier preparations for war. Achievement of these goals will require an extraordinary level of commitment from everyone concerned. Officials at the Pentagon and the Department of Energy must be supported in their efforts to comply with the environmental laws and to keep the public informed and involved in the process. Congress must be encouraged to incorporate this commitment into enduring national policy. The American people will have to muster the political will to pay more for an environmentally sound national defense. And individual citizens must work to stay informed and insist on compliance with the law. The opportunities and challenges are briefly outlined below.

Legislative Challenges

As we have seen, the domestic environmental laws provide a useful, familiar framework for evaluating the environmental impacts of national defense activities. We have considerable experience applying these laws in a variety of settings. However, while Congress has declared that these laws generally apply to national defense activities, it has failed to provide clear guidance on a number of critical issues. Here are some suggestions for improvements.

There is general agreement that pollution prevention is more desirable than either pollution control or waste management. Congress expressed this policy preference when it passed the Pollution Prevention Act of 1990,[2] but it failed to include a mandate to implement the policy. President Clinton took the next step in 1993 when he signed an executive order directing all federal agencies to drastically reduce their production of toxic wastes through source reduction and recycling, and to develop and test innovative pollution prevention technologies.[3] Even before the executive order was issued, both the Pentagon and DOE had launched major programs to reduce their production of all kinds of wastes. They have also begun to write contracts for goods and services that require suppliers to practice recycling and conservation and to comply with environmental rules. Congress should follow the President's lead by passing new legislation to require federal agencies, defense contractors, and others to reduce their production of dangerous chemical wastes. The new measure should include performance standards and require recycling and conservation of new materials wherever possible. Like the other environmental laws, it should also provide for citizen enforcement of its provisions.

Two major regulatory statutes should be amended to remove any doubt that they apply to the manufacture of nuclear weapons and cleanup afterwards. The Clean Water Act ought to be made expressly applicable to discharges of radionuclides from DOE facilities, reversing the 1976 Supreme Court decision that denied its applicability.[4] Procedures adopted by DOE and its predecessors under the Atomic Energy Act have repeatedly failed to protect the nation's waters from this threat. They have also excluded citizens and states from the monitoring and regulation of these discharges. The Resource Conservation and Recovery Act also should be amended to remove the exemption of substances and activities subject to regulation under the Atomic Energy Act, instead of relying on EPA regulations to govern their handling indirectly as "mixed" radioactive wastes.[5] Congress has already given EPA regulatory authority over disposal of radioactive wastes under the Nuclear Waste Policy Act[6] and over emissions of radionuclides under the Clean Air Act.[7] It makes sense to combine the environmental regulation of all defense nuclear materials in the same agency that administers most of the other environmental laws.

The major regulatory environmental statutes place heavy reliance on states for their application and enforcement. The states must adhere to minimum federal standards but otherwise may take account of local conditions in administering the laws. They may even apply more stringent rules than those prescribed by EPA. The most important exception to this pattern of federal delegation is found in CERCLA, which relegates states to an advisory role in the formulation and enforcement of interagency agreements for the cleanup of federal facilities on the National Priorities List.[8] As we saw in chapter 4, one federal appeals court has decided that a state may carry out its own RCRA cleanup program at a federal NPL site, applying its own more stringent state standards.[9] It would make more sense to make the states full partners in CERCLA remediations, not to slow the cleanup process, but to enable states to insist on a nondiscriminatory basis that contaminated defense sites be cleaned up more thoroughly than EPA otherwise would. In the same vein, EPA should be stripped of its ability to waive the imposition of state cleanup standards that are stricter than EPA requirements.

In the 1992 Federal Facility Compliance Act, Congress amended RCRA to waive the federal government's sovereign immunity, so that states could levy civil penalties for federal facility violations.[10] The same legislation authorized

EPA to issue administrative orders and file suit in federal court for enforcement of RCRA against federal facilities, refuting the Justice Department's claim of a "unitary executive." Congress should enact similar clarifying amendments for each of the other environmental regulatory statutes. These changes would enable the EPA and state regulators to ensure that federal facilities comply with the laws "in the same manner and to the same degree as any nongovernmental entity."[11]

One reason for the uneven application of the environmental laws to national defense activities has been the reluctance of courts to order a halt to defense-related violations.[12] Congress ought to remove that discretion from the courts by expressly requiring injunctions for all violations.

Congress needs to get on with the business of destroying United States chemical weapons inventories.[13] Over the last several years, it has closely monitored Army plans to incinerate the remaining weapons at eight sites around the country. Now Congress should abandon its insistence on the most cost-effective method of disposal and instead order the Army to employ the safest methods, possibly including alternatives to incineration, to finish the job as soon as possible. It is imperative to meet the deadline set out in the 1993 Chemical Weapons Convention for complete destruction of weapons stocks by the year 2005.

Finally, it is shockingly hypocritical for United States military forces to apply one set of environmental performance standards at home and another, laxer one everywhere else.[14] The Defense Department's recent decision to apply more stringent standards at active bases abroad than host country laws might require is a step in the right direction. Congress needs to take the initiative and declare unequivocally that all of the environmental statutes, including NEPA, apply to United States military activities abroad, unless host country laws or applicable status of forces agreements are more stringent. Congress should also make clear that it is as important to protect the environment in wartime as it is in time of peace.

It must be stressed that none of the changes suggested here would in any way hamper the efforts of the United States military and the Department of Energy to defend the nation, since each of the major environmental regulatory statutes provides for waiver of its requirements when it is in the "paramount interest of the United States."[15] If the President decides that a certain environmental sacrifice is required, he is entitled to order it, provided only that he explains his decision to Congress and the American people.

The current anti-regulatory mood in the Republican-controlled 104th Congress makes it seem unlikely that any of these changes will be enacted into law anytime soon. One measure currently under discussion would make enforcement of the various environmental laws turn on the outcome of an extensive analysis of the resulting economic impacts.[16] Some fear that such a requirement could paralyze the regulatory process and even lead to a dismantling of the entire federal environmental protection apparatus. However, proposed changes that would weaken the environmental laws are designed to provide financial relief for regulated industries and remove restrictions on the development of private land. As taxpayers, of course, we want the Pentagon and Department of Energy to operate as efficiently as possible. But these agencies do not measure their success in dollars-per-share, the way private corporations do. Their mission is to defend the nation and its citizens, present and future. Curtailing enforcement of the current health-based environmental laws against the military and DOE on economic grounds would be tantamount to financing America's defense establishment by increasing death and morbidity among its citizens. If the proposed changes are to be adopted for the benefit of private industries and landowners, the existing rules should nevertheless be retained and strengthened for federal agencies.

Administrative Challenges

Even without the legislative reforms just suggested, the Departments of Energy and Defense have ample authority under existing law to prevent most defense-related environmental damage. In fact, defense officials have begun to make environmental protection a central component of their overall defense mission.[17] A genuine change in attitude is evident among the staffs of both agencies. This new way of thinking must be encouraged and rewarded.

Both DOE and the Pentagon have done a much better job recently of informing the public about actions affecting local environments. The two defense agencies have begun publishing annual five-year plans for environmental compliance that make it easier for citizens to understand what is going on. They have organized representative interest groups at the national and local levels and invited those groups to become involved in planning and decision-making. Local installation newsletters and open houses have helped to foster a feeling of cooperation and have boosted public confidence in the good faith efforts of federal officials. On a national level, the declassification of millions of secret agency records, especially by the Department of Energy, and new procedures to discourage the unnecessary use of the "classified" stamp, mark a sharp change from earlier practice. This new openness is important in our democracy, because under our system it is finally the people who are supposed to make policy choices about how to defend the nation, the people who pay for implementation of those policies, and the people who must be satisfied that the policies are being implemented.

Both defense agencies are working hard to monitor their own performances and to improve internal communications. They now have large staffs devoted to environmental compliance and restoration. And they have developed extensive programs for training agency employees about their environmental responsibilities. Military and DOE officials have also begun writing environmental performance standards into contracts for goods and services, including contracts for operation of government facilities, and they have begun to monitor contractor activities more closely than in the past. To protect the government from open-ended liability for cleanup costs and environmental damages, both fixed-price and cost-reimbursable contracts should stipulate that the contractor will indemnify the government for any such claims.

The Department of Defense needs to incorporate coherent and consistent environmental analysis into its planning for armed conflict. Operation plans and rules of engagement should be routinely vetted for conformity to environmental performance standards, just as they are currently reviewed for compliance with the law of war. In the same vein, combat training and conventions for the use of particular weapons and tactics need to reflect an awareness of the environmental consequences.

The Department of Energy and the Pentagon need strong support from other parts of the executive branch. The Environmental Protection Agency must continue its efforts to improve communications with the defense establishment. New regulations are needed, for example, to provide guidance for cleanup of defense facilities on the National Priorities List. For its part, the Justice Department should formally renounce its so-called "unitary executive" policy adopted early in the Reagan administration, so EPA can enforce the law against federal facilities the same way it does against everyone else.[18] DOE should also vigorously and consistently enforce the criminal sanctions of the environmental statutes against officials of federal agencies and defense contractors. The 1989 criminal conviction of three Army civilian employees for RCRA violations at the Aberdeen Proving Grounds may have done more to capture the imaginations of defense officials than any development before or since. On the other hand, the Justice Department's refusal to enforce RCRA's criminal provisions against agency and contractor officials at DOE's Rocky Flats nuclear weapons plant has only bred cynicism and disrespect for the law.[19]

Financial Challenges

Cleaning up after the Cold War is going to be enormously expensive, although the exact cost cannot be accurately predicted. For example, we will have to pay more for environmental remediation at DOE's nuclear weapons complex than we have already paid to manufacture the weapons. The precise nature and extent of environmental contamination from all defense activities is still unknown, and the technology needed to carry out some of the cleanup has yet to be invented. Nor have we found a procedure or identified a place for safe disposal of all the radioactive waste and left-over weapons grade materials. Still, we must devote our best efforts to repairing a half-century of environmental damages as quickly as possible.

It may be difficult for us to sustain the political will to make the necessary expenditures. No one doubts that future federal budgets will be tight. Defense-related environmental protection will have to compete for funding with public education, health care, and myriad other discretionary programs, including of course military operations, hardware, and personnel. Following the Republican party's takeover of Congress in November 1994, for example, President Clinton announced that he would finance proposed middle-class tax cuts by reductions in federal spending that would take some $4.4 billion from DOE funding for environmental remediation over the next five years.[20] Two leading Republican members of the Senate Armed Services Committee have proposed cutting $5 billion from Defense Department environmental spending between 1995 and 1999.[21] The resulting delays in cleanup would almost certainly allow the wider dissemination of dangerous contaminants into the environment. When the cleanup was eventually completed, it would be more complicated and more expensive. The additional cost, as well as the additional risk of exposure, would have to be borne by future generations of Americans. If we decide not to clean up after ourselves now, members of the current generation may be glad they will not be around when the bill finally must be paid.

Diplomatic Challenges

Despite reductions in the nuclear arsenals of the United States and the former Soviet Union, nuclear war remains by far the gravest threat to the environment of the whole earth. The United States must redouble its efforts to reduce the numbers of such weapons and prevent their further spread among developing nations. With the recent unconditional extension of the Nuclear Non-Proliferation Treaty,[22] it is important now to press emerging nuclear powers such as Israel, Pakistan and India to join the treaty.

The current moratorium on nuclear weapons testing provides an effective bar to the development of new weapons. It also sends an important signal to other nations that the United States has withdrawn from the arms race. The United States should spare no effort to negotiate and conclude a Comprehensive Test Ban Treaty, which would include China, while conditions seem favorable.

United States renunciation of its long-standing policy of willingness to use nuclear weapons first in a conventional conflict would go far to reduce the danger of a nuclear war.[23] The United States should likewise abandon its development of anti-ballistic missile defenses (the Strategic Defense Initiative or "Star Wars" of the 1980s), the deployment of which would almost certainly violate the 1972 Anti-Ballistic Missile Treaty.[24] Such defensive weapons could also rekindle the arms race by encouraging a potential adversary to build large numbers of offensive nuclear weapons in order to deter a United States preemptive first strike.

Reductions in nuclear weapons stocks might be speeded by additional research into the environmental risks of nuclear war. So far, the Defense Department has refused to disclose whether or how such risks have figured in United States strategic planning.[25] Publication of such information would enable members of the public to participate in policy choices about whether it makes sense to continue manufacturing and testing nuclear weapons or to plan for their use in war.

The United States should quickly ratify and implement the 1993 Chemical Weapons Convention.[26] If we do not develop the technology and spend the money necessary to destroy our own chemical arsenals, we can hardly insist that other nations do so. Moreover, the former Soviet Union will almost surely need both technical and financial assistance from the United States in disposing of its chemical weapons inventories.

Finally, the United States can reduce the level of indiscriminate environmental destruction in wartime by halting its export of weapons of mass destruction, including land mines, cluster bombs, and missiles, and by putting pressure on other nations to do the same. Even if we could assume that transferee nations would use such weapons only in accordance with United States policy, experience has shown time and again that Arms Export Control Act limitations on retransfers by those nations[27] are ineffective. Despite the importance of preserving markets for domestic defense contractors when military budgets are shrinking, the United States must not continue to wage war indirectly on the people and environments of other nations. Nor can we ignore the possibility that such weapons might be turned against us if war ever returns to the American homeland. Our loss of control over Stinger antiaircraft missiles transferred to the Mujahadeen in Afghanistan provides a terrifying reminder of the potential danger.

Challenges to Individual Citizens

In the end, the responsibility for an environmentally responsible national defense rests with an active, informed electorate. The purpose of all military activities, of course, is to protect both the nation and its citizens. The American citizen has the most to lose from an ineffectual national defense. The same citizen will bear the cost of defense-related environmental excesses.

One need not be a chemist or a lawyer or an expert in military strategy to play a significant role in the debate about national defense and environmental protection. What is necessary is to recognize the importance of the issues, to invest the time and labor required to find out what is going on, and to have the courage to participate. Those who work in the defense establishment must, of course, faithfully execute the law, including environmental statutes and regulations. The rest of us must monitor the efforts of DOE and the Pentagon to defend the country, insisting on compliance with the environmental laws, applauding them when they comply. We all must use our imaginations to offer suggestions for improvement. And we must be willing to pay the bills, rather than passing the costs along to our children and grandchildren. It is part of our civic duty, whether we teach school, work in the home, operate a dairy farm, or design nuclear weapons for the Department of Energy. We are all in this together, and it is going to take all our best efforts to keep America strong, free, healthy, and beautiful.

Appendix A

Environmental Cases Involving National Defense Activities

Reported judicial decisions raising national defense and environmental issues are listed alphabetically by name for each of the major federal environmental statutes. Some unreported cases are also included. In general, when more than one decision has been reported for a single case, as often happens on appeal, only the most recent decision is included. A single case may be listed under more than one statute.

Contents

Base Closure and Realignment Acts
Pub. L. No. 100-526, 102 Stat. 2627 (1988)
Pub. L. No. 101-510, 104 Stat. 1808 (1990)

Dalton v. Specter, 112 S. Ct. 1719 (1994) (closure of Philadelphia Naval Shipyard).

Cohen v. Rice, 992 F.2d 376, 380 (1st Cir. 1993) (closure of Loring Air Force Base in Maine).

County of Seneca v. Cheney, 806 F. Supp. 387 (W.D. N.Y. 1992), preliminary injunction vacated, 12 F.3d 8 (2d Cir. 1994) (elimination and transfer of functions from Seneca Army Depot).

Greenwood v. Dalton, 1993 U.S. Dist. LEXIS 15369, 15374 (E.D. Pa.) (characterization of Naval Air Warfare Center—Warminster as military installation subject to 1990 Base Closure Act).

People ex rel. Hartigan v. Cheney, 726 F. Supp. 219 (C.D. Ill. 1989) (closure of Chanute Air Force Base and Army's Fort Sheridan).

Clean Air Act
42 U.S.C. §§7401–7671q (1988 & Supp. V 1993)

Alabama v. Seeber, 502 F.2d 1238 (5th Cir. 1974) (Army and TVA must obtain state air pollution permits).

Alabama ex rel. Graddick v. Veterans Administration, 648 F. Supp. 1208 (M.D. Ala. 1986) (state may impose civil penalties for violations at VA hospital).

Barcelo v. Brown, 478 F. Supp. 646, 672–674 (D.P.R. 1979) (discharges of ordnance and dust from Navy roadbuilding do not violate EPA or local standards).

California ex rel. State Air Resources Board v. Department of the Navy, 431 F. Supp. 1271 (N.D. Cal. 1977) (Navy jet engine test cells must conform to state and local standards under state implementation plan; however, no state civil penalties for violations).

Conservation Law Foundation, Inc. v. Department of the Air Force, No. 1:92-CV-156-M (D.N.H. 1992) (compliance with conformity provision upon closure of Pease Air Force Base).

Hancock v. Train, 426 U.S. 167 (1976) (Army, TVA, and Atomic Energy Commission facilities not required to obtain permit from state with federally approved implementation plan).

New York State Department of Environmental Conservation v. United States Department of Energy, 772 F. Supp. 91 (N.D.N.Y. 1991) (state may recover environmental regulatory costs for federal facilities).

Ohio ex rel. Celebreeze v. United States Department of the Air Force, 1987 WL 110399, 17 *Environmental Law Reporter* (Environmental Law Institute) 21,210 (S.D. Ohio 1987) (state may impose civil penalties for violations at two Air Force facilities in Ohio).

Idaho Department of Health and Welfare v. United States, 959 F.2d 149 (9th Cir. 1992) (shipments of spent nuclear fuel to DOE facility not enjoined as violation of state air quality standards).

United States v. New Mexico, 1992 WL 437,983 (D.N.M.) (state may regulate radioactive air emissions at Los Alamos National Laboratory).

United States v. South Coast Air Quality Management District, 748 F. Supp. 732 (C.D. Cal. 1990) (state may recover environmental regulatory fees from federal military installations in California).

Clean Water Act
33 U.S.C. §§1251–1387 (1988 & Supp. V 1993)

California v. United States Dept. of the Navy, 845 F.2d 222 (9th Cir. 1987) (only EPA, not the state, authorized to seek civil penalties for Navy's discharge of improperly treated waste into San Francisco Bay).

Friends of the Earth v. Hall, 693 F. Supp. 904 (W.D. Wash. 1988) (dredge and fill permit issued to Navy for construction of Everett, Washington, homeport was arbitrary and capricious).

Heart of America Northwest v. Westinghouse Hanford Co., 830 F. Supp. 1265 (E.D. Wash. 1993) (action for Clean Water Act violation at DOE's Hanford Reservation barred by ongoing CERCLA cleanup).

Legal Environmental Assistance Foundation, Inc. v. Hodel, 586 F. Supp. 1163 (E.D. Tenn. 1984) (DOE required to obtain permit for discharges at Oak Ridge nuclear weapons plant).

McClellan Ecological Seepage Situation v. Weinberger, 655 F. Supp. 601, 604 (E.D. Cal. 1986) (no state penalties for violations at McClellan Air Force Base in California).

McClellan Ecological Seepage Situation v. Weinberger, 707 F. Supp. 1182 (E.D. Cal. 1988) (violations of discharge permit limitations at McClellan Air Force Base in California).

McClellan Ecological Seepage Situation v. Weinberger, 1990 WL 117920 (E.D. Cal.) (no discharge permit required for hazardous waste leaking from burial pit into underlying aquifer).

Metropolitan Sanitary District of Greater Chicago v. United States Department of the Navy, 737 F. Supp. 51 (N.D. Ill. 1990) (no civil penalties for violation of agreement concerning indirect discharge to municipal sewage treatment plant by Glenview Naval Air Station).

Natural Resources Defense Council v. Watkins, 954 F.2d 974 (4th Cir. 1992) (action to prevent restart of K-reactor at DOE's Savannah River plant until cooling tower completed).

New York v. United States, 620 F. Supp. 374 (E.D.N.Y. 1985) (citizen suit not available for past groundwater contamination resulting in violation of water quality standards at former Suffolk County Air Force Base).

New York State Dept. of Environmental Conservation v. United States Dept. of Energy, 772 F. Supp. 91 (N.D.N.Y. 1991) (testing imposition of state regulatory charges on DOD and DOE facilities).

Ohio v. United States Department of Energy, 503 U.S. 607 (1992) (United States sovereign immunity from liablity for state penalties not waived by either federal facility or citizen suit provision).

Public Interest Research Group of New Jersey, Inc. v. Rice, 774 F. Supp. 317 (D.N.J. 1991) (discharges at McGuire Air Force Base in excess of permit limitations enjoined in citizen's suit).

Train v. Colorado Public Interest Research Group, 426 U.S. 1 (1976) (no Clean Water Act permit required for discharges of radionuclides from DOE's Rocky Flats nuclear weapons plant).

United States v. Curtis, 988 F.2d 946 (9th Cir. 1993) (federal official criminally liable for violation).

United States v. Pennsylvania Department of Environmental Resources, 778 F. Supp. 1328 (M.D. Pa. 1991) (discharge permit required for contaminated storm water runoff at Navy Ships Parts Control Center).

United States v. Pennsylvania Environmental Hearing Board, 584 F.2d 1273 (3rd Cir. 1978) (private corporation operating government ammunition plant not shielded from state penalties for violating state water pollution statutes).

United States v. Puerto Rico, 721 F.2d 832 (1983) (federal court may hear appeal of denial of water quality certification by Puerto Rico concerning Navy's application for discharge permit at its Vieques Island training facility).

Weinberger v. Romero-Barcelo, 456 U.S. 305 (1982) (Navy must apply for permit to discharge ordnance into water at Vieques Island, Puerto Rico, training facility).

Werlein v. United States, 746 F. Supp. 887 (D. Minn. 1990) (action for Clean Water Act violation at Twin Cities Army Ammunition Plant barred by ongoing CERCLA cleanup).

Coastal Zone Management Act
16 U.S.C. §§1451–1464 (1988 & Supp. V 1993)

Barcelo v. Brown, 478 F. Supp. 646, 680–688 (D.P.R. 1979) (Navy training activities on federal lands and waters on Vieques Island are not subject to Act, and do not significantly affect other coastal areas).

City and County of San Francisco v. United States, 615 F.2d 498 (9th Cir. 1980) (lease of Hunter's Point Naval Shipyard to private company).

Friends of the Earth v. United States Navy, 841 F.2d 927 (9th Cir. 1988) (Navy enjoined from dredging at Everett, Washington aircraft carrier homeport without state permit); 850 F.2d 599 (9th Cir. 1988) (injunction dissolved upon issuance of state permit).

Comprehensive Environmental Response, Compensation, and Liability Act (CERCLA)
42 U.S.C. §§9601–9675 (1988 & Supp. V 1993)

Clark v. United States, 16 *Environmental Law Reporter* (Environmental Law Institute) 20057 (Cl. Ct. 1985) (contamination of aquifer at McChord Air Force Base in Washington).

Colorado v. United States Dept. of the Army, 707 F. Supp. 1562 (D. Colo. 1989) (Colorado may enforce state RCRA cleanup plan for CERCLA site not on NPL at Army's Rocky Mountain Arsenal).

Conservation Law Foundation, Inc. v. Department of the Air Force, No. 1:92-CV-156-M (D.N.H. 1992) (transfer of contaminated lands at closing Pease Air Force Base in New Hampshire).

Conservation Law Foundation of New England, Inc. v. Reilly, 950 F.2d 38 (1st Cir. 1991) (no organizational standing to test EPA actions at federal facilities nationwide).

FMC Corp. v. United States Department of Commerce, 786 F. Supp. 471 (E.D. Pa. 1992), affirmed, 1994 WL 314814 (3d Cir.) (government jointly liable for cleanup costs at site of World War II defense plant).

Hanford Downwinders Coalition, Inc. v. Dowdle, 841 F. Supp. 1050 (E.D. Wash. 1993) (court lacked jurisdiction to order medical surveillance and other health studies during ongoing CERCLA cleanup at DOE's Hanford Reservation).

Heart of America Northwest v. Westinghouse Hanford Co., 820 F. Supp. 1265 (E.D. Wash. 1993) (citizen suit to abate Clean Water Act and RCRA violations barred by ongoing CERCLA cleanup at DOE's Hanford Reservation).

In re Hanford Nuclear Reservation Litigation, 780 F. Supp. 1551 (E.D. Wash. 1991) (claims for abatement and remediation of health hazards barred by ongoing CERCLA cleanup).

Key Tronic Corp. v. United States, 114 S. Ct. 1960 (1994) (Air Force liability for response costs at private disposal site did not include some attorney fees).

Maine v. Department of the Navy, 973 F.2d 1007 (1st Cir. 1992) (no waiver of sovereign immunity from state penalties for violations at Navy shipyard in Kittery, Maine).

New York v. United States, 620 F. Supp. 374 (E.D.N.Y. 1985) (Air Force may be responsible for cleanup costs at former Suffolk County Air Force Base).

Redland Soccer Club, Inc. v. Department of the Army, 801 F. Supp. 1432 (M.D. Pa. 1992) (state cleanup laws apply only to sites currently owned by federal government).

Rospatch Jessco Corp. v. Chrysler Corp., 829 F. Supp. 224 (W.D. Mich. 1993) (state cleanup laws apply only to sites currently owned by federal government).

Town of Bedford v. Raytheon Company, 755 F. Supp. 469 (D. Mass. 1991) (natural resource damages available to "state" are not recoverable by town for Air Force and Navy pollution of town's principal drinking water source).

United States v. Allied Signal Corp., 736 F. Supp. 1553 (N.D. Cal. 1990) (former owner of Concord Naval Station property responsible for cleanup costs).

United States v. Carr, 880 F.2d 1550 (2nd Cir. 1989) (relatively low-ranking civilian employee at Army's Fort Drum required to report release as person "in charge" of facility).

United States v. Colorado, 990 F.2d 1565 (10th Cir. 1993), cert. denied, 114 S. Ct. 922 (1994) (state RCRA corrective action may proceed in parallel with CERCLA cleanup of NPL site at Army's Rocky Mountain Arsenal).

United States v. Commissioner of Pennsylvania Department of Environmental Resources, 778 F. Supp. 1328 (M.D. Pa. 1991) (Pennsylvania DER may regulate cleanup of contaminated drainway at Navy Ships Parts Control Center).

United States v. Hooker Chemical and Plastics Corp., 136 F.R.D. 559 (W.D.N.Y. 1991) (discovery concerning Army's contribution to contamination of Love Canal).

United States v. Shell Oil Co., 605 F. Supp. 1064 (D. Colo. 1985) (oil company liable for cleanup costs at Army's Rocky Mountain Arsenal).

United States v. Werlein, 746 F. Supp. 887 (D. Minn. 1990) (no citizen suit challenge to ongoing CERCLA cleanup at Twin Cities Army Ammunition Plant; town may not recover natural resource damages for pollution of drinking water source).

Endangered Species Act
16 U.S.C. §§1531–1544 (1988 & Supp. V 1993)

Allied-Signal Inc. v. Lujan, 736 F. Supp. 1558 (N.D. Calif. 1990) (plaintiff lacked standing to challenge Navy and Department of the Interior drainage of marsh harboring endangered salt-marsh harvest mice).

Animal Lovers Volunteer Association, Inc. v. Weinberger, 765 F.2d 937 (9th Cir. 1985) (citizens lacked standing to challenge Navy's shooting of feral goats on its property).

Catholic Action of Hawaii/Peace Education Project v. Brown, 643 F.2d 569 (9th Cir. 1980), reversed on other grounds, 454 U.S. 139 (1981) (application of Act to be based on information in environmental impact statement for storage of nuclear weapons on Oahu).

Pyramid Lake Paiute Tribe of Indians v. United States Department of the Navy, 898 F.2d 1410 (9th Cir. 1990) (Navy justified in relying on Fish & Wildlife Service biological opinion when it leased land with water rights).

Romero-Barcelo v. Brown, 643 F.2d 835, 856–858 (1st Cir. 1981) (Navy violated Act when it failed to obtain a biological opinion from Fish & Wildlife Service for training on Vieques Island).

Stop H-3 Association v. Lewis, 538 F. Supp. 149 (D. Hawaii 1982) (construction of highway to connect Kaneohe Marine Corps Air Station to Pearl Harbor Naval Base and Hickham Air Force Base, Hawaii).

Federal Insecticide, Fungicide and Rodenticide Act (FIFRA) 7 U.S.C. §§136–139y (1988 & Supp. V 1993)

No reported cases found.

Federal Tort Claims Act 28 U.S.C. §§1346(b), 2671–2680 (1988)

Allen v. United States, 816 F.2d 1417 (10th Cir. 1987) (radiation injuries from atmospheric testing of nuclear weapons).

Barnson v. United States, 816 F.2d 549 (10th Cir.), cert. denied, 484 U.S. 896 (1987) (radiation injuries to uranium miners).

Begay v. United States, 768 F.2d 1059 (9th Cir. 1985) (radiation injuries to uranium miners).

Clark v. United States, 660 F. Supp. 1164 (W.D. Wash. 1987), affirmed, 856 F.2d 1433 (9th Cir. 1988) (contamination of aquifer at McChord Air Force Base in Washington).

Clark v. United States, 19 Cl. Ct. 220 (1990) (same).

Daigle v. Shell Oil Company, 972 F.2d 1527 (10th Cir. 1992) (air pollution from cleanup of Basin F at Army's Rocky Mountain Arsenal).

Dickerson, Inc. v. United States, 875 F.2d 1577 (11th Cir. 1989) (conveyance of waste oil contaminated with PCBs).

Duff v. United States, 999 F.2d 1280 (8th Cir. 1993) (fumes from painting military housing unit).

In re Agent Orange Product Liability Litigation, 818 F.2d 145 (2d Cir. 1987) (injuries from spraying of herbicides in Vietnam).

In re Consolidated United States Atmospheric Testing Litigation, 820 F.2d 982 (9th Cir. 1987) (radiation injuries from atmospheric testing of nuclear weapons).

Jones v. United States, 698 F. Supp. 826 (D. Hawaii 1988) (injuries from termite control pesticide).

Kirchmann v. United States, 8 F.3d 1273, 1274 (8th Cir. 1993) (pollution of groundwater from contruction of missile facility).

McKay v. United States, 703 F.2d 464 (10th Cir. 1983) (plutonium contamination of land near DOE's Rocky Flats nuclear weapons plant).

Nevin v. United States, 696 F.2d 1229 (9th Cir. 1983) (death resulting from Army's 1950 simulated biological warfare attack on San Francisco).

Prescott v. United States, 973 F.2d 696 (9th Cir. 1992) (radiation injuries to 220 workers at Nevada Test Site).

Roberts v. United States, 887 F.2d 899 (9th Cir. 1989) (radioactive fallout from venting of underground nuclear test).

Santa Fe Pacific Realty v. United States, 780 F. Supp. 687 (E.D. Cal. 1991) (sale of hazardous chemicals without proper packaging and handling).

Starrett v. United States, 847 F.2d 539 (9th Cir. 1988) (groundwater pollution from Navy submarine base).

Hazardous Materials Transportation Act
49 U.S.C. App. §§1801–1819 (1988 & Supp. V 1993)

City of New York v. United States Department of Transportation, 715 F.2d 732 (2d Cir. 1983) (regulation for highway shipment of radioactive materials from DOE's Brookhaven National Laboratory).

City of New York v. United States Department of Transportation, 700 F. Supp. 1294 (S.D.N.Y. 1988) (same).

Marine Mammal Protection Act
16 U.S.C. §§1361–1384 (1988 & Supp. V 1993)

Citizens to End Animal Suffering and Exploitation v. The New England Aquarium, 836 F. Supp. 45 (D. Mass. 1993) (citizens lacked standing to challenge transfer of dolphin from aquarium to Navy research facility).

Progressive Animal Welfare League v. Department of the Navy, 725 F. Supp. 475 (W.D. Wash. 1989) (military use of Atlantic bottlenose dolphins at Bangor, Washington Trident submarine base).

Marine Protection, Research and Sanctuaries Act (Ocean Dumping Act)
33 U.S.C. §§1401–1445 (1988 & Supp. V 1993)

Barcelo v. Brown, 478 F. Supp. 646, 667 (D.P.R. 1979) (discharge of ordnance into water at Navy's Vieques Island training area not covered by Act).

National Environmental Policy Act (NEPA)
42 U.S.C. §§4331–4361 (1988)

Aleut League v. Atomic Energy Commission, 337 F. Supp. 534 (D. Alaska 1971) (underground nuclear weapons test on Amchitka Island, Alaska).

Aluli v. Brown, 437 F. Supp. 602 (D. Hawaii 1977) (Navy bombing range on Hawaiian island).

Animal Lovers Volunteer Association v. Weinberger, 765 F.2d 937 (9th Cir. 1985) (shooting feral goats on Navy property).

Barcelo v. Brown, 478 F. Supp. 646 (D.P.R. 1979) (Navy bombing practice on Vieques Island, Puerto Rico).

Bargen v. Department of Defense, 623 F. Supp. 290 (D. Nev. 1985) (bombing range on public lands in Nevada).

Breckinridge v. Rumsfeld, 537 F.2d 864, 867 (6th Cir. 1976) (closing Lexington-Bluegrass Army depot and transferring 18 military, 2,630 civilian employees).

Citizens for Reid Park v. Laird, 336 F. Supp. 783 (D. Me. 1972) (mock amphibious landings on state park beach in Maine).

City and County of San Francisco v. United States, 615 F.2d 498 (9th Cir. 1980) (lease of Hunter's Point Naval Shipyard to private company).

City of New York v. United States Dept. of Energy, 715 F.2d 732 (2d Cir. 1983) (Department of Transportation regulations for shipment of radioactive materials from Brookhaven National Laboratory).

Committee for Nuclear Responsibility, Inc. v. Seaborg, 463 F.2d 796 (D.C. Cir. 1971) (underground nuclear weapons test on Amchitka Island, Alaska).

Concerned Citizens for the 442nd T.A.W. v. Bodycombe, 538 F. Supp. 184 (W.D. Mo. 1982) (deactivation of Air Force Reserve unit in Missouri).

Conservation Law Foundation, Inc. v. Department of the Air Force, No. 1:92-CV-156-M (D.N.H. 1992) (transfer of property following closure of Pease Air Force Base in New Hampshire).

Conservation Law Foundation of New England, Inc. v. General Service Administration, 707 F.2d 626 (1st Cir. 1983) (leasing and sale of former Navy facilities in Rhode Island).

Conservation Law Foundation of New England, Inc. v. United States Department of the Air Force, 1987 WL 46,370 (D. Mass.) (construction of radio towers to transmit instructions for launch of nuclear weapons).

Concerned About Trident v. Rumsfeld, 555 F.2d 817 (D.C. Cir. 1977) (construction of Trident submarine base in Bangor, Washington).

County of Seneca v. Cheney, 806 F. Supp. 387 (W.D.N.Y. 1992), preliminary injunction vacated, 12 F.3d 8 (1994) (realignment of Seneca Army Depot in New York).

Davidson v. Department of Defense, 560 F. Supp. 1019 (S.D. Ohio 1982) (use of Rickenbacker Air National Guard base in Ohio for civilian air cargo facility).

Don't Ruin Our Park v. Stone, 1992 WL 220000 (M.D. Pa. 1992) (relocation of a national guard training facility near Pennsylvania state park).

Drug Policy Foundation v. Bennett, No. C-90-2278 FMS (N.D. Cal. 1991) (Army involvement in domestic war on drugs).

Dunn v. United States, 842 F.2d 1420 (3d Cir. 1988) (DOE cleanup of Pennsylvania uranium ore processing plant).

Foundation on Economic Trends v. Carlucci, 1988 WL 60333 (D.D.C.) (electromagnetic pulse from simulated nuclear weapon test in North Carolina).

Foundation on Economic Trends v. Weinberger, 610 F. Supp. 829 (D.D.C. 1985) (toxin weapons testing at Dugway Proving Ground, Utah).

Friends of the Earth v. Hall, 693 F. Supp. 904 (W.D. Wash. 1988) (Navy plan to dredge and fill at Everett, Washington homeport).

Friends of the Earth v. Weinberger, 562 F. Supp. 265 (D.D.C. 1983) (siting of MX missile).

Greenpeace USA v. Stone, 748 F. Supp. 749 (D. Hawaii 1990), dismissed in part as moot, 924 F.2d 175 (9th Cir. 1991) (disposal of chemical weapons shipped from Germany to Johnston Island).

Hudson River Sloop Clearwater, Inc. v. Department of the Navy, 891 F.2d 414 (2nd Cir. 1989) (construction of Staten Island homeport for battleship group alleged to carry nuclear weapons).

Image of Greater San Antonio v. Brown, 570 F.2d 517 (5th Cir. 1978) (reduction in force at Kelly Air Force Base).

Jackson County, Mo. v. Jones, 571 F.2d 1004 (8th Cir. 1978) (movement of 2,000 military and 850 civilian workers from Richards-Gebaur Air Force Base near Kansas City, Missouri, to Scott Air Force Base near St. Louis).

Laine v. Weinberger, 541 F. Supp. 599 (C.D. Calif. 1982) (possible storage of nuclear weapons at Seal Beach Naval Weapons Station in California).

Maryland-National Capital Park and Planning Commission v. Martin, 447 F. Supp. 350 (D.D.C. 1978) (consolidation of two military mapping facilities in Maryland).

McQueary v. Laird, 449 F.2d 608 (10th Cir. 1971) (Army storage of chemical and biological weapons at Rocky Mountain Arsenal).

National Association of Government Employees v. Rumsfeld, 413 F. Supp. 1224 (D.D.C. 1976) (loss of 1800 civilian jobs due to realignment of Pueblo Army Depot in Colorado).

National Association of Government Employees v. Rumsfeld, 418 F. Supp. 1302 (E.D. Pa. 1976) (loss of 3,500 civilian jobs due to closure of Frankford Arsenal in Philadelphia).

National Wildlife Federation v. Adams, 629 F.2d 587 (9th Cir. 1980) (highway for Trident submarine base in Bangor, Washington).

Natural Resources Defense Council, Inc. v. Callaway, 524 F.2d 79 (2d Cir. 1975) (Navy dumping of polluted dredged spoil in Long Island Sound).

Natural Resources Defense Council, Inc. v. Marsh, 568 F. Supp. 1387 (E.D.N.Y. 1983) (Navy use of recreation area on Staten Island for offices).

Natural Resources Defense Council, Inc. v. Nuclear Regulatory Commission, 606 F.2d 1261 (D.C. Cir. 1979) (construction of storage tanks for high-level radioactive wastes at Hanford Reservation).

Natural Resources Defense Council, Inc. v. Seamans, 606 F.2d 1261 (D.C. Cir. 1979) (DOE construction of storage tanks for high-level radioactive waste from nuclear weapons program at Hanford and Savannah River facilities).

Natural Resources Defense Council, Inc. v. Vaughn, 566 F. Supp. 1472 (D.D.C. 1983) (restart of L-reactor at Savannah River weapons plant).

NEPA Coalition of Japan v. Aspin, 837 F. Supp. 466 (D.D.C. 1993) (U.S. military operations in Japan).

Nielson v. Seaborg, 348 F. Supp. 1369 (D. Utah 1972) (underground nuclear weapons tests in Utah).

No Gwen Alliance of Lane County, Inc. v. Aldridge, 855 F.2d 1380 (9th Cir. 1988) (construction of radio towers to transmit instructions for launch of nuclear weapons).

North Carolina v. Federal Aviation Admin., 957 F.2d 1125 (4th Cir. 1992) (designation of restricted air space for Navy bombing practice).

People of Enewetak v. Laird, 353 F. Supp. 811 (D. Hawaii 1973) (simulated nuclear explosion on Pacific island).

Prince George's County, Maryland v. Holloway, 404 F. Supp. 1181 (D.D.C. 1975) (transfer of Naval Oceanographic Program civilian personnel from Suitland, Maryland to Bay St. Louis, Mississippi).

Progressive Animal Welfare Society v. Department of the Navy, 725 F. Supp. 475 (D. Wash. 1989) (capture of dolphins for use at Trident submarine base).

Protect Key West, Inc. v. Cheney, 795 F. Supp. 1552 (S.D. Fla. 1992) (Navy housing project).

Protect Key West, Inc. v. Cheney, 1992 WL 219999 (S.D. Fla.) (Navy housing project).

Public Service Co. of Colorado v. Andrus, 825 F. Supp. 1483, as modified, 1993 WL 388312 (D. Idaho 1993) (shipment, processing, and storage of spent nuclear fuel at Idaho National Engineering Laboratory).

Pyramid Lake Paiute Tribe v. United States Department of the Navy, 898 F.2d 1410 (9th Cir. 1990) (leasing land and water rights said to jeopardize endangered species of fish in Nevada).

Romer v. Carlucci, 847 F.2d 445 (8th Cir. 1988) (en banc) (deployment of MX missles in Minuteman silos).

Shiffler v. Schlesinger, 548 F.2d 96 (3d Cir. 1977) (transfer of 2,485 employees and $2 million payroll from Ft. Monmouth, New Jersey, to Ft. Gordon, Georgia).

Sierra Club v. United States Department of Energy, 1991 WL 270190, 33 ERC 1720, 22 *Environmental Law Reporter* 20076 (D. Colo. 1991) (incineration of hazardous wastes at Rocky Flats nuclear weapons plant).

Smith v. Schlesinger, 371 F. Supp. 559 (C.D. Cal. 1974) (realignment of Long Beach Naval Shipyard, California).

Society for Animal Rights, Inc. v. Schlesinger, 512 F.2d 915 (D.C. Cir. 1975) (DOD destruction of blackbirds on military bases in Tennessee and Kentucky).

Stop H-3 Association v. Lewis, 538 F. Supp. 149 (D. Hawaii 1982) (construction of highway to connect Kaneohe Marine Corps Air Station to Pearl Harbor Naval Base and Hickham Air Force Base, Hawaii).

Tongass Conservation Society v. Cheney, 924 F.2d 1137 (D.C. Cir. 1991) (establishment of submarine acoustic testing range in Alaska).

Town of Groton v. Laird, 353 F. Supp. 344 (D. Conn. 1972) (construction of housing for Navy submarine base).

United States v. 45,149.98 Acres of Land, 455 F. Supp. 192 (E.D. N. Car. 1978) (condemnation of land for bombing range in North Carolina).

Valley Citizens for a Safe Environment v. Aldridge, 969 F.2d 1315 (1st Cir. 1992) (noise from C-5A cargo planes at Westover Air Force Base during Persian Gulf War).

Valley Citizens for a Safe Environment v. Aldridge, 886 F.2d 458 (1st Cir. 1989) (transfer of C-5A cargo planes from Dover Air Force Base, Deleware, to Westover Air Force Base, Massachusetts).

Weinberger v. Catholic Action of Hawaii/Peace Education Project, 454 U.S. 139 (1981) (possible storage of nuclear weapons on Oahu).

Westside Property Owners v. Schlesinger, 597 F.2d 1214 (9th Cir. 1979) (relocation of F-15 fighter aircraft to Luke Air Force Base in Arizona).

Wisconsin v. Weinberger, 745 F.2d 412 (7th Cir. 1984) (construction of low frequency submarine communications system in Wisconsin and Michigan's Upper Peninsula).

National Historic Preservation Act
16 U.S.C. §§470–470x-6 (1988 & Supp. V 1993)

Aluli v. Brown, 437 F. Supp. 602 (D. Hawaii 1977), reversed in part on other grounds, 602 F.2d 876 (9th Cir. 1979) (Navy ordered to protect candidate sites from bombing practice on small island).

Catholic Action of Hawaii/Peace Education Project v. Brown, 643 F.2d 569 (9th Cir. 1980), reversed on other grounds, 454 U.S. 139 (1981) (application of Act to be based on information in environmental impact statement for storage of nuclear weapons on Oahu).

Protect Key West, Inc. v. Cheney, 795 F. Supp. 1552 (S.D. Fla. 1992) (Navy housing project enjoined for failing to consider effect on adjoining historic district).

Romero-Barcelo v. Brown, 643 F.2d 835 (1st Cir. 1981) (Navy survey of candidate sites on Vieques Island not sufficiently thorough).

Noise Control Act
42 U.S.C. §§4901–4918 (1988 & Supp. V 1993)

Romero-Barcelo v. Brown, 643 F.2d 835 (1st Cir. 1981) (Act could not be used at Navy's Vieques Island training facility to enforce Puerto Rico's criminal nuisance statute for shock waves and excessive noise).

Westside Property Owners v. Schlesinger, 597 F.2d 1214 (9th Cir. 1979) (Federal Aviation Administration not authorized to control noise pollution from military aircraft at Air Force base in Arizona).

Nuclear Waste Policy Act
42 U.S.C. §§10101–10270 (1988 & Supp. V 1993)

County of Esmeralda v. United States Department of Energy, 925 F.2d 1216 (9th Cir. 1991) (standing for local governments as parties in Yucca Mountain proposal).

Maine v. Herrington, 790 F.2d 8 (1st Cir. 1986) (DOE refusal to extend comment period for possible high-level waste repository not reviewable by court).

Natural Resources Defense Council, Inc. v. United States Environmental Protection Agency, 824 F.2d 1258 (1st Cir. 1987) (EPA regulations for Nuclear Waste Policy Act must conform to EPA obligations under Safe Drinking Water Act).

Nevada v. Burford, 918 F.2d 854 (9th Cir. 1990), cert. denied sub nom. *Nevada v. Jamison*, 500 U.S. 932 (1991) (right-of-way for characterization studies at Yucca Mountain).

Nevada v. Herrington, 777 F.2d 529 (9th Cir. 1985) (Nevada entitled to DOE funding of characterization activities at proposed Yucca Mountain nuclear waste repository).

Nevada v. Herrington, 827 F.2d 1394 (9th Cir. 1987) (states not entitled to federal funding of judicial review under Act).

Nevada v. Watkins, 943 F.2d 1080 (9th Cir. 1991) (no need to prepare environmental assessment for characterization at Yucca Mountain).

Nevada v. Watkins, 939 F.2d 710 (9th Cir. 1991) (DOE guidelines for characterization of Yucca Mountain site not reviewable).

Nevada v. Watkins, 914 F.2d 1545 (9th Cir. 1990), cert. denied, 499 U.S. 906 (1991) (designation of federal lands in Nevada only for possible repository).

Public Service Co. of Colorado v. Andrus, 1991 WL 87511 (D. Idaho 1991) (shipment of spent nuclear fuel from Colorado to Idaho.)

Tennessee v. Herrington, 626 F. Supp. 1345 (M.D. Tennessee 1986) (Secretary of DOE violated consultation and cooperation requirements of the Nuclear Waste Policy Act).

Refuse Act of 1899
33 U.S.C. §407 (1988)

Crawford v. National Lead Co., 784 F. Supp. 439 (S.D. Ohio 1989) (violation bars assertion of "government contractor" defense at DOE's Fernald Feed Materials plant).

Romero-Barcelo v. Brown, 643 F.2d 835, 847–851 (1st Cir. 1981) (no private cause of action for Navy's alleged violations at Vieques Island training area).

Resource Conservation and Recovery Act (RCRA) 42 U.S.C. §§6901–6992k (1988 & Supp. V 1993)

Barcelo v. Brown, 478 F. Supp. 646, 667 (D.P.R. 1979) (discharge of ordnance into water at Navy's Vieques Island training area not covered by Act).

California v. United States Department of Defense, 878 F.2d 386 (9th Cir. 1989) (no waiver of sovereign immunity from state penalties for violations at Mare Island Naval Shipyard).

California v. Walters, 751 F.2d 977 (9th Cir. 1985) (federal official not criminally liable for violation).

Colorado v. United States Department of the Army, 707 F. Supp 1562 (D. Colo. 1989) (Army's ongoing CERCLA remediation at Rocky Mountain Arsenal Basin F did not preclude concurrent state RCRA corrective action).

Florida Department of Environmental Regulation v. Silvex Corp., 606 F. Supp. 159 (M.D. Fla. 1985) (Navy not responsible for cleanup costs and natural resource damages from spill of hazardous wastes sent to commercial disposal facility).

Heart of America Northwest v. Westinghouse Hanford Co., 830 F. Supp. 1265 (E.D. Wash. 1993) (action for RCRA violation at DOE's Hanford Reservation barred by ongoing CERCLA cleanup).

Legal Environmental Assistance Foundation, Inc. v. Hodel, 586 F. Supp. 1163 (E.D. Tenn. 1984) (DOE must obtain RCRA permit for hazardous waste disposal at Oak Ridge nuclear weapons plant).

Maine v. Department of Navy, 973 F.2d 1007 (1st Cir. 1992) (no waiver of sovereign immunity from state penalties for violations at Navy shipyard in Kittery, Maine).

McClellan Ecological Seepage Situation v. Weinberger, 655 F. Supp. 601 (E.D. Cal. 1986) (no state penalties for violations at McClellan Air Force Base).

McClellan Ecological Seepage Situation v. Weinberger, 707 F. Supp. 1182 (E.D. Cal. 1988) (no enforcement of state hazardous waste laws without EPA-approved state RCRA program).

McClellan Ecological Seepage Situation v. Cheney, 763 F. Supp. 431 (E.D. Cal. 1989) (hazardous waste disposed of in unlined pits at McClellan Air Force Base not "stored" for RCRA permit purposes).

Mitzelfelt v. Department of the Air Force, 903 F.2d 1293 (10th Cir. 1990) (no waiver of sovereign immunity from state penalities for violations at Canon Air Force Base in New Mexico).

New Mexico v. Watkins, 783 F. Supp. 628 (D.D.C. 1991), affirmed on other grounds, 969 F.2d 1122 (D.C. Cir. 1992) (DOE must obtain state RCRA permit for disposal of transuranic wastes at WIPP facility).

Ohio v. United States Department of Energy, 503 U.S. 607 (1992) (United States sovereign immunity from liablity for state penalties not waived by either federal facility or citizen suit provision).

Parola v. Weinberger, 848 F.2d 956 (9th Cir. 1988) (military installations in Monterrey, California, bound by local exclusive garbage collection franchise).

Sierra Club v. United States Department of Energy, 1991 WL 270190, 33 ERC 1720, 22 ELR 20076 (D. Colo. 1991) (DOE must obtain permit for hazardous waste incinerator at Rocky Flats).

Sierra Club v. United States Department of Energy, 770 F. Supp. 578 (D. Colo. 1991) (DOE may not store hazardous wastes mixed with plutonium at Rocky Flats facility without a RCRA permit).

Sierra Club v. United States Department of Energy, 734 F. Supp. 946 (D. Colo. 1990) (hazardous waste mixed with plutonium at DOE Rocky Flats facility subject to RCRA regulation).

Solano Garbage Co. v. Cheney, 779 F. Supp. 477 (E.D. Cal. 1991) (Travis Air Force Base in California subject to municipality's exclusive garbage collection franchise agreement).

United States v. Colorado, 990 F.2d 1565 (10th Cir. 1993), cert. denied, 114 S. Ct. 922 (1994) (Army's ongoing CERCLA remediation at Rocky Mountain Arsenal Basin F did not preclude concurrent state RCRA corrective action).

United States v. Dee, 912 F.2d 741 (4th Cir. 1990) (criminal conviction of Army civilian employees for improper storage and disposal of chemical weapons wastes at Aberdeen Proving Grounds).

United States v. New Mexico, 32 F.3d 494 (10th Cir. 1994) (state permit for hazardous waste incinerator at DOE's Los Alamos National Laboratory may include conditions to limit emissions of radionuclides).

United States v. Pennsylvania Department of Environmental Resources, 778 F. Supp. 1328 (M.D. Pa. 1991) (Navy must comply with state order to clean up contaminated drainage way at Navy Ships Parts Control Center).

United States v. Washington, 872 F.2d 874 (9th Cir. 1989) (no waiver of sovereign immunity from state penalties for violations at DOE's Hanford Nuclear Reservation).

Werlein v. United States, 746 F. Supp. 887 (D. Minn. 1990) (action for RCRA violation at Twin Cities Army Ammunition Plant barred by ongoing CERCLA cleanup).

Safe Drinking Water Act
42 U.S.C. §§300f–300j-11 (1988 & Supp. V 1993)

Natural Resources Defense Council, Inc. v. United States Environmental Protection Agency, 824 F.2d 1258 (1st Cir. 1987) (EPA regulations for Nuclear Waste Policy Act must conform to EPA obligations under Safe Drinking Water Act).

Toxic Substances Control Act
15 U.S.C. §§2601–2629 (1988 & Supp. V 1993)

No reported cases found.

Uranium Mill Tailings Radiation Control Act
42 U.S.C. §§7901–7942 (1988 & Supp. V 1993)

Dunn v. United States, 842 F.2d 1420 (3d Cir. 1988) (DOE compliance with public participation provisions of Act in cleaning up Pennsylvania ore processing mill).

Appendix B

Government Agencies Concerned with National Defense and the Environment

Department of Defense

Army Corps of Engineers, Pulaski Building, 20 Massachusetts Ave. NW, Washington, DC 20314-1000.

Assistant Deputy Under Secretary of Defense for Environmental Security/Cleanup, 3310 Defense Pentagon, Washington, DC 20301-3310.

Defense Environmental Restoration Program, 400 Army Navy Drive, Arlington, VA 22202.

Defense Logistics Agency, Cameron Station, 5100 Duke Street, Alexandria, VA 22304-6100.

Defense Nuclear Agency, 6801 Telegraph Road, Alexandria, VA 22310-3393.

Department of Defense Inspector General, 400 Army Navy Drive, Arlington, VA 22202-2884. Copies of reports: (703) 614-8542.

Department of the Army, Office of the Deputy Assistant Secretary of the Army for Environment, Safety, and Occupational Health, 110 Army Pentagon, Washington, DC 20310-0110.

Department of the Navy, Office of the Deputy Assistant Secretary of the Navy for

Environment and Safety, Crystal Plaza #5, 2211 Jefferson Davis Highway, Arlington, VA 22244-5110.

Department of the Air Force, Office of the Deputy Assistant Secretary of the Air Force for Environment, Safety, and Occupational Health, 1660 Air Force Pentagon, Washington, DC 20330-1660.

Department of Energy

Center for Environmental Management Information, P.O. Box 23769, Washington, D.C. To order publications: (800) 736-3282.

Office of the Secretary, Environment, Safety, and Health, 1000 Independence Ave. SW, Washington, DC 20585.

Environmental Protection Agency

401 M Street SW, Washington, DC 20460.

General Accounting Office

P.O. Box 6015, Gaithersburg, MD 20884-6015. To order reports: phone (202) 512-6000, fax (301) 258-4066.

Superintendent of Documents

Government Printing Office, Washington, DC 20402-9325. To order documents: (202) 512-1800.

Appendix C

Public Interest Organizations Concerned with National Defense and the Environment

Center for Defense Information, 1500 Massachusetts Avenue NW, Washington, DC 20005, (202) 862-0700.

Conservation Foundation, 1255 23rd Street NW, Washington, DC 20037, (202) 293-4800.

Conservation Law Foundation, 62 Summer Street, Boston, MA 02110-1008, (617) 350-0990; *and* 21 East State Street, Montpelier, VT 05602-2152, (802) 223-2152; *and* 119 Tillson Avenue, Rockland, ME 04841-3632, (207) 594-8107.

Environmental Defense Fund, 257 Park Avenue South, New York, NY 10010, (212) 505-2100.

Foreign Bases Project, P.O. Box 150753, Brooklyn, NY 11215, (718) 788-6071.

Friends of the Earth, 218 D Street SE, Washington, DC 20003, (202) 544-2600.

Government Accountability Project, 810 First Street NE, Suite 630, Washington, DC 20002, (202) 408-0034; *and* 1402 Third Avenue, Suite 1215, Seattle, WA 98101, (206) 292-2850.

Greenpeace, 1611 Connecticut Avenue NW, Washington, DC 20009, (202) 462-1177.

Military Production Network, 2000 T Street NW, Washington, DC 20530, (202) 833-4668.

Military Toxics Network, RR1, Box 220, Litchfield, ME 04350, (207) 268-4071

National Toxics Campaign Fund, 1168 Commonwealth Avenue, Boston, MA 02134, (617) 232-0327.

Natural Resources Defense Council, 40 West 20th Street, New York, NY 10011, (212) 727-2700; *and* 1350 New York Avenue NW, Washington, DC 20005, (202) 783-7800.

Nuclear Free America, 325 East 25th Street, Baltimore, MD 21218, (410) 235-3575.

Physicians for Social Responsibility, 1000 16th Street NW, #810, Washington, DC 20036, (202) 898-0150.

Sierra Club, 408 C Street NE, Washington, DC 20002, (202) 547-1141.

Notes

Preface (pp. xiii–xv)

1. *Inaugural Addresses of the Presidents of the United States* 305 (Washington, D.C.: U.S. Government Printing Office, 1989), quoted in Department of Energy, *Fact Sheets: Environmental Restoration and Waste Management* (1991–1992).
2. See ch. 4, "Cleaning Up DOE's Weapons Complex," and ch. 8, "Recovery of Damages in Defense-Related Cases."

1. National Defense vs. Environmental Protection (pp. 1–10)

1. Laurent Hourcle, "Overview, Master Environmental Lawyer's Edition," 31 *Air Force Law Review* (1989): 1, 3–5.
2. Address to the Defense and Environment Initiative Forum, Washington, D.C., Sept. 3, 1990.
3. Memorandum from the Secretary of Defense to Secretaries of Military Departments re: Environmental Management Policy, Oct. 10, 1989.
4. *National Security Strategy of the United States* (White House, Aug. 1991): 3. The term "environmental security" was also appropriated by former Soviet President Mikhail S. Gorbachev. Michael Wines, "Capitol Stirs to Mr. Green, Né Gorbachev," *New York Times*, Apr. 16, 1993. See also Jessica Tuchman Mathews, "Redefining Security," *Foreign Affairs* (spring 1989): 162.
5. Joseph J. Romm, *Defining National Security: The Nonmilitary Aspects* (New York: Council on Foreign Relations Press, 1993), 85.
6. Department of Defense, *Defense and the Environment: A Commitment Made* (1991): 1. See also Michael Renner, *National Security: The Economic and Environmental Dimensions* (Worldwatch Institute, 1989), 6, 29–38; Peter H. Gleick, "Environment and Security: The Clear Connections," *Bulletin of the Atomic Scientists* (April 1991): 17. One writer argues that the environment deserves our protection because it is inherently valuable: Whether we regard the natural environment as the manifestation of a divine presence, or value it for the "beauty and sense of order and balance" we perceive in it, we ought to guard it the same way we do works of art and other evidence of our cultural heritage. Merritt P. Drucker, "The Military Commander's Responsibility for the Environment," 11 *Environmental Ethics* (1989): 135, 136–140.
7. See L. R. Hourcle, "DOD's Budgeting Plans to Meet Environmental Challenges," 1 *Federal Facilities Environmental Journal* (1990): 109.
8. For example, no more W-88 nuclear warheads, designed to fit on the submarine-launched Trident II missile, will be made. They were the last in active production. "Bomb for a Submarine Will Be Discontinued," *New York Times*, Jan. 26, 1992. Even the production of tritium, used to boost the yield of existing weapons, has now stopped.

9. See ch. 5.
10. See *Drug Policy Foundation v. Bennett*, No. C-90–2278 FMS (N.D. Cal. Dec. 20, 1991).
11. See 16 U.S.C. §§670a–670f (1988 & Supp. V 1993), known as the "Sikes Act." See generally George H. Siehl, *Natural Resource Issues in National Defense Programs* (Washington, D.C.: Congressional Research Service, 1991). See also Dick Wingerson, "Scenic Vistas, Sonic Booms," *New York Times*, Mar. 28, 1992.
12. Keith Schneider, "U.S. Shares Blame in Abuses at A-Plant," *New York Times*, Mar. 27, 1992.
13. See generally ch. 4, "Disposal of Defense Nuclear Wastes."
14. See ch. 4, "Cleaning Up DOE's Weapons Complex," and case study, ch. 2, "Buttoning Up for a Nuclear Winter."
15. U.S. Army Corps of Engineers, *Commander's Guide to Environmental Management* (1990): unnumbered. See also John L. Fugh, Scott P. Isaacson and Lawrence E. Rouse, "The Commander and Environmental Compliance," *The Army Lawyer* (May 1990): 3.
16. 42 U.S.C. §§4331–4361 (1988).
17. 16 U.S.C. §§1531–1544 (1988 & Supp. V 1993).
18. *No GWEN Alliance of Lane County, Inc. v. Aldridge*, 855 F.2d 1380 (9th Cir. 1988). This case is described in further detail in case study, ch. 2, "Command and Control for a Nuclear War."
19. *Hudson River Sloop Clearwater, Inc. v. Department of the Navy*, 745 F.2d 414 (2d Cir. 1989); see ch. 2, "Secret Environmental Impact Statements."
20. 33 U.S.C. §§1251–1387 (1988 & Supp. V 1993).
21. 42 U.S.C. §§9601–9675 (1988 & Supp. V 1993).
22. *Weinberger v. Romero-Barcelo*, 456 U.S. 305 (1982). See ch. 3.
23. *United States v. Colorado*, 990 F.2d 1565 (10th Cir. 1993), cert. denied, 114 S. Ct. 922 (1994); see case study, ch. 4, "Struggle for Control at Basin F."
24. *Ohio v. United States Department of Energy*, 503 U.S. 607 (1992).
25. *Conservation Law Foundation v. Department of the Air Force*, No. 1:92-CV-156-M (D. N.H. 1992). See case study, ch. 5, "New Management at Pease Air Force Base."
26. See, e.g., *Little v. Barreme*, 6 U.S. (2 Cranch) 170 (1804) (limiting the use of naval forces in the quasi-war with France); *Youngstown Sheet & Tube Co. v. Sawyer*, 343 U.S. 579 (1952) (nullifying the president's unilateral seizure of steel mills during the Korean War).
27. *Weinberger v. Romero-Barcelo*, 456 U.S. 305, 310 (1982).
28. *Crawford v. National Lead Co.*, 784 F. Supp. 439 (S.D. Ohio 1989). See case study, ch. 8, "Flawed Fiat at Fernald."

2. Environmental Planning for National Defense (pp. 11–35)

1. 42 U.S.C. §§4331–4361 (1988).
2. Citations herein are to NEPA's section numbers as originally enacted, as well as to the United States Code. In this section we address only those issues that are essential to an understanding of NEPA's application to national security activities. Other

aspects of NEPA law are surveyed in John Henry Davidson and Orlando E. Delogu, *Federal Environmental Regulation*, vol. 1 (Salem, N.H.: Butterworth, 1993), ch. 1; Daniel R. Mandelker, *NEPA Law and Litigation*, 2d ed. (Deerfield, Ill.: Callaghan & Co. 1992); Sheldon M. Novick, ed., *Law of Environmental Protection*, vol. 1 (Deerfield, Ill.: Clark, Boardman, Callaghan 1994), ch. 9; *NEPA Deskbook* (Washington, D.C.: Environmental Law Institute, 1989). See also Karin P. Sheldon, "NEPA in the Supreme Court," 25 *Land and Water Law Review* (1990): 83.

3. 40 C.F.R. §1500.1(a).
4. CEQ is also directed to assist the President in preparing an annual report to Congress on the condition of the environment, along with recommendations for improvements.
5. §101(b), 42 U.S.C. §4331(b).
6. *Marsh v. Oregon Natural Resources Council*, 490 U.S. 360, 371 (1989).
7. *Robertson v. Methow Valley Citizens Council*, 490 U.S. 332, 350 (1989) (citations omitted).
8. *Vermont Yankee Nuclear Power Corporation v. Natural Resources Defense Council, Inc.*, 435 U.S. 519, 558 (1978).
9. *Kleppe v. Sierra Club*, 427 U.S. 390, 410 n. 21 (1976). See also *Strycker's Bay Neighborhood Council, Inc. v. Karlen*, 444 U.S. 223, 227–228 (1980).
10. 40 C.F.R. pt. 1500.
11. See, e.g., 32 C.F.R. pt. 214 (DOD); 32 C.F.R. pts. 650, 651, as modified by 53 Fed. Reg. 46,322 (1988) (Army); 32 C.F.R. pt. 989 (Air Force); 32 C.F.R. pt. 775 (Navy); 44 Fed. Reg. 45,431 (1979) (Central Intelligence Agency).
12. §102(2)(C), 42 U.S.C. §4332(2)(C).
13. 40 C.F.R. §1502.16.
14. 40 C.F.R. §§1502.14, 1502.16, 1508.25(b). The discussion of alternatives is meant to influence not only agency decision-makers, but also the President and members of Congress "for the guidance of these ultimate decision-makers . . . for their consideration along with the various other elements of the public interest." *Natural Resources Defense Council, Inc. v. Morton*, 458 F.2d 827, 835 (D.C. Cir. 1972).
15. *Robertson v. Methow Valley Citizens Council*, 490 U.S. 332, 351–352 (1989); 40 C.F.R. §§1502.14(f), 1502.16(h), 1505.2(c), 1508.25(b).
16. 40 C.F.R. §§1507.3(b)(2)(ii), 1508.4; 32 C.F.R. pt. 651, App. A.
17. 40 C.F.R. §§1501.3; 1501.4(b), (c) & (e); 1508.9; 1508.13.
18. Environmental Protection Agency, *Facts About the National Environmental Policy Act* (1989), 3.
19. §102(2)(C), 42 U.S.C. §4332(2)(C); 40 C.F.R. §§1501.7, 1502.19, 1503.1(a), 1508.25. See generally, Environmental Protection Agency, "Policy and Procedures for the Review of Federal Actions Impacting the Environment" (1984), reproduced in *NEPA Deskbook* (Environmental Law Institute, 1989), 291.
20. 40 C.F.R. §§1500.1(b), 1500.2(d), 1502.9(b), 1503.1(a)(4), 1503.4, 1506.6. See *California v. Block*, 690 F.2d 753, 773–774 (9th Cir. 1982).
21. 40 C.F.R. §1505.2. While an agency must discuss reasonable mitigation measures in the EIS, under a recent Supreme Court decision it need not commit itself in the ROD to implementation of any of them. *Robertson v. Methow Valley Citizens*

Council, 490 U.S. 332, 352–353 (1989). However, the agency is bound to implement any mitigation measures or other conditions it adopts. 40 C.F.R. §1505.3; Council on Environmental Quality, *Forty Most Asked Questions Concerning CEQ's National Environmental Policy Act Regulations*, 46 Fed. Reg. 18,026, 18,037 (1981), as amended, 51 Fed. Reg. 15,618 (1986).

22. 40 C.F.R. §1502.4. The leading case is *Kleppe v. Sierra Club*, 427 U.S. 390 (1976). By the same token, an agency may not divide a project into several small "segments," so that the cumulative impact of the related parts is ignored. 40 C.F.R. §§1502.4(a), 1508.25(a)(1).
23. 40 C.F.R. §1508.25(a)(1).
24. *Hudson River Sloop Clearwater, Inc. v. Department of the Navy*, 836 F.2d 760 (2d Cir. 1988).
25. 40 C.F.R. §1502.9(c).
26. 40 C.F.R. §§1506.8, 1508.17. Although the legislative EIS could provide meaningful environmental review at nearly the earliest point in the planning process, and in a very public forum, very few have been prepared. See Joseph Mendelson III and Andrew Kimbrell, "The Legislative Environmental Impact Statement: An Analysis of Public Citizen v. Office of the U.S. Trade Representative," 23 *Environmental Law Reporter* (Environmental Law Institute) (1993): 10,653; Ian M. Kirschner, Note, "NEPA's Forgotten Clause: Impact Statements for Legislative Proposals," 58 *Boston University Law Review* (1978): 560.
27. 40 C.F.R. §1508.27.
28. *Marsh v. Oregon Natural Resources Commission*, 490 U.S. 360, 374–375, 378 (1989). While the *Marsh* case involved the need for a supplemental EIS, the Court suggested that similar considerations govern the preparation of an initial EIS.
29. Nicholas C. Yost and James W. Rubin, "The National Environmental Policy Act," in *NEPA Deskbook* (Environmental Law Institute, 1989), 23.
30. Clean Air Act §309, 42 U.S.C. §7609 (1988).
31. 40 C.F.R. §1504.2.
32. 40 C.F.R. §1504.3. The same regulation enables agencies other than EPA to refer objections to CEQ as well. See Council on Environmental Quality, *15th Annual Report* (1985): 523–524; Joseph A. Wellington, "A Primer on Environmental Law for the Naval Services," 38 *Naval Law Review* (1989): 5, 37.
33. For example, NEPA practice in the Air Force is described in Julie K. Fegley, "The National Environmental Policy Act: The Underused, Much-Abused Compliance Tool," 31 *Air Force Law Review* (1989): 153.
34. §§101(a), 102(2); 42 U.S.C. §§4331(a), 4332(2).
35. See, e.g., Pub. L. No. 98–94, §110, 97 Stat. 614 (1983) (procurement and deployment of the MX missile); Pub. L. No. 100-256, §204(c), 102 Stat. 2623 (1988) (military base closures and realignments).
36. §§101(b), 102, 42 U.S.C. §§4331(b), 4332 (emphasis supplied).
37. Cf. *Valley Citizens for a Safe Environment v. Vest*, 1991 WL 330963, 22 *Environmental Law Reporter* 20,335 (D. Mass.). In *Nielson v. Seaborg*, 348 F. Supp. 1369, 1372 (D. Utah 1972), the court emphasized the distinction between substantive and procedural duties under NEPA, declaring that once the AEC had complied with NEPA's procedural requirements by preparing environmental impact

statements for underground nuclear weapons tests, it was free to proceed with the testing, since "compliance with the general policy of NEPA [was] subject to its discretionary weighing of other 'essential considerations.'"

38. Section 102(2)(F), 42 U.S.C. §4332(2)(F) requires all federal agencies to:

> recognize the worldwide and long-range character of environmental problems and, where consistent with the foreign policy of the United States, lend appropriate support to initiatives, resolutions, and programs designed to maximize international cooperation in anticipating and preventing a decline in the quality of mankind's world environment.

Commenting on this provision, Senator Jackson indicated that such international cooperation is "possible because the problems of the environment do not, for the most part, raise questions related to ideology, national security and the balance of world power." 115 Congressional Record 40,417 (1969). Use of the terms "foreign policy" and "national security" here is an indication, though hardly a strong one, that Congress was mindful of the potential for conflict between national defense and environmental protection, yet decided not to carve out an exception that would have avoided the conflict. The terms do not appear elsewhere in the statute or in the committee reports on its passage.

39. 40 C.F.R. §1500.3.
40. Executive Order No. 11,991, 42 Federal Register 26,967 (1977).
41. 32 C.F.R. pt. 214, Encl. 1, §E.5. Other DOD regulations for emergencies are found at 32 C.F.R. pt. 197, Encl. 1, §C.3; and 32 C.F.R. pt. 197, Encl. 2, §C.3.a. The subject of emergency waivers of NEPA's procedural requirements is examined in ch. 6, "Environmental Law Waivers in Wartime."
42. See, e.g., 32 C.F.R. pt. 197, Encl. 2, §C.3.a.(3) (DOD NEPA regulations for environmental effects abroad). The applicability of NEPA to warmaking is addressed in chapter 6.
43. See Cary Ichter, Comment, "'Beyond Judicial Scrutiny': Military Compliance with NEPA," 18 *Georgia Law Review* (1984): 639, 647–653.
44. *Concerned About Trident v. Rumsfeld*, 555 F.2d 817, 823 (D.C. Cir. 1977).
45. Ibid. (citations omitted).
46. See *No GWEN Alliance v. Aldridge*, 855 F.2d 1380, 1384 (9th Cir. 1988) ("There is no 'national defense' exception to NEPA."); *Weinberger v. Catholic Action of Hawaii*, 454 U.S. 139, 147 (1981) (Blackmun, J., concurring) ("If the Navy proposes to engage in a major action that will have a significant environmental effect, it must prepare an environmental impact statement (EIS) addressing the consequences of the proposed activity."); *Citizens for Reid Park v. Laird*, 336 F. Supp. 783, 786 (D. Me. 1972). Cf. *McQueary v. Laird*, 449 F.2d 608, 612 (10th Cir. 1971) ("the Federal Government has traditionally exercised unfettered control with respect to the internal management and operations of federal military establishments.").
47. Reported cases applying NEPA to national security activities are listed in Appendix A.
48. Charles A. Brothers, "The Judge Advocate and the Action-Forcing Provisions of

the National Environmental Policy Act," 18 *Air Force Law Review* (1976): 1; Lance D. Wood, "A Prescriptive Analysis of the U.S. Navy's Program to Implement the National Environmental Policy Act," 5 *Environmental Law Reporter* (Environmental Law Institute) (1975): 50,049.

49. Environmental Law Institute, *NEPA in Action: Environmental Offices in Nineteen Federal Agencies* (1981): 101.
50. Department of Defense Inspector General, *Environmental Consequence Analyses of Major Defense Acquistion Programs* (Rep. No. 94-020) (1993): 15, 77–83, 117–126 (hereinafter *DOD IG Report*).
51. Ibid.
52. See *Protect Key West, Inc. v. Cheney*, 795 F. Supp. 1552 (S.D. Fla. 1992) (environmental analysis prepared after decision to go ahead with Navy housing project).
53. See *DOD IG Report*, p. 16.
54. James I. Mangi, "Window on the Pentagon: EIA and Defense Decision Making," 5 *Environmental Impact Assessment Review* (1985): 3, 4–7.
55. This account is drawn from the opinion of the court in *Concerned About Trident v. Rumsfeld*, 555 F.2d 817 (D.C. Cir. 1977).
56. §101(b)(3), 42 U.S.C. §4331(b)(3).
57. NEPA calls on federal agencies to evaluate the environmental consequences of actions "which *may* have an impact on man's environment." §102(2)(A) (emphasis added). See *Louisiana v. Lee*, 758 F.2d 1081, 1084 (5th Cir. 1985), cert. denied sub nom. *Dravo Basic Materials Co. v. Louisiana*, 475 U.S. 1044 (1986).
58. *Carolina Environmental Study Group v. United States*, 510 F.2d 796, 799 (D.C. Cir. 1975) (chance of "Class 9" nuclear reactor accident somewhere between one in one hundred thousand and one in one billion per plant per year).
59. *Scientists Institute for Public Information v. Atomic Energy Commission*, 481 F.2d 1079, 1092 (D.C. Cir. 1973). In that case the court required the AEC to prepare an impact statement on incompletely understood long-term environmental consequences of the liquid metal fast breeder reactor program. See 40 C.F.R. §§1508.27(b)(5), 1508.25(c), 1508.8(b).
60. 43 Fed. Reg. 55997 (1978).
61. 40 C.F.R. §1502.22, 51 Federal Register 15,625 (1986). The rationale for the amendment is set forth succinctly in Vicki O'Meara Masterman, "Worst Case Analysis: The Final Chapter?" 19 *Environmental Law Reporter* (1989): 10,026.
62. *Robertson v. Methow Valley Citizens Council*, 490 U.S. 332, 354–356 (1989).
63. 40 C.F.R. §1502.22(b).
64. *No GWEN Alliance of Lane County, Inc. v. Aldridge*, 855 F.2d 1380 (9th Cir. 1988).
65. The plaintiffs, who lived near one of the towers, did not ask the Air Force to address their heightened fears attributable to such primacy in targeting. In *Metropolitan Edison Co. v. People Against Nuclear Energy* (*PANE*), 460 U.S. 766 (1983), the Supreme Court held that an EIS need not include an analysis of such psychological harm, even when the fear and anxiety are caused by an attendant risk of physical harm.
66. 855 F.2d at 1386; Appellant's Brief at 13.

67. 855 F.2d at 1386. In another NEPA case concerned with GWEN, the court found, without explanation, that there is no "proximate causal relationship" between the installation of GWEN and nuclear war. *Conservation Law Foundation of New England v. United States Department of the Air Force*, 1987 WL 46,370 (D. Mass.).
68. 855 F.2d at 1386.
69. Pub. L. No. 100–526, 102 Stat. 2627 (1988).
70. Pub. L. No. 98–94, §110, 97 Stat. 614, 621 (1983).
71. See *Romer v. Carlucci*, 847 F.2d 445 (8th Cir. 1988) (en banc). The litigation is analyzed in Douglas E. Baker, "Anticipating Lamm v. Weinberger (Romer v. Carlucci): The Political Question Doctrine, MX Missiles, and NEPA's Environmental Impact Statement," 21 *Creighton Law Review* 1245 (1987–1988).
72. 847 F.2d at 448 (emphasis supplied).
73. Ibid., p. 464.
74. NEPA §101(b); 42 U.S.C. §4331(b).
75. FEMA, *U.S. Crisis Relocation Planning* (P&P-7 1981).
76. See Jeannie Peterson, ed., *The Aftermath: The Human and Ecological Consequences of Nuclear War* (New York: Pantheon, 1983).
77. A. Barrie Pittock et al., *Environmental Consequences of Nuclear War* (Chichester & New York: John Wiley & Sons, 1986) (the SCOPE Report). See also John W. Birks and Anne H. Ehrlich, "If Deterrence Fails: Nuclear Winter and Ultraviolet Spring," in *Hidden Dangers: Environmental Consequences of Preparing for War*, ed. Anne H. Ehrlich and John W. Birks (San Francisco: Sierra Club Books, 1990), 119; National Research Council, *The Effects on the Atmosphere of a Major Nuclear Exchange* (Washington, D.C.: National Academy Press, 1985).
78. Starley L. Thompson and Stephen H. Schneider, "Nuclear Winter Reappraised," *Foreign Affairs* (summer 1986): 981. See also Edwin Teller, "Widespread Aftereffects of Nuclear War," *Nature*, Aug. 23, 1984, 621; Malcolm Browne, "Nuclear Winter Theorists Pull Back," *New York Times*, Jan. 23, 1990.
79. Pub. L. No. 98–525, §1107(b)(1), 98 Stat. 2583 (1984).
80. Caspar Weinberger, *The Potential Effects of Nuclear War on the Climate: A Report to the United States Congress* (1985).
81. Department of Defense, *1986 Report on Nuclear Winter* (1986).
82. Pub. L. No. 99–661, §1371, 100 Stat. 4004–4005 (1986). See letter from Secretary of Defense Caspar Weinberger to Senator Sam Nunn, Jan. 11, 1988.
83. 40 C.F.R. §1502.22(b).
84. *DOD IG Report*, p. 14.
85. §102(2)(C), 42 U.S.C. §4332(2)(C).
86. 5 U.S.C. §552 (1988). FOIA §552(b)(1) provides that agencies may withhold information that is "specifically authorized under criteria established by an Executive order to be kept secret in the interest of national defense or foreign policy and [is] in fact properly classified pursuant to such Executive order." The order currently applicable is Executive Order No. 12,958, 60 Fed. Reg. 19,825 (1995). FOIA §552(b)(3) also permits the withholding of records "specifically exempted from disclosure" by another statute. For example, the Atomic Energy Act provides for the protection of "all data concerning design, manufacture, or utilization

of atomic weapons." 42 U.S.C. §2014(y) (1988). Operation of both FOIA exemptions is explained in Center for National Security Studies, *Litigation Under the Federal Open Government Laws*, 18th ed., ed. Allan Robert Adler (Washington: American Civil Liberties Foundation, 1993); Stephen Dycus, Arthur L. Berney, William C. Banks and Peter Raven-Hansen, *National Security Law* (Boston: Little, Brown 1990), 543–571; James T. O'Reilley, *Federal Information Disclosure*, 2nd ed. (Colorado Springs, CO: Shepard's/McGraw-Hill, 1994); *Guidebook to the Freedom of Information and Privacy Acts*, 2nd ed., eds. Robert F. Bouchard and Justin D. Franklin (Deerfield, Ill.: Clark Boardman Callaghan, 1994); United States Department of Justice, *Freedom of Information Act Guide and Privacy Overview*, ed. Pamela Maida (Washington, D.C.: Superintendent of Documents, 1993).

87. The CEQ regulation is 40 C.F.R. §1507.3(c). See also 32 C.F.R. pt. 214, encl. 1, §D.10. (Department of Defense).
88. *Weinberger v. Catholic Action of Hawaii/Peace Education Project*, 454 U.S. 139, 145 (1981).
89. Id. p. 146.
90. This case study is based on *Weinberger v. Catholic Action of Hawaii/Peace Education Project*, 454 U.S. 139 (1981). The case is reviewed in Cary Ichter, Comment, "'Beyond Judicial Scrutiny': Military Compliance With NEPA," 18 *Georgia Law Review* (1984): 639; F. L. McChesney, Comment, "Nuclear Weapons and 'Secret' Impact Statements: High Court Applies FOIA Exemptions to EIS Disclosure Rules," 12 *Environmental Law Reporter* (Environmental Law Institute) (1982): 10,007.
91. This same distinction was used to limit the scope of NEPA review for a biological warfare test facility at Dugway Proving Ground in Utah. *Foundation on Economic Trends v. Weinberger*, 610 F. Supp. 829, 839–840 (D.D.C. 1985).
92. *Catholic Action*, 454 U.S. at 146. The Court also invoked its holding in a Civil War-era case that "public policy forbids the maintenance of any suit in a court of justice, the trial of which would inevitably lead to the disclosure of matters which the law itself regards as confidential, and respecting which it will not allow the confidence to be violated." Id. at 146–147, citing *Totten v. United States*, 92 U.S. 105, 107 (1876). In *Totten*, a Union spy sought compensation for his services pursuant to a secret contract with President Lincoln. Without reaching the merits of his claim, the Court decided that the subject of the suit itself was too sensitive to try in court, thus leaving the claimant without redress. Contract claims with national security implications have similarly been rejected in several recent cases. See, e.g., *Guong v. United States*, 860 F.2d 1063 (Fed. Cir. 1988). This dangerous rationale, rejected by two concurring Justices in *Catholic Action*, could allow courts in other cases simply to avoid ruling on the government's compliance with the law.
93. Several commentators have misinterpreted this language as altogether excusing the Navy from complying with NEPA. See, e.g., Amy J. Sauber, Comment, "The Application of NEPA to Nuclear Weapons Production, Storage, and Testing: Weinberger v. Catholic Action of Hawaii/Peace Education Project," 11 *Boston College Environmental Affairs Law Review* (1984): 805, 806, 828, 830; Com-

ment, "Weinberger v. Catholic Action of Hawaii/Peace Education Project: Assessing the Environmental Impact of Nuclear Weapons Storage," 3 *Virginia Journal of Natural Resources Law* (1984): 335, 335. But the Court made plain that "[t]he Navy must consider environmental consequences in its decisionmaking process, even if it is unable to meet NEPA's public disclosure goals." *Catholic Action*, 454 U.S. at 146. The fact that the Court would not rule on the Navy's compliance with NEPA in no way diminished the Navy's obligation to uphold and abide by the law. Indeed, the Navy's own regulations require preparation of a classified EIS under circumstances like those in *Catholic Action*. 32 C.F.R. §775.5.

94. *Concerned About Trident v. Rumsfeld*, 555 F.2d 817, 825 (D.C. Cir. 1977). See case study, "Finding a Home for the Trident Submarine."
95. *Laine v. Weinberger*, 541 F. Supp. 599 (C.D. Calif. 1982).
96. *Hudson River Sloop Clearwater, Inc. v. Department of the Navy*, 745 F.2d 414 (2d Cir. 1989). These decisions are analyzed in Stephen Dycus, "NEPA Secrets," 2 *New York University Environmental Law Journal* (1993): 300.
97. See generally Harry H. Almond, "The Extraterritorial Reach of United States Regulatory Authority Over the Environmental Impacts of Its Activities," 44 *Albany Law Review* (1980): 739; Nicholas A. Robinson, "Extraterritorial Environmental Protection Obligations of the Foreign Affairs Agencies: The Unfulfilled Mandate of NEPA," 7 *NYU Journal of International Law and Policy* (1974): 257; Joan A. Goldfarb, "Extraterritorial Compliance with NEPA Amid the Current Wave of Environmental Alarm," 18 *Boston College Environmental Affairs Law Review* (1991): 543; R. David Kitchen, Comment, "NEPA's Overseas Myopia: Real or Imagined?" 71 *Georgetown Law Journal* (1983): 1201; Comment, "NEPA's Role in Protecting the World Environment," 131 *University of Pennsylvania Law Review* (1982): 353.
98. §2, 42 U.S.C. §4321.
99. §102(2)(F), 42 U.S.C. §4332(2)(F).
100. The House Report on NEPA indicates:

> Implicit in this section [sec. 101] is the understanding that the international implications of our current activities will also be considered, inseparable as they are from the purely national consequences of our actions.

H. Rep. No. 91–378, *U.S. Code Congressional & Administrative News* (1969): 2751, 2759, quoted in *People of Enewetak v. Laird*, 353 F. Supp. 811, 817 (D. Hawaii 1973). Sen. Jackson, NEPA's principal sponsor, remarked that we "do not intend, as a government or as a people, to initiate actions which endanger the continued existence or the health of mankind." 115 Cong. Rec. 40,416 (1969), also quoted in *People of Enewetak v. Laird*, 353 F. Supp. at 817. See S. Report No. 352, 101st Cong., 2d Sess. 10–12 (1990); Goldfarb, *Extraterritorial Compliance with NEPA*, pp. 556–557. One court's observation that "NEPA's legislative history illuminates nothing in regard to extraterritorial application" seems unwarranted. *Natural Resources Defense Council, Inc. v. Nuclear Regulatory Commission*, 647 F.2d 1345, 1366 (D.C. Cir. 1981).

101. *Sierra Club v. Adams*, 578 F.2d 389 (D.C. Cir. 1978); *Wilderness Society v.*

Morton, 463 F.2d 1261 (D.C. Cir. 1972); *National Organization for Reform of Marijuana Laws (NORML) v. United States Dept. of State*, 452 F. Supp. 1226 (D.D.C. 1978); *People of Enewetak v. Laird*, 353 F. Supp. 811 (D. Hawaii 1973); *People of Saipan v. Department of the Interior*, 356 F. Supp. 645 (D. Hawaii 1973), aff'd as modified 502 F.2d 90 (9th Cir. 1974), cert. denied, 420 U.S. 1003 (1975).

102. *People of Enewetak v. Laird*, 353 F. Supp. at 818.
103. *Natural Resources Defense Council, Inc. v. Nuclear Regulatory Commission*, 647 F.2d 1345 (D.C. Cir. 1981). The same concern for foreign policy was echoed in a recent case, *Greenpeace USA v. Stone*, 748 F. Supp. 749 (D. Hawaii 1990), described in the case study below.
104. 42 U.S.C. §§2011–2296 (1988).
105. 22 U.S.C. §§3201–3282 (1988).
106. 647 F.2d at 1366.
107. Exec. Order No. 12,114, 44 Fed. Reg. 1957 (1979). The executive order is implemented by the military in *Environmental Effects Abroad of Major Department of Defense Actions*, DOD Dir. 6050.7 (Mar. 31, 1979). See generally Sanford E. Gaines, "'Environmental Effects Abroad of Major Federal Actions': An Executive Order Ordains a National Policy," 3 *Harvard Environmental Law Review* (1979): 136; Glenn Pincus, Comment, "The 'NEPA-Abroad' Controversy: Unresolved by an Executive Order," 30 *Buffalo Law Review* (1981): 611.
108. This provision reflects a reluctance to interfere in any way with the sovereignty of another nation that has voluntarily acceded to an environmentally damaging U.S. activity. See Office of General Counsel, Department of Defense, *The Application of the National Environmental Policy Act to Major Federal Actions with Environmental Impacts Outside the United States*, reprinted in 124 *Congressional Record* (Oct. 14, 1978), 37,773, 37,792. However, concern for the sovereignty of other nations does not extend to all U.S. actions. Sec. 2–3(c) and (d) of the executive order require formal evaluation of the environmental effects on foreign soil of toxic or radioactive materials that are prohibited or strictly regulated in the United States, and of risks to resources of recognized global importance, regardless of the wishes of the nation in question.
109. Exec. Order No. 12,114, §2-5(a). "Armed conflict" is defined in Army regulations implementing Executive Order No. 12,114 as "hostilities in which Congress has declared war or enacted a specific authorization for the use of armed forces; hostilities . . . in which a report is prescribed by section 4(a)(1) of the War Powers Resolution . . . ; and other actions by the armed forces that involve defensive use or introduction of weapons in situations where hostilities occur or are expected." Army Regulation 200-2, Incl. 2, §C.3.a.(3) (1988).
110. Exec. Order No. 12,114, §§2-5(b) and (c), 3-1.
111. Pub. L. No. 99–145, §1412(a), 99 Stat. 583 (1986) (codified as amended, 50 U.S.C. §1521 (1988 & Supp. V 1993)).
112. *Greenpeace USA v. Stone*, 748 F. Supp. 749 (D. Hawaii 1990), dismissed in part as moot, 924 F.2d 175 (9th Cir. 1991).
113. 748 F. Supp. at 760.
114. Ibid., p. 762.

115. *Environmental Defense Fund, Inc. v. Massey*, 986 F.2d 528 (D.C. Cir. 1993).
116. *NEPA Coalition of Japan v. Aspin*, 837 F. Supp. 466, 467 n. 5 (D.D.C. 1993). See Robert Quan Lee, Note, "The Presumption Against Extraterritorial Application of the National Environmental Policy Act: Has the Iron Curtain Been Lifted?" 19 *Land and Water Law Review* (1994): 533.
117. The leading case, *Flint Ridge Development Co. v. Scenic Rivers Association of Oklahoma*, 426 U.S. 776, 778 (1976), addresses a conflict with other *statutory* duties, but the principle might be extended to treaty obligations, as well.
118. *NEPA Coalition of Japan*, 837 F. Supp. at 468.
119. S. 1278, 102d Cong., 1st Sess. §1 (1991).
120. George H. Brauchler, Jr., "United States Environmental Policy and the United States Army in Western Europe," 5 *Colorado Journal of International Law and Policy* (1994): 479.
121. 16 U.S.C. §§668dd–668ee (1988 & Supp. V 1993).
122. 16 U.S.C., §§470aa–470mm (1988 & Supp. V 1993).
123. 16 U.S.C. §§1531–1544 (1988 & Supp. V 1993). References herein are to section numbers of the act as originally passed, as well as to the United States Code. The act is described in greater detail in John Henry Davidson and Orlando E. Delogu, *Federal Environmental Regulation*, vol. 3 (Salem, N.H.: Butterworth, 1993), ch. 15. See also Craig E. Teller, "Effective Installation Compliance with the Endangered Species Act," *Army Lawyer*, June 1993, 5; David D. Joy, Jr., "A Tennessee Snail Darter at Grafenwoehr? The Application of the Endangered Species Act to Military Actions Abroad," *Army Lawyer*, Dec. 1991, 23.
124. §§3(6), (20); 16 U.S.C. §§1532(6), (20).
125. §4, 16 U.S.C. §1533.
126. §4(f), 16 U.S.C. §1533(f).
127. §§9, 11(a), (b), (g)(1); 16 U.S.C. §§1538, 1540(a), (b), (g)(1).
128. §7(a)(2), 16 U.S.C. §1536(a)(2).
129. *Thomas v. Peterson*, 753 F.2d 754, 763 (9th Cir. 1985) (citations omitted).
130. §§7(h), (j); 16 U.S.C. §§1536(h), (j).
131. *Pyramid Lake Paiute Tribe v. United States Department of the Navy*, 898 F.2d 1410 (9th Cir. 1990).
132. The decision is criticized in Judith Luck, "Diversions of Nevada's Truckee River Foreshadow Doom for Endangered Species," 31 *Natural Resources Journal* (1991): 931.
133. *Romero-Barcelo v. Brown*, 643 F.2d 835, 856–858 (1st Cir. 1981), reversed on other grounds sub nom. *Weinberger v. Romero-Barcelo*, 456 U.S. 305 (1982).
134. *Animal Lovers Volunteer Association, Inc., v. Weinberger*, 765 F.2d 937 (9th Cir. 1985); *Allied-Signal, Inc. v. Lujan*, 736 F. Supp. 1558 (N.D. Cal. 1990).
135. Department of Defense, *1994 Report on Environmental Compliance for Fiscal Years 1995–1999* (1994): 2–10.
136. 16 U.S.C. §§1361–1384 (1988 & Supp. V 1993). See generally John Henry Davidson and Orlando E. Delogu, *Federal Environmental Regulation*, vol. 3 (Salem, N.H.: Butterworth, 1993), ch. 11.
137. *Progressive Animal Welfare League v. Department of the Navy*, 725 F. Supp. 475 (W.D. Wash. 1989).

138. 10 U.S.C. §7524 (1988).
139. *Citizens to End Animal Suffering and Exploitation, Inc. v. The New England Aquarium*, 836 F. Supp. 45 (D. Mass. 1993).
140. 16 U.S.C. §§1451–1464 (1988 & Supp. V 1993). See generally John Henry Davidson and Orlando E. Delogu, *Federal Environmental Regulation*, vol. 3 (Salem, N.H.: Butterworth, 1993), ch. 13.
141. 16 U.S.C. §1456(c).
142. See Richard Lee Kuersteiner, Paul M. Sullivan and David Block Temin, "Protecting Our Coastal Interests: A Policy Proposal for Coordinating Coastal Zone Management, National Defense, and the Federal Supremacy Doctrine," 8 *Boston College Environmental Affairs Law Review* (1980): 705.
143. 16 U.S.C. §§1456(c)(3)(A), 1456(c)(3)(B)(iii), and 1456(d).
144. *Friends of the Earth v. United States Navy*, 841 F.2d 927 (9th Cir. 1988).
145. See ch. 3, "Clean Water Act."
146. A permit was subsequently granted, and the court dissolved its injunction. *Friends of the Earth v. United States Navy*, 850 F.2d 599 (9th Cir. 1988).
147. 16 U.S.C. §§470–470x-6 (1988 & Supp. V 1993). See John Henry Davidson and Orlando E. Delogu, *Federal Environmental Regulation*, vol. 1 (Salem, N.H.: Butterworth, 1993), ch. 3.
148. 16 U.S.C. §§470(b)(2), 470a(a)(1)(A), 470f, 470h-2(f).
149. Exec. Order No. 11,593, 36 Fed. Reg. 8921 (1971).
150. 16 U.S.C. §§470-h2(j), 470v.
151. *Romero-Barcelo v. Brown*, 643 F.2d 835, 858–860 (1st Cir. 1981).
152. *Aluli v. Brown*, 437 F. Supp. 602 (D. Hawaii 1977), reversed in part on other grounds, 602 F.2d 876 (9th Cir. 1979).
153. See also *Protect Key West, Inc. v. Cheney*, 795 F. Supp. 1552 (S.D. Fla. 1992) (Navy housing project enjoined for failure to properly consider, among other environmental impacts, effect on adjoining Historic District listed on National Register).

3. Environmental Regulation of the Defense Establishment (pp. 36–79)

1. 42 U.S.C. §§6901–6992k (1988 & Supp. V 1993).
2. 33 U.S.C. §§1251–1387 (1988 & Supp. V 1993).
3. 15 U.S.C. §§2601–2671 (1988 & Supp. V 1993).
4. 42 U.S.C. §§9601–9675 (1988 & Supp. V 1993).
5. Clean Water Act §304(e), 33 U.S.C. §1314(e) (1988).
6. RCRA §§1003(b), 3002(b); 42 U.S.C. §§6902(b), 6922(b) (1988). See also §3005(h), 42 U.S.C. §6925(h).
7. RCRA §6002, 42 U.S.C. §6962. See also §§5001–5006, 42 U.S.C. §§6951–6956 (Secretary of Commerce to encourage commercialization of recycling technology); §§8001–8007, 42 U.S.C. §§6981–6987 (research and development of waste reduction and recycling). See Robert S. Lingo, "Something New From Something Old: Federal Procurement of Recycled Products," 31 *Air Force Law Review* (1989): 269.

8. 15 U.S.C. §§2601–2629 (1988 & Supp. V 1993).
9. 7 U.S.C. §§136–136y (1988 & Supp. V 1993).
10. 42 U.S.C. §§13,101–13,109 (Supp. V 1993). The act is described in E. Lynn Grayson, "The Pollution Prevention Act of 1990: Emergence of a New Environmental Policy," 22 *Environmental Law Reporter* (Environmental Law Institute) (1992): 10,392.
11. 42 U.S.C. §13,101(a)(4) and (b).
12. EPA's strategy set a voluntary goal for industry to reduce environmental releases of fifteen to twenty toxic chemicals by one-third by the end of 1992, and by 50 percent before the end of 1995. 56 Fed. Reg. 7849 (1991).
13. 42 U.S.C. §13,106.
14. *Federal Compliance with Right-to-Know Laws and Pollution Prevention Requirements*, Exec. Order No. 12,856, 58 Fed. Reg. 41,981 (1993).
15. §313, 42 U.S.C. §11,023 (1988).
16. Exec. Order No. 12,856, §§3-302(d), 5-508, 6-601, 902.
17. *Federal Acquisition, Recycling, and Waste Prevention*, Exec. Order No. 12,873, 58 Fed. Reg. 54,911 (1993).
18. See Department of Energy, *Environmental Management 1994* (DOE/EM-0119) (1994), 5; Department of Energy, *Pollution Prevention Program: Technology Summary* (DOE/EM-0137P) (1994).
19. See Department of Defense, *Report on Environmental Requirements and Priorities* (1992), 1-16; Department of Defense, *Environmental Considerations During Weapons Systems Acquisitions* (1991).
20. Office of the Inspector General, Department of Defense, *Audit Report: Environmental Consequence Analysis of Major Defense Acquisition Programs* (Rep. No. 94–020) (Dec. 20, 1993).
21. Office of the Inspector General, Department of Defense, *Hazardous Waste Minimization Within the Department of Defense* (Rep. No. 93-INS-06) (Dec. 28, 1992).
22. 33 U.S.C. §1323(a) (1988).
23. Exec. Order No. 12,088, 43 Fed. Reg. 47,707 (1978). See also Exec. Order No. 12,580, 52 Fed. Reg. 2923 (1987), addressing compliance with the Comprehensive Environmental Response, Compensation and Liability Act (CERCLA), 42 U.S.C. §§9601–9675 (1988 & Supp. V 1993).
24. Department of Defense, *Report on Environmental Compliance for Fiscal Years 1995–1999* (1994), 1-4.
25. Department of Energy, *Environmental Management 1994* (DOE/EM-0119) (1994), 99.
26. Environmental Protection Agency, *Federal Facilities Compliance Strategy* (EPA 130/4–89/003) (1988) [hereinafter *Yellow Book*]. See James R. Edward, "Implementing the New EPA Federal Facilities Compliance Strategy: An EPA View of the Legal and Policy Issues," 31 *Air Force Law Review* (1989): 237.
27. *Yellow Book*, at II-1.
28. *Ohio v. United States Dept. of Energy*, 503 U.S. 607 (1992).
29. Federal Facility Compliance Act, Pub. L. No. 102-386, §102(a)(3), 106 Stat. 1505 (1992). See generally Laurent R. Hourcle and William J. McGowan, "Federal

Facility Compliance Act of 1992: Its Provisions and Consequences," *Fed. Facilities Environmental Journal* (winter, 1992–93): 359; Nelson D. Cary, Note, "A Primer on Federal Facility Compliance with Environmental Laws: Where Do We Go From Here?" 50 *Washington & Lee Law Review* (1993): 801.

30. The quoted language, from 42 U.S.C. §300j-6(a) (Safe Drinking Water Act), is typical.
31. See generally Nelson D. Cary, "Primer on Federal Facility Compliance"; Stan Millan, "Federal Facilities and Environmental Compliance: Toward a Solution," 36 *Loyola Law Review* (1990): 319, 340–387; Michael W. Steinberg, "Can EPA Sue Other Agencies?" 17 *Ecology Law Quarterly* (1990): 317; *Yellow Book*, VI-1 to VI-23.
32. Exec. Order No. 12,088, 43 Fed. Reg. 47,707 (1978); Exec. Order No. 12,146, 44 Fed. Reg. 42,657 (1979).
33. See ch. 7, "Enforcement by EPA."
34. Pub. L. No. 102–386, §§102(b), 103, 106 Stat. 1506, 1507 (1992); amending RCRA §§6001, 1004(15), 42 U.S.C. §§6961, 6903(15).
35. Memo from Steven Herman, EPA Assistant Administrator for Enforcement, *EPA Enforcement Policy for Private Contractor Operators at Government-Owned/Contractor-Operated (GOCO) Facilities*, Jan. 7, 1994; *Yellow Book*, at VI-14. See also Robert T. Lee and Scott E. Slaughter, "Government Contractors and Environmental Litigation," 19 *Environment Reporter* (BNA) (1989): 2138; Stanley Millan, "Environmental Dilemma of Defense Contractors—Cheshire Grin or Smile of Immunity?" 2 *Tulane Environmental Law Journal* (1989): 15.
36. *United States v. Dee*, 912 F.2d 741 (4th Cir. 1990).
37. 33 U.S.C. §1323(a) (Clean Water Act); 42 U.S.C. §7418 (Clean Air Act); 42 U.S.C. §300j-6 (Safe Drinking Water Act); 42 U.S.C. §6961 (Resource Conservation and Recovery Act); 7 U.S.C. §136p (Federal Insecticide, Fungicide and Rodenticide Act); 15 U.S.C. §2621 (Toxic Substances Control Act); 16 U.S.C. §1456(c)(3)(A) (Coastal Zone Management Act); 42 U.S.C. §4903 (Noise Control Act); 16 U.S.C. §1536(j) (Endangered Species Act); 42 U.S.C. §9620(j)(1) (Comprehensive Environmental Response, Compensation, and Liability Act).
38. Exec. Order No. 12,244, 45 Fed. Reg. 66,443 (1980), extended for one year, Exec. Order No. 12,327, 46 Fed. Reg. 48,893 (1981), waiving application of the Clean Water Act, Clean Air Act, RCRA, and Noise Control Act. See *Commonwealth of Puerto Rico v. Muskie*, 507 F. Supp. 1035 (D.P.R. 1981).
39. 13 *Environment Reporter* (BNA) (1982): 91.
40. 40 C.F.R. §§122.4 (monitoring), 122.48, 122.44(i)(2) (reporting).
41. See *Yellow Book* at V-3, V-11 to V-13.
42. EPA's environmental auditing policy is set forth at 51 Fed. Reg. 25,004 (1986), reproduced in *Yellow Book*, App. D. The Defense Department's environmental audit program is described in J. Michael Abbott, "Environmental Audits: Pandora's Box or Aladdin's Lamp?" 31 *Air Force Law Review* (1989): 225.
43. See *Yellow Book*, ch. V and App. F.
44. 42 U.S.C. §300j-9(i).
45. Pub. L. No. 101–12, 103 Stat. 16 (1989), amending 5 U.S.C. §2302 (1988).

46. OMB Circular No. A-106, Dec. 31, 1974, reproduced in *Yellow Book*, App. G. The A-106 process is described in *Yellow Book*, V-6 to V-10.
47. 33 U.S.C. §§1251–1387 (1988 & Supp. V 1993). The act is the product of sweeping amendments to the Federal Water Pollution Control Act in 1972, Pub. L. No. 92–500, 86 Stat. 816 (1972), and several amendments since that time. Citations herein are to section numbers of the Clean Water Act as originally enacted, as well as to sections of the United States Code. For a more detailed description of the Act, see John Henry Davidson and Orlando E. Delogu, *Federal Environmental Regulation*, vol. 1 (Salem, N.H.: Butterworth, 1993), ch. 2; William H. Rodgers, Jr., *Environmental Law: Air and Water*, vol. 2 (St. Paul: West Publishing Co., 1992), ch. 4; Sheldon M. Novick, ed., *Law of Environmental Protection*, vol. 2 (Deerfield, Ill.: Clark, Boardman, Callaghan, 1994), ch. 12; *Clean Water Deskbook* (Washington, D.C.: Environmental Law Institute, 1991).
48. §101(a), 33 U.S.C. §1251(a).
49. §303, 33 U.S.C. §1313.
50. Interaction between the states and the EPA in this process is described in *Mississippi Commission on Natural Resources v. Costle*, 625 F.2d 1269 (5th Cir. 1980). See Jeffrey M. Gaba, "Federal Supervision of State Water Quality Standards Under the Clean Water Act," 36 *Vanderbilt Law Review* (1983): 1167.
51. §303(c)(2)(A), 33 U.S.C. §1313(c)(2)(A); 40 C.F.R. §131.12.
52. §§301, 402, 502(6); 33 U.S.C. §§1311, 1342, 1362(6).
53. 40 C.F.R. §457.12.
54. §§301(b)(1)(A), 304(b)(1); 33 U.S.C. §§1301(b)(1)(A), 1314(b)(1).
55. §§301(b)(2)(E), 304(a)(4), 304(b)(4); 33 U.S.C. §§1311(b)(2)(E), 1314(a)(4), 1314(b)(4).
56. Toxic pollutants are those which "upon exposure, ingestion, inhalation, or assimilation by any organism, either directly from the environment or indirectly by ingestion through the food chains, will . . . cause death, disease, behavioral abnormalities, cancer, genetic mutations, physiological dysfunctions . . . or physical deformations, in such organisms or their offspring." §502(13), 33 U.S.C. §1362(13). EPA has identified sixty-five categories of toxic pollutants. 40 C.F.R. §401.15. See Oliver Houck, "The Regulation of Toxic Pollutants Under the Clean Water Act," 21 *Environmental Law Reporter* (Environmental Law Institute) (1991): 10,529.
57. §§301(b)(2)(A), 304(b)(2); 33 U.S.C. §§1311(b)(2)(A), 1314(b)(2). In determining what is "best achievable technology" (BAT), EPA must take into account the cost of the technology but need not compare such costs with the benefits of the effluent reduction. See *Rybacheck v. Environmental Protection Agency*, 904 F.2d 1276, 1290–1291 (9th Cir. 1990). A discharger may receive an "economic variance" from the BAT standard if a lower standard represents "the maximum use of technology within the economic ability" of the discharger, and will result in "reasonable progress" toward limiting further pollution discharges. §301(c), 33 U.S.C. §1311(c). The economic variance is not available for toxic pollutants, however. §301(l), 33 U.S.C. §1311(l). See also §301(g), 33 U.S.C. §1311(g).
58. §306, 33 U.S.C. §1316.

59. §307(b), 33 U.S.C. §1317(b).
60. §§301(b)(1)(C), 302; 33 U.S.C. §§1311(b)(1)(C), 1312.
61. §301(g), 33 U.S.C. §1311(g).
62. See General Accounting Office, *Water Pollution, Greater EPA Leadership Needed to Reduce Nonpoint Source Pollution* (GAO/RCED-91–10) (1990).
63. §§304(f), 319; 33 U.S.C. §§1314(f), 1329.
64. §§303(h), 311, 316, 403; 33 U.S.C. §§1313(h), 1311, 1326, 1343.
65. §304, 33 U.S.C. §1344.
66. §304(c), 33 U.S.C. §1344(c). See *Bersani v. United States Environmental Protection Agency*, 674 F. Supp. 405, 408–409, affirmed, 850 F.2d 36, 40 (2d Cir. 1988).
67. §309, 33 U.S.C. §1319.
68. §505, 33 U.S.C. §1365.
69. §313(a), 33 U.S.C. §1323(a). Federal agencies also must comply with all state laws regarding the discharge of dredge or fill material in intrastate waters. §404(t), 33 U.S.C. §1344(t). Federal vessels must comply with EPA regulations for marine sanitation devices as well, unless the Secretary of Defense finds that compliance "would not be in the interest of national security." §312(d), 33 U.S.C. §1322(d). DOD exemptions for vessels engaged in training and readiness exercises are described in *Barcelo v. Brown*, 478 F. Supp. 646, 663 n. 41 (D.P.R. 1979).
70. *United States v. Pennsylvania Department of Environmental Resources*, 778 F. Supp. 1328 (M.D. Pa. 1991).
71. General Accounting Office, *Stronger Enforcement Needed to Improve Compliance at Federal Facilities* (GAO/RCED-89–13) 3–4 (1988).
72. §313(a), 33 U.S.C. §1323(a).
73. *Natural Resources Defense Council, Inc. v. Watkins*, 954 F.2d 974, 982–983 (4th Cir. 1992).
74. *Romero-Barcelo v. Brown*, 478 F. Supp. 646, 667, 706–708 (D.P.R. 1979).
75. *Weinberger v. Romero-Barcelo*, 456 U.S. 305, 323 (1982).
76. §401(a), 33 U.S.C. §1341(a).
77. See *United States v. Puerto Rico*, 721 F.2d 832 (1st Cir. 1983); Joseph A. Wellington, "A Primer on Environmental Law for the Naval Services," 38 *Naval Law Review* (1989): 5, 46.
78. §§301(f), 502(6); 33 U.S.C. §§1311(f), 1362(6).
79. 42 U.S.C. §§2011–2296 (1988 & Supp. V 1993).
80. *Train v. Colorado Public Interest Research Group*, 426 U.S. 1 (1976).
81. See Department of Energy, *Report of the Task Group on Operation of Department of Energy Tritium Facilities* (DOE/EH-0198P) (1991); General Accounting Office, *Nuclear Waste: Impact of Savannah River Plant's Radioactive Waste Management Practices* (GAO/RCED-86–143) (1986).
82. Energy Secretary James D. Watkins, press conference, Augusta, Ga., Jan. 8, 1992.
83. See ch. 7, "Enforcement by EPA."
84. *Public Interest Research Group of New Jersey, Inc. v. Rice*, 774 F. Supp. 317 (D. N.J. 1991).
85. *United States Department of Energy v. Ohio*, 503 U.S. 607 (1992).
86. *California v. United States Dept. of the Navy*, 845 F.2d 222 (9th Cir. 1987). Legislation introduced in the 103d Congress, H.R. 2580 (1993), would amend the

Clean Water Act to permit states to collect civil penalties and EPA to issue administrative orders for federal facility violations.

87. *New York v. United States*, 620 F. Supp. 374 (E.D. N.Y. 1985).
88. *United States v. Pennsylvania Environmental Hearing Board*, 584 F.2d 1273 (3d Cir. 1978).
89. 42 U.S.C. §§7401–7671q (1988 & Supp. V 1993). References in these notes are to section numbers of the act as originally enacted, as well as to the United States Code.
90. §101(b)(1); 42 U.S.C. §7401(b)(1); see also 40 C.F.R. §50.1(e).
91. A fuller description of the Clean Air Act may be found in John Henry Davidson and Orlando E. Delogu, *Federal Environmental Regulation*, vol. 1 (Salem, N.H.: Butterworth, 1993), ch. 3; Theodore L. Garrett and Sonya D. Winner, "A Clean Air Act Primer" (pts. I, II, & III), 22 *Environmental Law Reporter* (Environmental Law Institute) (1992): 10,159, 10,235, and 10,301; William H. Rodgers, Jr., *Environmental Law: Air and Water*, vol. 1 (St. Paul: West Publishing Co., 1992), ch. 3; Sheldon M. Novick, ed., *Law of Environmental Protection*, vol. 2, (Deerfield, Ill.: Clark, Boardman, Callaghan, 1994), ch. 11; *Clean Air Deskbook* (Washington, D.C.: Environmental Law Institute, 1992).
92. §§108(a)(1)(A), 109; 42 U.S.C. §§7408(a)(1)(A), 7409; S. Rep. No. 228, 101st Cong., 2d Sess., at 5, reprinted at 1990 U.S.C.C.A.N. 3385, 3389.
93. §109(b)(2), 42 U.S.C. §7409(b)(2).
94. See 40 C.F.R. pt. 50.
95. §112, 42 U.S.C. §7412. See 40 C.F.R. pt. 61.
96. §112(b), 42 U.S.C. §7412(b). Between 1970 and 1990, the EPA had developed standards for only seven hazardous air pollutants: asbestos, beryllium, mercury, vinyl chloride, radionuclides, inorganic arsenic, and benzene. 40 C.F.R. pt. 61 (1991).
97. §§112(d)(2) and (f); 42 U.S.C. §§7412(d)(2) and (f).
98. §110, 42 U.S.C. §7410. A state must demonstrate that it has the requisite resources and authority to implement its plan, and it must faithfully implement and enforce the plan once it is approved by EPA. §§110(a)(2) and (e); 42 U.S.C. §§7410(a)(2) and (e).
99. §§110(a)(2), 116, 160–169B, 171–193; 42 U.S.C. §§7410(a)(2), 7416, 7470–7492, 7501–7515.
100. §176, 42 U.S.C. §7506.
101. §§111(c), 112(l), 42 U.S.C. §§7411(c), 7412(l).
102. §§501–506, 42 U.S.C. §§7661–7661f.
103. §§110(a)(2)(C), 113, 304; 42 U.S.C. §§7410(a)(2)(C), 7413, 7604.
104. §112(i)(4), 42 U.S.C. §7412(i)(4).
105. §118(a), 42 U.S.C. §7418(a).
106. *Ohio ex rel. Celebreeze v. United States Department of the Air Force*, 1987 WL 110399, 17 *Environmental Law Reporter* (Environmental Law Inst.) 21,210 (S.D. Ohio 1987).
107. *New York State Department of Environmental Conservation v. United States Department of Energy*, 772 F. Supp. 91 (N.D.N.Y. 1991); *United States v. South Coast Air Quality Management District*, 748 F. Supp. 732 (C.D. Cal. 1990).

108. *California ex rel. State Air Resources Board v. Department of the Navy*, 431 F. Supp. 1271 (N.D. Cal. 1977).
109. 54 Fed. Reg. 51,654 (1989); 40 C.F.R. pt. 61, subpt. H.
110. See ch. 7, "Protecting Military Secrets."
111. Background information for this study was drawn from Department of Energy, *Environmental Restoration and Waste Management Five-Year Plan: Fiscal Years 1994–1998* (1993), at II-22, and from interviews with officials at the Department of Energy and the New Mexico Attorney General's office.
112. N.M. Stat. Ann. §§74-4 and -4.2(C) (Michie 1978 & Supp. 1993).
113. RCRA §6001, 42 U.S.C. §6961, makes federal facilities subject to "all federal, State, interstate, and local requirements, both substantive and procedural," concerning solid waste disposal.
114. *United States v. New Mexico*, 1992 WL 437983 (D.N.M.), affirmed, 32 F.3d 494 (10th Cir. 1994).
115. §322, 42 U.S.C. §7622.
116. *Nochumson v. Los Alamos National Laboratory* (No. 92-CAA-1) (U.S. Department of Labor Office of Admin. L. Judges) (Sept. 22, 1994).
117. *Concerned Citizens for Nuclear Safety, Inc. v. United States Department of Energy*, Civ. No. 94–1039M (D.N.M. filed Sept. 13, 1994).
118. §118(b), 42 U.S.C. §7418(b). No exemption is available for new sources. Exemption from rules covering hazardous air pollutants is possible only when technology needed for compliance is not available.
119. John S. Hannah, "Chlorofluorocarbons: A Scientific, Environmental, and Regulatory Assessment," 31 *Air Force Law Review* (1989): 85, 102.
120. Montreal Protocol on Substances that Deplete the Ozone Layer, reprinted in 26 *International Legal Materials* (1987): 1616, entered into force Sept. 22, 1988.
121. §§601–618, 42 U.S.C. §§7671–7671q.
122. Copenhagen Amendments to the Montreal Protocol, reprinted in 32 *International Legal Materials* (1993): 74. See also London Amendments, reprinted in 30 *International Legal Materials* (1991): 537.
123. 58 Fed. Reg. 65018 (1992), announcing regulations to be published at 40 C.F.R. pt. 82.
124. §604, 42 U.S.C. §7671c.
125. Pub. L. No. 101–189, §356, 103 Stat. 1425 (1989). See also Pub. L. No. 101-510, §345, 104 Stat. 1538 (1990), and Pub. L. No. 102-484, §325, 106 Stat. 2367 (1992) (ordering evaluation of military uses and alternatives to various ozone-depleting chemicals).
126. Ibid., §326.
127. Department of Defense, *Report on Environmental Requirements and Priorities* (1992), 1–7.
128. "Air Pollution: Defense Secretary Orders Military to Revise Specifications for Ozone-Depleting Chemicals," 22 *Environment Reporter* (BNA) (1992): 2466.
129. §118(a), 42 U.S.C. §7418(a).
130. §304(a), 42 U.S.C. §7604(a).
131. See *United States v. Air Pollution Control Board of Tennessee Department of Health and the Environment*, No. 3:88–1030 (M.D. Tenn. March 2, 1990); *Ohio*

ex rel. Celebreeze v. United States Department of the Air Force, 1987 WL 110,399, 17 *Environmental Law Reporter* (Environmental Law Institute) 21,210, 21,212 (S.D. Ohio 1987); *Alabama ex rel. Graddick v. Veterans Administration*, 648 F. Supp. 1208 (M.D. Ala. 1986). Cf. *California ex rel. State Air Resources Board v. Department of the Navy*, 431 F. Supp. 1271 (N.D. Cal. 1977).

132. *Ohio v. United States Dept. of Energy*, 503 U.S. 607 (1992). The subject of sovereign immunity is analyzed in greater depth in chapter 7.
133. 42 U.S.C. §§300f to 300j-11 (1988 & Supp. V 1993); H.R. Rep. No. 1185, 93d Cong., 2d Sess, reprinted in *U.S. Code Congressional & Administrative News* (1974): 6454.
134. The Safe Drinking Water Act and its operation generally are described in John Henry Davidson and Orlando E. Delogu, *Federal Environmental Regulation*, vol. 2 (Salem, N.H.: Butterworth, 1993), ch. 9; Sheldon M. Novick, ed., *Law of Environmental Protection*, vol. 3 (Deerfield, Ill.: Clark, Boardman, Callaghan, 1994), ch. 16; Kenneth Gray, "The Safe Drinking Water Amendments of 1986: Now a Tougher Act to Follow," 16 *Environmental Law Reporter* (Environmental Law Institute) (1986): 10,338.
135. 42 U.S.C. §300f(4).
136. Ibid., §300g-1. The regulations are found at 40 C.F.R. pt. 141.
137. 42 U.S.C. §300g-1(b); 40 C.F.R. §§141.50–141.63.
138. 40 C.F.R. §141.61. See *Natural Resources Defense Council, Inc. v. Environmental Protection Agency*, 812 F.2d 721 (D.C. Cir. 1987) (upholding EPA's MCL for fluoride). See also Seth Shulman, *The Threat at Home: Confronting the Toxic Legacy of the U.S. Military* (Boston: Beacon Press, 1992), 24–25, 86–87.
139. 40 C.F.R. §§141.15 and .16.
140. 42 U.S.C. §§300g-3(e), 300j-6(a).
141. 40 C.F.R. §264.94 (Resource Conservation and Recovery Act, described in the next section); 40 C.F.R. §300.430(e)(2)(B) (Comprehensive Environmental Response, Compensation, and Liability Act, addressed in chapter 4); 40 C.F.R. §144.12 (Safe Drinking Water Act UIC program, described below).
142. 42 U.S.C. §300f(2); 40 C.F.R. pt. 143.
143. See 42 U.S.C. §300g-3(d); 40 C.F.R. §143.1.
144. 42 U.S.C. §§300h to 300h-5; 40 C.F.R. pts. 144, 146. See generally William H. Rodgers, Jr., *Environmental Law: Hazardous Wastes and Substances*, vol. 4 (St. Paul: West Publishing Co. 1992), 210.
145. 42 U.S.C. §§300h(b)(1) and (d)(1).
146. The operation of RCRA is described below.
147. 42 U.S.C §§300h(b)(1), (d)(2).
148. RCRA §§3004(d)–(i), 42 U.S.C. §§6924(d)–(i). On the other hand, EPA has decided that it can exempt an aquifer from protection of the UIC program if the aquifer is already "so contaminated that it would be economically or technologically impracticable to render that water fit for human consumption." 40 C.F.R. §146.4(b)(3). See *Western Nebraska Resource Council v. Environmental Protection Agency*, 793 F.2d 194 (8th Cir. 1986). This regulation appears to reward those who have contaminated groundwater in the past by allowing them to continue polluting.

149. 42 U.S.C. §§300h-1 and -2.
150. 42 U.S.C. §300h-3; 40 C.F.R. pt. 149.
151. 42 U.S.C. §300h-3(a)(1). In addition, 1986 amendments to the Safe Drinking Water Act established a sole source aquifer demonstration program for the protection of "critical aquifer protection areas." Id. §300h-6(a). States were invited to develop comprehensive management plans for such critical areas to "maintain the quality of the ground water . . . in a manner reasonably expected to protect human health, the environment and ground water resources." Id. §300h-6(f).
152. Id. §300h-3(b)(3).
153. Id. §300h-7.
154. Id. §300j-8.
155. Id. §300i(a).
156. See generally, Paul R. Smith, "The Impact of the Safe Drinking Water Act Amendments of 1986 on Military Installations: How Real is the Encroachment Threat?" 38 *Naval Law Review* (1989): 49.
157. 42 U.S.C. §300j-6(a).
158. Id. §300j-6(b).
159. Id. §300h-7(h).
160. 42 U.S.C. §§10,101–10,270 (1988 & Supp. V 1993).
161. Id. §10141(a).
162. 50 Fed. Reg. 38,066 (1985). According to Nuclear Waste Policy Act §10,141(a), the regulations are to be based on EPA's existing authority under the Atomic Energy Act, 42 U.S.C. §2201(b).
163. See *Natural Resources Defense Council, Inc. v. United States Environmental Protection Agency*, 824 F.2d 1258 (1st Cir. 1987).
164. 824 F.2d at 1270–1271.
165. 824 F.2d at 1276. The litigation is described in Charles H. Montange, "Federal Nuclear Waste Disposal Policy," 27 *Natural Resources Journal* (1987): 309, 380–389.
166. Waste Isolation Pilot Plant Land Withdrawal Act, Pub. L. No. 102–579, 106 Stat. 4777 (1992).
167. 58 Fed. Reg. 66,398 (1993), to be codified at 40 C.F.R. pt. 191.
168. Regulations for the proposed Yucca Mountain repository are called for in the Energy Policy Act of 1992, Pub. L. No. 102–486, §801(a)(1), 106 Stat. 2921 (1992).
169. 40 C.F.R. §§144.6(d), 144.13.
170. 42 U.S.C. §300j-8(a)(1).
171. Id. §300j-6(a).
172. 42 U.S.C. §§6901–6992k (1988 & Supp. V 1993). Citations in these notes are to RCRA's section numbers as originally enacted, as well as to the United States Code.
173. §1004(27), 42 U.S.C. §6903(27).
174. Some of these activities are described in Department of Defense, *Defense Environmental Cleanup Program: Annual Report to Congress for Fiscal Year 1993* (1994); Seth Shulman, *The Threat at Home: Confronting the Toxic Legacy of the*

U.S. Military (Boston: Beacon Press, 1992); Office of Technology Assessment, *Complex Cleanup: The Environmental Legacy of Nuclear Weapons Production* (OTA-0–484) (1991).

175. RCRA is described in greater detail in John Henry Davidson and Orlando E. Delogu, *Federal Environmental Regulation*, vol. 2 (Salem, N.H.: Butterworth, 1993), ch. 4; William H. Rodgers, Jr., *Environmental Law: Hazardous Wastes and Substances* (St. Paul: West Publishing Co.), vol. 3 (1988): 509–568, and vol. 4 (1992): 1–467; Donald W. Stever, *Law of Chemical Regulation and Hazardous Wastes*, vol. 1 (Deerfield, Ill.: Clark, Boardman, Callaghan, 1994), ch. 5; Sheldon M. Novick, ed., *Law of Environmental Protection*, vol. 3 (Deerfield, Ill.: Clark, Boardman, Callaghan, 1994), ch. 13; *RCRA Deskbook* (Washington, D.C.: Environmental Law Institute, 1991); and Randolph L. Hill, "An Overview of RCRA: The 'Mind-Numbing' Provisions of the Most Complicated Environmental Statute," 21 *Environmental Law Reporter* (Environmental Law Institute) (1991): 10,254.
176. §§4001–4010; 42 U.S.C. §§6941–6949a.
177. 53 Fed. Reg. 33,314 (1988), 40 C.F.R. pt. 258.
178. §4005(c)(2), 42 U.S.C. §6945(c)(2).
179. §§3002(b), 3005(h), 6002; 42 U.S.C. §§6922(b), 6925(h), 6962.
180. §§3001–3020, 42 U.S.C. §§6921–6939b.
181. §§1004(5), (27); 42 U.S.C. §§6903(5), (27); 40 C.F.R. §§261.2, 261.3.
182. §§3001(a), (b); 42 U.S.C. §§6921(a), (b).
183. 40 C.F.R. pt. 261, subpts. B, C & D.
184. 40 C.F.R. §261.3(a)(2)(iv).
185. §1004(27), 42 U.S.C. §6903(27).
186. 51 Fed. Reg. 24,504 (1986); 52 Fed. Reg. 15,937 (1987); 53 Fed. Reg. 37,045 (1988). See 10 C.F.R. §962.3(b). See also *Sierra Club v. Department of Energy*, 734 F. Supp. 946, 949 (D. Colo. 1990).
187. §§3002, 3003, 3005, 3006; 42 U.S.C. §§6922, 6923, 6925, 6926; 40 C.F.R. pts. 262–263, 270.
188. In order to obtain interim status, a facility must either have been in existence on November 18, 1980, or on the date of a statutory or regulatory change requiring the facility to obtain a hazardous waste permit, and must also submit a permit application. §3005(e), 42 U.S.C. §6925(e); 40 C.F.R. pt. 270, subpt. G.
189. §§3004(d)(1), (e)(1), (g)(5); 42 U.S.C. §§6924(d)(1), (e)(1), (g)(5). See 40 C.F.R. §268.6 (land disposal generally), 40 C.F.R. pt. 148 (underground injection wells).
190. See generally §3004, 42 U.S.C. §6924; 40 C.F.R. pts. 264–270.
191. 40 C.F.R. §264.94.
192. §3007, 42 U.S.C. §6927. Only trade secrets and confidential personal information are insulated from public view.
193. §§9001–9010; 42 U.S.C. §§6991–6991i; 40 C.F.R. pt. 280. See Laura J. Nagle, "RCRA Subtitle I: The Federal Underground Storage Tank Program," 24 *Environmental Law Reporter* (Environmental Law Institute) (1994): 10,057.
194. 40 C.F.R. §270.60.
195. 40 C.F.R. pt. 264, subpt. O.
196. §§3004(u), 3004(v), 3008(h); 42 U.S.C. §§6924(u), 6924(v), 6928(h); 40 C.F.R. pt. 264, subpt. F.

197. 40 C.F.R. pt. 264, subpt. G; 40 C.F.R. pt. 265, subpt. G.
198. §§3008, 7002; 42 U.S.C. §§6928, 6972.
199. §§7002(a)(1)(B), 7003(a); 42 U.S.C. §§6972(a)(1)(B), 6973(a).
200. §6001; 42 U.S.C. §6961.
201. Ibid.
202. SECNAV memo to Chief of Naval Operations and Commandant of the Marine Corps, re: Resources in Support of Environmental Compliance (Apr. 29, 1988), quoted in Joseph A. Wellington, "A Primer on Environmental Law for the Naval Services," 38 *Naval Law Review* (1989): 5, 31.
203. See H. Rep. No. 111, 102d Cong., 2d Sess. 4 (1992), reprinted in *U.S. Code Congressional and Administrative News* (1992): 1287, 1290.
204. Department of Defense, *Report on Environmental Compliance for Fiscal Years 1995–1999* (1994), 2–10.
205. See Dan W. Reicher and S. Jacob Scherr, "Laying Waste to the Environment," *Bulletin of the Atomic Scientists* (January/February 1988): 32–33.
206. See Matthew L. Wald, "Atomic-Unit Leaks Cited in Tennessee," *New York Times*, May 9, 1990.
207. *Legal Environmental Assistance Foundation v. Hodel*, 586 F. Supp. 1163 (E.D. Tenn. 1984). The litigation is analyzed in Barbara Finamore, "Regulating Hazardous and Mixed Waste at DOE Nuclear Weapons Facilities: Reversing Decades of Environmental Neglect," 9 *Harvard Environmental Law Review* (1985): 83.
208. §1006(a), 42 U.S.C. §6905(a).
209. §1004(27); 42 U.S.C. §6903(27).
210. 586 F. Supp. at 1167.
211. Ibid., p. 1169.
212. 50 Fed. Reg. 45,736 (1985).
213. In 1986, the General Accounting Office reported that the Hanford facility was "seriously out of compliance with RCRA." General Accounting Office, *Nuclear Waste: Unresolved Issues Concerning Hanford's Waste Management Practice* (GAO/RCED-87–30) (1986). See case study, ch. 4, "DOE's Cleanup at Hanford."
214. 51 Fed. Reg. 24,504 (1986); 52 Fed. Reg. 15,940 (1987). See also *New Mexico v. Watkins*, 969 F.2d 1122, 1130–1132 (D.C. Cir. 1992).
215. *Sierra Club v. United States Department of Energy*, 734 F. Supp. 946, 949 (D. Colo. 1990) (emphasis in original).
216. *Sierra Club v. United States Dept. of Energy*, 770 F. Supp. 578 (D. Colo. 1991).
217. *Environmental Defense Fund v. Watkins*, 783 F. Supp. 633 (D.D.C. 1992).
218. 43 U.S.C. §§1701–1784 (1988).
219. *New Mexico v. Watkins*, 969 F.2d 1122 (D.C. Cir. 1992).
220. 40 C.F.R. pt. 264, subpt. O.
221. Background information for this case study was drawn from David A. Koplow, "How Do We Get Rid of These Things?: Dismantling Excess Weapons While Protecting the Environment," 89 *Northwestern University Law Review* (1995): 445; National Research Council, *Alternative Technologies for the Destruction of Chemical Agents and Munitions* (Washington, D.C.: National Academy Press, 1993); Office of Technology Assessment, *Disposal of Chemical Weapons: Alter-*

native Strategies—Background Paper (1992); Michael Renner, "Cleaning Up After the Arms Race," in Lester R. Brown et al., *State of the World 1994* (New York: W.W. Norton & Co., 1994), 145–150; Lawrence E. Rouse, "The Disposition of the Current Stockpile of Chemical Munitions and Agents," 121 *Military Law Review* (1988): 17; Amy E. Smithson, "Chemical Destruction: The Work Begins," *Bulletin of the Atomic Scientists*, (April 1993): 38.

222. Pub. L. No. 99–145, §1412(a), 99 Stat. 583 (1986). See also R. Jeffrey Smith, "U.S. Ushers In New Era of Chemical Weapons," *Washington Post*, Jan. 15, 1989.
223. Pub. L. No. 102-484, §171, 106 Stat. 2315 (1992).
224. See Keith Schneider, "U.S. Plan to Burn Chemical Weapons Stirs Public Fear," *New York Times*, Apr. 29, 1991; *Citizen Attitudes on the Destruction and Disposal of Chemical Weapon Stockpiles* (Washington, D.C.: Lawyers Alliance for World Security/Committee on National Security, 1994).
225. R. Jeffrey Smith, "Army Poison Gas Stockpile Raises Worries in Kentucky," *Washington Post*, Jan. 22, 1989.
226. See 50 U.S.C. §1512 (1988) (imposing restrictions on transportation of chemical warfare agents between military installations); and Pub. L. No. 103-139, §8075A(a), 107 Stat. 1418 (1992) (prohibiting the Army from further study of relocation options).
227. Pub. L. No. 100-180, §125, 101 Stat. 1043 (1987).
228. Dean Hill Rivkin, "Programmatic Environmental Impact Statements, Tiering, and the NEPA Process: A Case Study of the U.S. Army's Chemical Stockpile Disposal Program," unpublished paper, 1988.
229. See National Research Council, *Alternative Technologies for the Destruction of Chemical Agents and Munitions* (Washington, D.C.: National Academy Press, 1993); Office of Technology Assessment, *Disposal of Chemical Weapons: Alternative Technologies* (1992).
230. Pub. L. No. 102-484, §173, 106 Stat. 2442 (1992).
231. National Research Council, *Recommendations for the Disposal of Chemical Agents and Munitions* (Washington, D.C.: National Academy Press, 1994).
232. Program Manager for Chemical Demilitarization, Department of the Army, *U.S. Army's Alternative Demilitarization Technology Report for Congress* (April 11, 1994).
233. Convention on the Prohibition of the Development, Stockpiling and Use of Chemical Weapons and On Their Destruction, Jan. 13, 1993, 32 *International Legal Materials* 800, reproduced with extensive explanation in *Chemical Weapons Convention*, S. Treaty Doc. No. 21, 103d Cong., 1st Sess. (1993).
234. Ibid., Art. IV., par. 10.
235. Ibid., Annex on Implementation and Verification, Part IV(a), par. 13.
236. The unitary executive theory is described in ch. 7, "Enforcement by EPA."
237. *Ohio v. United States Dept. of Energy*, 503 U.S. 607 (1992); sovereign immunity is discussed in ch. 7, "Enforcement by the States."
238. Pub. L. No. 102–386, §102(b), 106 Stat. 1506 (1992). EPA's new policy for enforcing RCRA at federal facilities is set forth at 58 Fed. Reg. 49044 (1993). See Hourcle and McGowan, "Federal Facility Compliance Act."
239. "Navy Air Station Hit With Fine Under New Federal Law Provisions," 24 *Environment Reporter* (BNA) (1993): 190.

240. 23 *Environmental Law Reporter* (Environmental Law Institute) (1993): 10,519.
241. Pub. L. No. 102–386, §102(a), 106 Stat. 1505. Special provisions for storage of mixed wastes at DOE facilities are described in ch. 4, "Cleaning Up DOE's Weapons Complex."
242. 23 *Environmental Law Reporter* (Environmental Law Institute) (1993): 10,411.
243. See, e.g., *Sierra Club v. United States Department of Energy*, 770 F. Supp. 578 (D. Colo. 1991).
244. 15 U.S.C. §2601–2629 (1988 & Supp. V 1993). References in these notes are to section numbers of the act as originally enacted, as well as to the United States Code. For a detailed discussion of TSCA, see John Henry Davidson and Orlando E. Delogu, *Federal Environmental Regulation*, vol. 2 (Salem, N.H.: Butterworth, 1993), ch. 5; William H. Rodgers, Jr., *Environmental Law: Pesticides and Toxic Substances*, vol. 3 (St. Paul: West Publishing Co. 1988), 370–503; Donald W. Stever, *Law of Chemical Regulation and Hazardous Wastes*, vol. 1 (Deerfield, Ill.: Clark, Boardman, Callaghan, 1994), ch. 2; Sheldon M. Novick, ed., *Law of Environmental Protection*, vol. 1 (Deerfield, Ill.: Clark, Boardman, Callaghan, 1994), ch. 15.
245. §8(b), 15 U.S.C. §2607(b); FIFRA is described in the next section.
246. §6, 15 U.S.C. §2605; 40 C.F.R. pts. 747, 761, 762, and 763.
247. §§4, 5, 7; 15 U.S.C. §§2603, 2604(f), 2606.
248. §6(e), 15 U.S.C. §2605(e). Comprehensive rules for PCBs are set out at 40 C.F.R. pt. 761. See Marc W. Trost, "The Regulation of Polychlorinated Biphenyls Under the Toxic Substances Control Act," 31 *Air Force Law Review* (1989): 117.
249. Department of Defense, *Defense Environmental Restoration Program: Annual Report to Congress for Fiscal Year 1992* (1993), 16, 57.
250. §28, 15 U.S.C. §2627.
251. §20, 15 U.S.C. §2619.
252. §22, 15 U.S.C. §2621.
253. 7 U.S.C. §§136–136y (1988 & Supp. V 1993). References in these notes are to section numbers of the act as originally enacted, as well as to the United States Code. For a more detailed treatment of FIFRA, see John Henry Davidson and Orlando E. Delogu, *Federal Environmental Regulation*, vol. 2 (Salem, N.H.: Butterworth, 1993), ch. 8; William H. Rodgers, Jr., *Environmental Law: Pesticides and Toxic Substances* vol. 3, (St. Paul: West Publishing Co., 1988), ix–369; Donald W. Stever, *Law of Chemical Regulation and Hazardous Wastes*, vol. 1 (Deerfield, Ill.: Clark, Boardman, Callaghan, 1994), ch. 3; Sheldon M. Novick, ed., *Law of Environmental Protection*, vol. 3 (Deerfield, Ill.: Clark, Boardman, Callaghan, 1994), ch. 17.
254. See generally §3, 7 U.S.C. §136a.
255. §2(bb), 7 U.S.C. §136(bb).
256. §11, 7 U.S.C. §136i.
257. §12(a)(2)(G), 7 U.S.C. §136j(a)(2)(G).
258. §14, 7 U.S.C. §136l.
259. Department of Defense, *Report on Environmental Requirements and Priorities* (1992), 1–7.
260. See ch. 6, "Application of Domestic Environmental Laws."
261. 42 U.S.C. §§4901–4918 (1988 & Supp. V 1993). References herein are to section numbers of the act as originally enacted, as well as to the United States Code. See

generally Note, "The Noise Control Act of 1972," 58 *Minnesota Law Review* (1973): 273; John Henry Davidson and Orlando E. Delogu, *Federal Environmental Regulation*, vol. 2 (Salem, N.H.: Butterworth, 1993), ch. 7.

262. §§2(b), 3(3), 5(a), 6(a); 42 U.S.C. §§4901(b), 4902(3), 4904(a), 4905(a).
263. 40 C.F.R. pts. 204, 205; see Sidney A. Shapiro, "Lessons from a Public Policy Failure: EPA and Noise Abatement," 19 *Ecology Law Quarterly* (1992): 1.
264. §§10–12, 42 U.S.C. §§4909–4911.
265. §4(b), 42 U.S.C. §4903(b).
266. §10(b)(1), 42 U.S.C. §4909(b)(1).
267. 40 C.F.R. §205.5-2.
268. §4(b), 42 U.S.C. §4903(b).
269. *Romero-Barcelo v. Brown*, 643 F.2d 835, 852–856 (1st Cir. 1981).
270. §4(a), 42 U.S.C. §4903(a).
271. *Westside Property Owners v. Schlesinger*, 597 F.2d 1214 (9th Cir. 1979).
272. See Bernard K. Schafer, "The Air Installation Compatible Use Zone Program: The Science and the Law," 31 *Air Force Law Review* (1989): 165.
273. 49 U.S.C. App. §§1801–1819 (1988 & Supp. V 1993). See generally Stan Millan and Andrew J. Harrison, Jr., "A Primer on Hazardous Materials Transportation Law in the 1990's: The Awakening," 22 *Environmental Law Reporter* (Environmental Law Institute) (1992): 10,583; Bradley M. Marten, "Regulation of the Transportation of Hazardous Materials: A Critique and a Proposal," 5 *Harvard Environmental Law Review* (1982): 345.
274. 49 U.S.C. App. §§1803, 1804.
275. Id. §§1804(a)(4), 1811(a). See, e.g., *Colorado Public Utilities Commission v. Harmon*, 951 F.2d 1571 (10th Cir. 1991).
276. *City of New York v. United States Dept. of Transportation*, 715 F.2d 732 (1st Cir. 1983); *City of New York v. United States Department of Transportation*, 700 F. Supp. 1294 (S.D.N.Y. 1988).
277. 49 U.S.C. App. §1804(b).
278. Some of the hazards of defense-related rail and highway shipments are described in Gerald Jacob and Andrew Kirby, "On the Road to Ruin: The Transport of Military Cargoes," in *Hidden Dangers: Environmental Consequences of Preparing for War*, ed. Anne H. Ehrlich and John W. Birks (San Francisco: Sierra Club Books, 1990), 71.
279. "Trucking of Rocket Fuel Is Called Dangerous," *New York Times*, Oct. 11, 1987.
280. See generally 49 C.F.R. pt. 173.
281. John M. Broder, "Pollution 'Hot Spots' Taint Water Sources," *Los Angeles Times*, June 18, 1990.
282. John M. Broder, "U.S. Military Leaves Toxic Trail Overseas," *Los Angeles Times*, June 18, 1990.
283. Joseph Gerson and Bruce Birchard, eds., *The Sun Never Sets: Confronting the Network of Foreign U.S. Military Bases* (Boston: South End Press, 1991), 19–21.
284. General Accounting Office, *Hazardous Waste: Management Problems Continue at Overseas Military Bases* (GAO/NSIAD-91–231) (1991): 2, 3.
285. See *Equal Employment Opportunity Commission v. Arabian American Oil Co.*, 499 U.S. 244, 248 (1991).

286. *Amlon Metals, Inc. v. FMC Corp.*, 775 F. Supp. 668 (S.D.N.Y. 1991).
287. *Defenders of Wildlife v. Lujan*, 911 F.2d 117, 122–125 (8th Cir. 1990), reversed on other grounds, 504 U.S. 555 (1992).
288. *Environmental Defense Fund v. Massey*, 986 F.2d 528, 529 (D.C. Cir. 1993). Cf. *NEPA Coalition of Japan v. Aspin*, 837 F. Supp. 466, 467 (D.D.C. 1993), citing *Foley Bros., Inc. v. Filardo*, 336 U.S. 281, 285 (1942). See ch. 2, "Applying NEPA to Military Activities Abroad."
289. Exec. Order No. 12,088, §1-801, 43 Fed. Reg. 47,707 (1978).
290. Pub. L. No. 101-510, §342(b), 104 Stat. 1537 (1990).
291. Department of Defense, *Overseas Environmental Baseline Guidance Document* (Oct. 1992), 1-3.
292. Ibid., 1-8.
293. Ibid., 1-1.
294. The annual report, first published in 1994, is prescribed in 10 U.S.C. §2706(b). A more detailed description of military environmental programs appears in Department of Defense, *Report on Environmental Requirements and Priorities* (1992). Congress called for the earlier report in Pub. L. No. 101-189, §358, 103 Stat. 1427 (1989).
295. Department of Defense, *1994 Report on Environmental Compliance for Fiscal Years 1995–1999* (1994).
296. See, e.g., Department of Energy, *Environmental Restoration and Waste Management Five-Year Plan: Fiscal Years 1994–1998* (DOE/S-00097P) (1993).
297. Ibid., II-198 to -206.
298. Ibid., 1-2, 1-6.
299. Department of Energy, *Environmental Management 1994* (DOE/EM-0119) (1994), 2.
300. See case study, ch. 4, "DOE's Cleanup at Hanford."
301. The separate defense budget accounts for cleanups at operating and closing bases are described in chaps. 4 and 5.
302. Department of Energy, *Environmental Restoration and Waste Management Five-Year Plan: Fiscal Years 1994–1998* (DOE/S-00097P) (1993), at I-34 to -36.
303. See 57 Fed. Reg. 57,170 (1992).
304. See generally Laurent R. Hourcle, Robert S. Lingo and Francis H. Esposito, "Environmental Law in the Fourth Dimension: Issues of Responsibility and Indemnification with Government Owned-Contractor Operated Facilities," 31 *Air Force Law Review* (1989): 245; Stanley Millan, "Environmental Dilemma of Defense Contractors—Cheshire Grin or Smile of Immunity?" 2 *Tulane Environmental Law Journal* (1989): 15.
305. Federal Acquisition Regulation (FAR) 31.201–3(b). The FAR is set out at 48 C.F.R. ch. 1.
306. A tradition of DOD reimbursement of contractor operators at government facilities is described in Lenny Siegel, Gary Cohen and Ben Goldman, *The U.S. Military's Toxic Legacy* (Boston: National Toxic Campaign Fund, 1991), ch. 4.
307. FAR 50.403; FAR 31.205–15(a).
308. *In re Fernald Litigation*, 1989 WL 267039 (S.D. Ohio). See case study, ch. 8, "Flawed Fiat at Fernald."

309. FAR 23.103(b), 23.104(c).
310. FAR 9.406-2(a)(5), 9.407-2(a)(7).
311. Department of Energy, *Environmental Restoration and Waste Management Five-Year Plan: Fiscal Years 1994–1998* (1993), I-53.
312. John H. Cushman, Jr., "204 Secret Nuclear Tests By U.S. Are Made Public," *New York Times*, Dec. 8, 1993; Keith Schneider, "U.S. Reveals a Few More of Its Atomic Secrets," *New York Times*, June 28, 1994.
313. Department of Defense, *Defense Environmental Cleanup Program: Annual Report to Congress for Fiscal Year 1993* (1994): 33. Technical review committees are prescribed for cleanups at active military bases. 10 U.S.C. §2705(c).
314. The Defense Department's environmental audit program is described in J. Michael Abbott, "Environmental Audits: Pandora's Box or Aladdin's Lamp?" 31 *Air Force Law Review* (1989): 225. See also Department of Defense, *Report on Environmental Requirements and Priorities* (1992), 2–9.
315. Office of the Inspector General, Department of Defense, *Comprehensive Long-Term Environmental Action, Navy Contract* (Rep. No. 93–097) (1993).
316. Department of Defense, *1994 Report on Environmental Compliance for Fiscal Years 1995–1999* (1994), 2-10.
317. William J. Broad, "Energy Department Giving Critics More Voice," *New York Times*, Oct. 16, 1994.
318. United States Army Corps of Engineers, *Commander's Guide to Environmental Management* (1989). This guide was recently updated by Kerrin J. Dame, *Commander's Guide to Environmental Management* (Aberdeen Proving Ground, Md.: U.S. Army Environmental Center, 1993).

4. Dangerous Legacy: Cleaning Up After the Cold War (pp. 80–124)

1. See generally Kenneth P. Doyle, "Cleaning Up Federal Facilities: Controversy Over an Environmental Peace Dividend," 23 *Environment Reporter* (BNA) (1993): 2659; General Accounting Office, *Superfund: Backlog of Unevaluated Federal Facilities Slows Cleanup Efforts* (GAO/RCED-93–119) (1993); Seth Schulman, *The Threat at Home: Confronting the Toxic Legacy of the U.S. Military* (Boston: Beacon Press, 1992); Lenny Siegel, Gary Cohen and Ben Goldman, *The U.S. Military's Toxic Legacy: America's Worst Environmental Enemy* (Boston: National Toxic Campaign Fund, 1991); Office of Technology Assessment, *Complex Cleanup: The Environmental Legacy of Nuclear Weapons Production* (OTA-0–484) (1991); National Governors' Association and National Association of Attorneys General, *From Crisis to Commitment: Environmental Cleanup and Compliance at Federal Facilities* (1990); *Cleanup at Federal Facilities*, Hearing on H.R. 765 before the Subcommittee on Transportation and Hazardous Materials of the House Committee on Energy and Commerce, 101st Cong., 1st Sess. (1989).
2. Department of Defense, *Defense Environmental Cleanup Program: Annual Report to Congress for Fiscal Year 1993* (1994), p. 2.
3. Department of Defense, *Defense Environmental Restoration Program, Annual Report to Congress for Fiscal Year 1992* (1993), p. 27; General Accounting

Office, *Hazardous Waste: DOD Estimates for Cleaning Up Contaminated Sites Improved But Still Constrained* (GAO/NSIAD-92–37) (1991), pp. 2, 3, 10.

4. Keith Schneider, "Military Has New Strategic Goal in Cleanup of Vast Toxic Waste," *New York Times*, Aug. 5, 1991.
5. Department of Energy, *Environmental Management 1994* (DOE/EM-0119) (1994), p. 1.
6. The estimates have grown steadily. See General Accounting Office, *Dealing With Problem Areas in the Nuclear Defense Complex Expected to Cost Over $100 Billion* (GAO/RCED-88-53) (1988); General Accounting Office, *Nuclear Weapons Complex: Improving DOE's Management of the Environmental Cleanup* (GAO/T-RCED-92–43) (1992), 1; Congressional Budget Office, *Cleaning Up the Department of Energy's Nuclear Weapons Complex* (1994), 1.
7. See *Interim Report of the Federal Facilities Environmental Restoration Dialogue Committee* (1993).
8. See generally Doyle, "Cleaning Up Federal Facilities."
9. *Cleanup at Federal Facilities*, p. 35.
10. 42 U.S.C. §§6901–6992k (1988 & Supp. V 1993).
11. As amended by the Superfund Amendments and Reauthorization Act, 42 U.S.C. §9601–9675 (1988 & Supp. V 1993).
12. 42 U.S.C. §§6901–6992k (1988 & Supp. V 1993). References in the notes are to RCRA's section numbers as originally enacted, as well as to the United States Code.
13. RCRA is described in greater detail in John Henry Davidson and Orlando E. Delogu, *Federal Environmental Regulation*, vol. 2 (Salem, N.H.: Butterworth, 1993), ch. 4; William H. Rodgers, Jr., *Environmental Law: Hazardous Wastes and Substances* (St. Paul: West Publishing Co.), vol. 3 (1988): 509–568, and vol. 4 (1992): 1–467; Sheldon M. Novick, ed., *Law of Environmental Protection*, vol. 3 (Deerfield, Ill.: Clark, Boardman, Callaghan, 1994), ch. 13; and *RCRA Deskbook* (Washington, D.C.: Environmental Law Institute, 1991). See especially Richard G. Stoll, "The New RCRA Cleanup Regime: Comparisons and Contrasts with CERCLA," 44 *Southwestern Law Journal* (1991): 1299.
14. *United States v. Colorado*, 990 F.2d 1565 (10th Cir. 1993), cert. denied, 114 S. Ct. 922 (1994). The role of states in cleanups of federal facilities is examined in Laurie Morissette and Laurent R. Hourcle, "State Environmental Laws Redefine 'Substantial and Meaningful Involvement,'" 31 *Air Force Law Review* (1989): 137.
15. §3005, 42 U.S.C. §6925; 40 C.F.R. pt. 270.
16. §3004(u), 42 U.S.C. §6924(u); 40 C.F.R. §264.101.
17. §§3004(u), 3008(h); 42 U.S.C. §§6924(u), 6928(h); Environmental Protection Agency, *Federal Facilities Compliance Strategy* (EPA 130/4–89/003) (1988) [hereinafter *Yellow Book*], II-4. See 40 C.F.R. pt. 264, subpt. F. Comprehensive guidance for RCRA corrective actions may be found in EPA proposed regulations at 55 Fed. Reg. 30,796 (1990), and 58 Fed. Reg. 25,045 (1993), to be codified at 40 C.F.R. pt. 264 Subpt. S, and in special regulations for hazardous wastes generated by the cleanup process, 58 Fed. Reg. 8,658 (1993). See also Richard G. Stoll, "Corrective Action in RCRA Permits: An Emerging Rival to Superfund as

the Hot Area for Environmental Lawyers and Consultants," 21 *Environmental Law Reporter* (Environmental Law Institute) (1991): 10,666.

18. §3004(v), 42 U.S.C. §6924(v); 40 C.F.R. §264.100(e)(2). Cleanup beyond the facility boundary may be delayed, but not excused, if the facility owner or operator cannot get permission from the adjacent landowner to perform the cleanup.
19. 40 C.F.R. pt. 264, subpt. F.
20. 40 C.F.R. pt. 264, subpt. G and pt. 265, subpt. G.
21. §§3006(b) and (g)(2); 42 U.S.C. §§6926(b) and (g)(2); 40 C.F.R. §§271.3(b)(2) and (b)(3). See John C. Chambers, Jr. and Peter L. Gray, "EPA and State Roles in RCRA and CERCLA," 4 *Natural Resources & Environment* (Summer 1989): 7.
22. "Regulating Solid and Hazardous Wastes: Has Federal Regulation Lived Up to Its Mandate or Can the States Do a Better Job?" 22 *Environmental Law Reporter* (Environmental Law Institute) (1992): 10,038, 10,039.
23. §3008, 42 U.S.C. §6928.
24. §7003(a), 42 U.S.C. §6973(a). See Joel A. Mintz, "Abandoned Hazardous Waste Sites and the RCRA Imminent Hazardous Provision: Some Suggestions for a Sound Judicial Construction," 11 *Harvard Environmental Law Review* (1987): 247.
25. §7002(a)(1)(B), 42 U.S.C. §6972(a)(1)(B).
26. §6001; 42 U.S.C. §6961.
27. §3007(c); 42 U.S.C. §6927(c).
28. §3016; 42 U.S.C. §6937.
29. §3016(a); 42 U.S.C. §6937(a).
30. Pub. L. No. 102-386, 106 Stat. 1505 (1992).
31. *Florida Dept. of Environmental Regulation v. Silvex Corp.*, 606 F. Supp. 159 (M.D. Fla. 1985).
32. 23 *Environmental Law Reporter* (Environmental Law Institute) (1993): 10,519.
33. §§3008, 7002; 42 U.S.C. §§6928, 6972.
34. Pub. L. No. 96–510, 94 Stat. 2767, codified as amended at 42 U.S.C. §§9601–9675 (1988 & Supp. V 1993). References hereinafter are to section numbers of the statute as originally enacted, as well as to the United States Code.
35. Pub. L. No. 99-499, 100 Stat. 1613 (1986).
36. A detailed description of the Superfund law may be found in John Henry Davidson and Orlando E. Delogu, *Federal Environmental Regulation*, vol. 2 (Salem, N.H.: Butterworth, 1993), ch. 6; William H. Rodgers, Jr., *Environmental Law: Hazardous Wastes and Substances*, vol. 4 (St. Paul: West Publishing Co., 1992), ch. 8; Donald W. Stever, *Law of Chemical Regulation and Hazardous Wastes*, vol. 1 (Deerfield, Ill.: Clark, Boardman, Callaghan, 1994), 6-52 to 6-310; Sheldon M. Novick, ed., *Law of Environmental Protection*, vol. 3 (Deerfield, Ill.: Clark, Boardman, Callaghan, 1994), ch. 13; *Superfund Deskbook* (Washington, D.C.: Environmental Law Institute, 1992).
37. §101(22), 42 U.S.C. §9601(22). The CERCLA definition of "releases" excludes emissions from certain "nuclear incidents" specified in the Atomic Energy Act, 42 U.S.C. §§2011-2296 (1988). A nuclear incident is "any occurrence, including an

extraordinary nuclear occurrence, within the United States causing, within or outside the United States, bodily injury, sickness, disease, or death, or loss of or damage to property, or loss of use of property, arising out of or resulting from the radioactive, toxic, explosive, or other hazardous properties of source, special nuclear, or byproduct material." Id. §2014(q). Also excluded are releases at sites being cleaned up by the Department of Energy under the Uranium Mill Tailings Radiation Control Act of 1978, 42 U.S.C. §7901–7942 (1988). See Steven R. Miller, *The Applicability of CERCLA and SARA to Releases of Radioactive Materials*, 17 *Environmental Law Reporter* (Environmental Law Institute, 1987): 10071; Allan E. Curlee, *Regulation of Radiation and Radioactive Materials*, 31 *Air Force Law Review* (1989): 69.

38. §§101(14), 102; 42 U.S.C. §§9601(14), 9602; 40 C.F.R. pt. 302.
39. §§103(a) and (c); 42 U.S.C. §§9603(a) and (c).
40. §§103(b) and (c); 42 U.S.C. §§9603(b) and (c). Even a person of low rank who is in a position to detect, prevent, and abate a release may be punished for failure to report it. See *United States v. Carr*, 880 F.2d 1550 (2d Cir. 1989).
41. §104, 42 U.S.C. §9604.
42. §§101(23), (24); 42 U.S.C. §§9601 (23), (24).
43. §105(a), 42 U.S.C. §9605(a). The National Contingency Plan is reproduced at 40 C.F.R. pt. 300.
44. §106(a), 42 U.S.C. §9606(a).
45. §105(a)(8)(A), 42 U.S.C. §9605(a)(8)(A). EPA is directed to establish a Hazard Ranking System for assessing the "relative degree of risk to human health and the environment posed by sites and facilities subject to review." §105(c), 42 U.S.C. §9605(c). The System is used to assign numerical values to each site for comparison. EPA rules for ranking are set forth at 40 C.F.R. pt. 300, App. A.
46. §105(a)(8)(B), 42 U.S.C. §9605(a)(8)(B).
47. §121(b)(1), 42 U.S.C. §9621(b)(1).
48. §121(d)(1), 42 U.S.C. §9621(d)(1).
49. See, e.g., Casey Scott Padgett, "Selecting Remedies at Superfund Sites: How Should 'Clean' Be Determined?" 18 *Vermont Law Review* (1994): 361; E. Donald Elliot, "Superfund: EPA Success, National Debacle?" 6 *Natural Resources and Environment* (winter 1992): 11.
50. §121(d)(2)(A), 42 U.S.C. §9621(d)(2)(A). See Stephen Merrill Smith, "CERCLA Compliance with RCRA: The Labyrinth," 18 *Environmental Law Reporter* (Environmental Law Institute, 1988): 10,518, 10,530–10,531.
51. §121(d)(4), 42 U.S.C. §9621(d)(4). See *Colorado v. Idarado Mining Co.*, 916 F.2d 1486 (10th Cir. 1990); William D. Turkula, "Determining Cleanup Standards for Hazardous Waste Sites," 135 *Military Law Review* (1992): 167.
52. §121(f)(1), 42 U.S.C. §9621(f)(1).
53. §117, 42 U.S.C. §9617.
54. §§121(f)(2) and (3); 42 U.S.C. §§9621(f)(2) and (3).
55. 59 Fed. Reg. 27,989 (1994).
56. Doyle, "Cleaning Up Federal Facilities," p. 2664.
57. 55 Fed. Reg. 6163 (1990).

58. §107(a), 42 U.S.C. §9607(a).
59. §107(b), 42 U.S.C. §9607(b). One defense of special relevance here is described in G. Nelson Smith, III, "The 'Act of War' Defense Under CERCLA," 2 *Federal Facilities Environmental Journal* (1991): 267. There is also a limited "innocent landowner" defense for purchasers of contaminated property. §101(35)(A)(i), 42 U.S.C. §9601(35)(A)(i).
60. §107(j), 42 U.S.C. §9607(j).
61. §113(f), 42 U.S.C. §9613(f); *Key Tronic Corp. v. United States*, 766 F. Supp. 865 (E.D. Wash. 1991). However, the federal government is not responsible for attorney fees and related costs, except those expended to identify other PRPs. *Key Tronic Corp. v. United States*, 114 S. Ct. 1960 (1994).
62. *United States v. Allied-Signal Corp.*, 736 F. Supp. 1553 (N.D. Cal. 1990). See also *United States v. Shell Oil Co.*, 605 F. Supp. 1064 (D. Colo. 1985) (oil company liable for cleanup costs at Rocky Mountain Arsenal).
63. Letter from Undersecretary of Defense John M. Deutch to Senator Robert C. Byrd, July 26, 1993, with enclosure, reproduced in Department of Defense Inspector General, *Environmental Consequence Analyses of Major Defense Acquisition Programs* (Rep. No. 94–020) (1993), 84–89.
64. Department of Defense Inspector General, *Environmental Consequence Analyses*, pp. 45–59, 84–89.
65. *FMC Corp. v. United States Dept. of Commerce*, 786 F. Supp. 471 (E.D. Pa. 1992), affirmed, 1994 WL 314814 (3d Cir.). *See* Van S. Katzman, Note, "The Waste of War: Government CERCLA Liability at World War II Facilities," 79 *Virginia Law Review* (1993): 1191.
66. §104(a), 42 U.S.C. §9604(a). A responsible party who is capable but who refuses to clean up a site may be directed by administrative order or judicial injunction to do so. §106(a), 42 U.S.C. §9606(a).
67. 26 U.S.C. §9507 (1988) (part of the Internal Revenue Code); CERCLA §111(a), 42 U.S.C. §9611(a).
68. 26 U.S.C. §9507(b); CERCLA §111(p), 42 U.S.C. §9611(p).
69. SARA introduced a "mixed funding" provision to facilitate cleanup of CERCLA sites. Under this provision, EPA may use Superfund to reimburse responsible parties for certain remedial costs paid by them. §122(b)(1), 42 U.S.C. §9622(b)(1). These costs usually include the liability of other recalcitrant or insolvent responsible parties ("orphan shares"). See Peter F. Sexton, Comment, "Superfund Settlements: The EPA's Role," 20 *Connecticut Law Review* (1988): 923, 938–939.
70. §§107(a)(4)(C), 107(f); 42 U.S.C. §§9607(a)(4)(C), 9607(f).
71. *Town of Bedford v. Raytheon Co.*, 755 F. Supp. 469 (D. Mass. 1991). See also *United States v. Werlein*, 746 F. Supp. 887 (D. Minn. 1990).
72. §§121(e)(2), 310; 42 U.S.C. §§9621(e)(2), 9659.
73. §113(h), 42 U.S.C. §9613(h).
74. *United States v. Werlein*, 746 F. Supp. 887 (D. Minn. 1990); *Heart of America Northwest v. Westinghouse Hanford Co.*, 820 F. Supp. 1265 (E.D. Wash. 1993). See also *In re Hanford Nuclear Reservation Litigation*, 780 F. Supp. 1551 (E.D.

Wash. 1991) (damage claims barred during pendency of CERCLA cleanup). But cf. *United States v. Colorado*, 990 F.2d 1565 (10th Cir. 1993), cert. denied, 114 S. Ct. 922 (1994).

75. Pub. L. No. 99-499, tit. III, 100 Stat. 1729 (1986), codified at 42 U.S.C. §§11,001–11,050.
76. §120(a)(1), 42 U.S.C. §9620(a)(1). See generally Andrew M. Gaydosh, "The Superfund Federal Facility Program: We Have Met the Enemy and It Is U.S.," 6 *Natural Resources and Environment* (winter 1992): 21.
77. Executive Order No. 12,580, 52 Fed. Reg. 2923 (1987); §120(a)(2), 42 U.S.C. §9620(a)(2).
78. See Laurent R. Hourcle, "Subpart K of the National Contingency Plan: The 'Missing Link' in the Federal Facilities Cleanup Plan," *Federal Facilities Environmental Journal* (1993–94): 401.
79. §120(j)(1), 42 U.S.C. §9620(j)(1).
80. §120(c), 42 U.S.C. §9620(c); RCRA §3016, 42 U.S.C. §6937.
81. CERCLA §120(j)(2), 42 U.S.C. §9620(j)(2).
82. §120(d), 42 U.S.C. §9620(d).
83. §§117, 120(e), 42 U.S.C. §§9617, 9620(e).
84. On the process of concluding an IAG, see E. David Hoard and Terrence M. Lyons, "Negotiating With Environmental Regulatory Agencies: Working Towards Harmony," 31 *Air Force Law Review* (1989): 201.
85. See 54 Fed. Reg. 10,520 (1989); §120(f), 42 U.S.C. §9620(f).
86. §120(e)(4)(A), 42 U.S.C. §9620(e)(4)(A).
87. §310(a)(1), 42 U.S.C. §9659(a)(1).
88. §122(l), 42 U.S.C. §9622(l).
89. §120(a)(4), 42 U.S.C. §9620(a)(4).
90. *United States v. Commissioner of Pennsylvania Department of Environmental Resources*, 778 F. Supp. 1328 (M.D. Pa. 1991).
91. *Rospatch Jessco Corp. v. Chrysler Corp.*, 829 F. Supp. 224 (W.D. Mich 1993); *Redland Soccer Club, Inc. v. Department of the Army*, 801 F. Supp. 1432 (M.D. Pa. 1992). Cf. *Tenaya Assoc. Ltd. Partnership v. United States Forest Service*, No. CV-F-92-5375 REC (E.D. Cal. May 18, 1993).
92. See *United States v. Colorado*, No. 98-C-1646 (D.C. Colo, Aug. 14, 1991), 22 *Environment Reporter* (BNA) (August 23, 1991): 1143, reversed, 990 F.2d 1565 (10th Cir. 1993), cert. ref. 114 S. Ct. 922 (1994). Cf. Kyle E. McSlarrow, "The Department of Defense Environmental Cleanup Program: Application of State Standards to Federal Facilities After SARA," 17 *Environmental Law Reporter* (Environmental Law Institute) (1987): 10,120, 10,123–10,125.
93. §114(a), 42 U.S.C. §9614(a). See James P. Young, Comment, "Expanding State Initiation and Enforcement Under Superfund," 57 *Chicago Law Review* (1990): 985.
94. §302(d), 42 U.S.C. §9652(d).
95. Environmental Protection Agency, *Listing Policy for Federal Facilities*, 54 Fed. Reg. 10,520 (1989).
96. *Apache Powder Co. v. United States*, 968 F.2d 66, 68–69 (D.C. Cir. 1992).
97. Air Force Environmental Restoration Division, HQ AF/CEVR, Bolling AFB, *FY*

1993–97 Defense Environmental Restoration Program Management Guidance for Non-Closure Bases (1993), p. 17.
98. See McSlarrow, "DOD Environmental Cleanup Program," pp. 10,125–10,127.
99. 49 Fed. Reg. 41,036 (1984).
100. *Colorado v. United States Dept. of the Army*, 707 F. Supp. 1562 (D. Colo. 1989).
101. Ibid., pp. 1569–1570.
102. *Daigle v. Shell Oil Co.*, 972 F.2d 1527 (10th Cir. 1992).
103. 54 Fed. Reg. 10,512, 10,515–10,516 (1989).
104. *United States v. Colorado*, No. 98-C-1646 (D. Colo., Aug. 14, 1991), 22 *Environment Reporter* (BNA) (August 23, 1991): 1143.
105. *United States v. Colorado*, 990 F.2d 1565 (10th Cir. 1993).
106. Citing §§114(a), 302(d), 42 U.S.C. §§9614(a), 9652(d).
107. §120(i), 42 U.S.C. §9620(i).
108. *United States v. Colorado*, 114 S. Ct. 922 (1994). The litigation is traced in Vicky L. Peters, Laura E. Perrault and Susan Mackay Smith, "Can States Enforce RCRA at Superfund Sites? The Rocky Mountain Arsenal Decision," 23 *Environmental Law Reporter* (Environmental Law Institute) (1993): 10,419; John F. Seymour, "The Rocky Mountain Arsenal Experience," 1 *Federal Facilities Environmental Journal* (1990): 117.
109. *United States v. Colorado*, No. 94-C-491 (D. Colo. 1994).
110. §120(e)(5), 42 U.S.C. §9620(e)(5). See, e.g., Department of Defense, *Defense Environmental Cleanup Program: Annual Report to Congress for Fiscal Year 1993* (1994); Department of Energy, *Environmental Management 1994* (DOE/EM-0119) (1994).
111. 59 Fed. Reg. 27,989 (1994).
112. §111(e)(3); 42 U.S.C. §9611(e)(3). See also *Yellow Book*, II-7.
113. 10 U.S.C. §2703 (1988).
114. See ch. 5, "Paying for Environmental Restoration."
115. U.S. Const. art. I, §9, cl. 7 prohibits the expenditure of unappropriated funds, while the Anti-Deficiency Act, 31 U.S.C. §§1341, 1350 (1988), criminalizes either the obligation or expenditure of unappropriated funds.
116. *Yellow Book*, VI-1.
117. §106(a), 42 U.S.C. §9606(a).
118. *Maine v. Department of the Navy*, 973 F.2d 1007 (1st Cir. 1992).
119. Unless otherwise noted, information in this section was drawn from the following sources: Department of Defense, *Defense Environmental Cleanup Program: Annual Report to Congress for Fiscal Year 1993* (1994); Kyle E. McSlarrow, "The Department of Defense Environmental Cleanup Program: Application of State Standards to Federal Facilities After SARA," 17 *Environmental Law Reporter* (Environmental Law Institute) (1987): 10,120. Seth Shulman, *The Threat at Home: Confronting the Toxic Legacy of the U.S. Military* (Boston: Beacon Press, 1992); Lenny Siegel, Gary Cohen and Ben Goldman, *The U.S. Military's Toxic Legacy* (Boston: National Toxic Campaign Fund, 1991).
120. 10 U.S.C. §§2701–2708 (1988 & Supp. V 1993).
121. Ibid., §2701(d).
122. Pub. L. No. 101-510, §2923(c), 104 Stat. 1821 (1990).

123. Department of Defense, *Report of the Defense Environmental Response Task Force* (1991).
124. See Department of Defense, *Defense Environmental Restoration Program: Annual Report to Congress for Fiscal Year 1992* (1993): 62–63.
125. Pub. L. No. 102–484, §323, 106 Stat. 2365 (1992).
126. See Department of Defense, *Revitalizing Base Closure Communities Announced by President Clinton* (July 2, 1993); Memorandum from Deputy Secretary of Defense re *Fast Track Cleanup at Closing Installations* (Sept. 9, 1993). Chapter 5 addresses the closure and realignment of military bases generally.
127. 10 U.S.C. §2703.
128. Pub. L. No. 100-526, §§204(a)(3), 207, 102 Stat. 2627 (1988); Pub. L. No. 101-510, §§2905(a), 2906, 104 Stat. 1808 (1990).
129. *McClellan Ecological Seepage Situation (MESS) v. Weinberger*, 655 F. Supp. 601 (E.D. Cal. 1986).
130. *McClellan Ecological Seepage Situation (MESS) v. Weinberger*, 707 F. Supp. 1182 (E.D. Cal. 1989).
131. *McClellan Ecological Seepage Situation (MESS) v. Cheney*, 763 F. Supp. 431 (E.D. Cal. 1989).
132. §121(e)(1), 42 U.S.C. §9621(e)(1).
133. *McClellan Ecological Seepage Situation (MESS) v. Cheney*, 1990 WL 117920 (E.D. Cal.).
134. The jurisdictional argument is based on §113(h) of CERCLA, described above.
135. *McClellan Ecological Seepage Situation (MESS) v. Perry*, 1995 WL 31863 (9th Cir.).
136. DOD, *Annual Report, 1993*, p. 14.
137. Department of Defense, *Defense Environmental Restoration Program: Annual Report to Congress for Fiscal Year 1992* (1993): 24; see also Warren G. Foote, "Operation Safe Removal: Cleanup of World War I Era Munitions in Washington, D.C.," *Army Lawyer*, Aug. 1994, 34.
138. See Shulman, *The Threat at Home*, pp. 3–9; Seigel et al., *U.S. Military's Toxic Legacy*, p. 28.
139. Pub. L. No. 102–386, §107, 106 Stat. 1513 (1992), amending RCRA §3004, 42 U.S.C. §6924.
140. See case study, ch. 3, "Regulating the Disposal of Chemical Weapons."
141. 10 U.S.C. §2706(a). See Department of Defense, *Defense Environmental Cleanup Program: Annual Report to Congress for Fiscal Year 1993* (1994).
142. 10 U.S.C. §2706(b). See Department of Defense, *1994 Report on Environmental Compliance for Fiscal Years 1995–1999* (1994).
143. Id. §2706(c).
144. Exec. Order No. 12,088, §1-801, 43 Fed. Reg. 47707 (1978).
145. Department of Defense, *Overseas Environmental Baseline Guidance Document* (1992). See ch. 3, "Applying Environmental Regulations Abroad."
146. See generally Joseph Gerson and Bruce Birchard, eds., *The Sun Never Sets . . . : Confronting the Network of Foreign U.S. Military Bases* (Boston: South End Press, 1991); John M. Broder, "U.S. Military Leaves Toxic Trail Overseas," *Los Angeles Times*, June 18, 1990.
147. Pub. L. No. 101-510, §342(b)(2), 104 Stat. 1537 (1990).

148. Memo from Secretary of Defense, *DOD Policy and Procedures for the Realignment of Overseas Sites*, Dec. 14, 1993.
149. See Philip Shabecoff, "Senator Urges Military Resources Be Turned to Environmental Battle," *New York Times*, June 29, 1990; Jessica Tuchman Matthews, "How Green the Pentagon," *Washington Post*, June 19, 1990.
150. Pub. L. No. 101-510, §344, 104 Stat. 1538 (1990); Pub. L. No. 103-160, §§1333–1335, 107 Stat. 1793–1804 (1993).
151. DOD, *Annual Report 1993*, p. 27.
152. Unless otherwise noted, information in this section was drawn from the following sources: F. G. Gosling and Terrence R. Fehner, *Closing the Circle: The Department of Energy and Environmental Management 1942–1994* (Washington, D.C.: Department of Energy 1994) (draft); Department of Energy, *Closing the Circle on the Splitting of the Atom: The Environmental Legacy of Nuclear Weapons Production in the United States and What the Department of Energy Is Doing About It* (1995); Department of Energy, *Environmental Management 1995* (DOE/EM-0228) (1995); Department of Energy, *Environmental Management Fact Sheets* (1994); Department of Energy, *Department of Energy 1977–1994: A Summary History* (DOE/HR-0098) (1994); Department of Energy, *Environmental Management 1994* (DOE/EM-0119) (1994); Department of Energy, Office of Technology Development, *Office of Research and Development/Office of Demonstration, Testing, and Evaluation: FY 1994 Program Summary* (DOE/EM-0216) (1994); Department of Energy, *Environmental Restoration and Waste Management Five-Year Plan (Fiscal Years 1994–1998)* (DOE/S-00097P) (3 vols.) (1993) (hereinafter *EM Five-Year Plan*); Dan W. Reicher, "Nuclear Energy and Weapons," in *Sustainable Environmental Law*, eds. Celia Campbell-Mohn, Barry Breen and J. William Futrell (St. Paul: West Publishing Co., 1993), 873; Department of Energy, *Environmental Restoration and Waste Management (EM) Program: An Introduction* (DOE/EM-0013P) (1992); Office of Technology Assessment, *Complex Cleanup: The Environmental Legacy of Nuclear Weapons Production* (OTA-O-484) (1991) (hereinafter *Complex Cleanup*); Advisory Committee on Nuclear Facility Safety, *Final Report on DOE Nuclear Facilities* (*Ahearne Report*) (1991); Anne H. Ehrlich and John W. Birks, eds., *Hidden Dangers: Environmental Consequences of Preparing for War* (San Francisco: Sierra Club Books, 1990); Department of Energy, *EM Progress* (quarterly newsletter of the Office of Environmental Restoration and Waste Management).
153. OTA, *Complex Cleanup*, p. 4.
154. Ibid., p. 15.
155. See, e.g., Keith Schneider, "U.S. Shares Blames in Abuses at A-Plant," *New York Times*, Mar. 27, 1992; H. Rep. No. 111, 102d Cong., 2d Sess. 4 (1992), reprinted in *U.S. Code Congressional and Administrative News* (1992): 1287, 1290; Kenneth B. Noble, "U.S., For Decades, Let Uranium Leak at Weapon Plant," *New York Times*, Oct. 15, 1988.
156. Senate Committee on Governmental Affairs, *Early Health Problems of the U.S. Nuclear Weapons Industry and Their Implications for Today*, S. Prt. No. 63, 101st Cong., 1st Sess. (1989); Matthew L. Wald, "Risks to A-Plant Workers Were Ignored, Study Says," *New York Times*, Dec. 19, 1989.
157. See, e.g., Keith Schneider, "U.S. Spread Radioactive Fallout In Secret Cold War

Weapon Tests," *New York Times*, Dec. 16, 1993; Keith Schneider, "New View of Peril from A-Plant Emissions," *New York Times*, Apr. 22, 1994.

158. See Dan W. Reicher and S. Jacob Scherr, "Laying Waste to the Environment," *Bulletin of the Atomic Scientists* (Jan./Feb. 1988): 31; Dan W. Reicher and S. Jacob Scherr, "The Bomb Factories: Out of Compliance and Out of Control," in *Hidden Dangers: Environmental Consequences of Preparing for War*, eds. Anne H. Ehrlich and John W. Birks (San Francisco: Sierra Club Books 1990), 35.
159. DOE, *Environmental Management 1994*, p. i.
160. DOE, *EM Five-Year Plan*, p. I-9.
161. Russell Watson, "America's Nuclear Secrets," *Newsweek*, Dec. 27, 1993, 14.
162. OTA, *Complex Cleanup*, pp. 9, 183–202.
163. Clyde Frank, DOE Deputy Assistant Secretary for Technology Development, quoted in "Cutting the Costs of Waste Disposal," *Initiatives in Environmental Technology Investment* (fall 1993): 8.
164. General Accounting Office, *Department of Energy: Management Changes Needed to Expand Use of Innovative Cleanup Technologies* (GAO/RCED-94–205) (1994): 3.
165. Matthew L. Wald, "Rusting Uranium in Storage Pools Is Troubling U.S.," *New York Times*, Dec. 8, 1993.
166. 42 U.S.C. §2201(i)(3) (1988).
167. See 42 U.S.C. §§2021 and 5812(d) (1988).
168. Environmental Protection Agency, *Environmental Radiation Protection Standards for Nuclear Power Operations*, 42 Fed. Reg. 2860 (1977), codified at 40 C.F.R. pt. 190. See also Clean Air Act regulations for radionuclides, 40 C.F.R. pt. 61. EPA was given authority to promulgate these standards in Reorganization Plan No. 3 of 1970, §§1–2, 84 Stat. 2086, 2086–2089 (1970).
169. Pub. L. No. 100-456, §1441, 102 Stat. 1918, 2076 (1988).
170. See letters to John Herrington, Secretary of Energy, from Dan W. Reicher, Natural Resources Defense Council (Dec. 14, 1988), and from Congressman Mike Synar and other members of Congress (Dec. 14, 1988). The environmental impact statement requirement is described in ch. 2.
171. *Natural Resources Defense Council, Inc. v. Department of Energy*, Civil Action No. 89-1835 (D.D.C. 1989).
172. DOE press release, Jan. 12, 1990.
173. 60 Fed. Reg. 4607 (1995).
174. See Department of Energy, *Implementation Plan Executive Summary* (DOE/EIS-0200) (1994).
175. The Five-Year Plans were subsequently prescribed by Congress. 42 U.S.C. §7274(g) (Supp. V 1993).
176. See, e.g., DOE, *EM Five-Year Plan.*
177. Pub. L. No. 103-160, §3153, 107 Stat. 1950 (1993).
178. RCRA §3004, 42 U.S.C. §6924; 40 C.F.R. §268.35(d) and (e).
179. Federal Facility Compliance Act, Pub. L. No. 102-386, §§102(c), 105, 106 Stat. 1506, 1508-1513 (1992), to be codified as 42 U.S.C. §§6961 note and 6939(c). See Laurent R. Hourcle and William J. McGowan, "Federal Facility Compliance Act of 1992: Its Provisions and Consequences," *Federal Facilities Environmental Journal* (winter 1992–93): 359.

180. Paradoxically, while facilities meeting these conditions can avoid state fines and penalties, injunctive relief for violation of the storage ban is still available under RCRA §6001, 42 U.S.C. §6961. H. Conf. Rep. No. 886, 102d Cong., 1st Sess. 21 (1992), reprinted in *U.S. Code Congressional and Administrative News* (1992): 1317, 1321.
181. Department of Energy, *Interim Mixed Waste Inventory Report: Waste Streams, Treatment Capacities, and Technologies* (DOE/NBM 1100) (Apr. 21, 1993).
182. See Department of Energy, *FY 1993 Program Summary: Office of Research and Development, Office of Demonstration, Testing and Evaluation* (DOE/EM-0109P) (1994); Department of Energy, *Innovation Investment Area: Technology Summary* (DOE/EM-0146P) (1994); Department of Energy, *Technology Catalogue* (DOE/EM-0138P) (1994); Department of Energy, *FY 1995 Technology Development Needs Summary* (DOE/EM-0147P) (1994).
183. General Accounting Office, *Department of Energy: Management Changes Needed.*
184. 42 U.S.C. §§7901–7942 (1988 & Supp. V 1993).
185. Cleanup standards are set forth at 40 C.F.R. pt. 192.
186. DOE Order No. 1230.2 (Apr. 1992).
187. Matthew L. Wald, "At an Old Atomic-Waste Site, The Only Sure Thing is Peril," *New York Times*, June 21, 1993. Hanford's history is documented in Michele Stenehjem Gerber, *On the Home Front: The Cold War Legacy of the Hanford Nuclear Site* (Lincoln, Neb.: University of Nebraska Press, 1992); and M. S. Gerber, *Legend and Legacy: Fifty Years of Defense Production at the Hanford Site* (Richland, Wash.: Westinghouse Hanford Co., 1992).
188. Keith Schneider, "Nuclear Complex Threatens Indians," *New York Times*, Sept. 3, 1990.
189. Matthew L. Wald, "Wider Peril Seen in Nuclear Waste From Bomb Making," *New York Times*, Mar. 28, 1991; Michael D'Antonio, *Atomic Harvest: Hanford and the Lethal Legacy of America's Nuclear Arsenal* (New York: Crown Publishers, 1993).
190. Matthew L. Wald, "At an Old Atomic-Waste Site, The Only Sure Thing Is Peril," *New York Times*, June 21, 1993.
191. Matthew L. Wald, "Nuclear Waste Tanks at Hanford Could Explode, U.S. Panel Warns," *New York Times*, July 31, 1990; "Better 'Urp' than 'Boom!'" *New York Times*, June 21, 1993.
192. DOE, *Environmental Management 1994*, p. 58.
193. Matthew L. Wald, "At an Old Atomic-Waste Site."
194. Matthew L. Wald, "U.S. Reaches Pact on Plant Cleanup," *New York Times*, Oct. 3, 1993.
195. OTA, *Complex Cleanup*, pp. 6–9, 75–127.
196. Office of Technology Assessment, *Hazards Ahead: Managing Cleanup Worker Health and Safety at the Nuclear Weapons Complex* (OTA-BP-O-85) (1993): iii.
197. Senators Hatfield (R-Ore) and Johnston (D-La), quoted in Sarah Pekkanen, "Committee Urges Faster, More Efficient Cleanup of Nuclear Weapons Sites," *States News Service*, July 29, 1993.
198. DOE, *EM Five-Year Plan*, p. iii.
199. Ibid., pp. v, I-35.

200. DOE, *Closing the Circle*, p. 64.
201. John H. Cushman, Jr., "204 Secret Nuclear Tests By U.S. Are Made Public," *New York Times*, Dec. 8, 1993.
202. Keith Schneider, "U.S. Reveals a Few More of Its Atomic Secrets," *New York Times*, June 28, 1994.
203. Department of Energy, *Fact Sheet: Environmental Restoration Activities at Nevada Field Office*, DOE/EM 0045P (1991).
204. John H. Cushman, Jr., "204 Secret Nuclear Tests By U.S. Are Made Public," *New York Times*, Dec. 8, 1993; Keith Schneider, "U.S. Reveals a Few More of Its Atomic Secrets," *New York Times*, June 28, 1994.
205. Michael B. Gerard, "Fear and Loathing in the Siting of Hazardous and Radioactive Waste Facilities: A Comprehensive Approach to a Misperceived Crisis," 68 *Tulane Law Review* (1994): 1047.
206. Department of Energy National Security and Military Applications of Nuclear Energy Authorization Act, Pub. L. No. 96-164, §213, 93 Stat. 1259, 1265 (1979).
207. 42 U.S.C. §§10,101–10,270 (1988 & Supp. V 1993). NWPA provisions for the "temporary" monitored retrievable storage of high-level wastes while a permanent repository is under development are described in Melinda Kassen, "Siting the MRS—A Lesson in How Even Bribes Don't Work," *Natural Resources and Environment Journal* (winter 1993): 16.
208. 42 U.S.C. §10,132(a). The guidelines may be found at 10 C.F.R. pt. 960. See Neill Edwards, Note, "Yucca Mountain or: How We Learn to Stop Worrying and Love the Department of Energy's High-Level Waste Disposal Guidelines," 12 *Virginia Environmental Law Journal* (1993): 271, 273.
209. 42 U.S.C. §10,132(a).
210. Ibid., §§10,133, 10,172.
211. Ibid., §§10,134(a), 10,135(b) and (c).
212. 42 U.S.C. §5842(4).
213. 42 U.S.C. §10,141(b)(1)(A).
214. Ibid., §§10,141(a) and (b).
215. 10 C.F.R. §60.113(a)(1)(ii).
216. 48 Fed. Reg. 38,085 (1985), promulgating 40 C.F.R. pt. 191.
217. *Natural Resources Defense Council, Inc. v. United States Environmental Protection Agency*, 824 F.2d 1258 (1st Cir. 1987). That litigation is described in case study, ch. 3, "Nuclear Weapons and Drinking Water."
218. 58 Fed. Reg. 66,398 (1993).
219. Energy Policy Act of 1992, Pub. L. No. 102-486, §801(a)(1), 106 Stat. 2921 (1992).
220. See Charles H. Montange, "Federal Nuclear Waste Disposal," 27 *Natural Resources Journal* (1987): 309, 391–395.
221. See 40 C.F.R. §§191.01, 191.03, 191.11.
222. Dan W. Reicher and S. Jacob Scherr, "The Bomb Factories: Out of Compliance and Out of Control," in *Hidden Dangers: Environmental Consequences of Preparing for War*, eds. Anne H. Erlich and John W. Birks, (San Francisco: Sierra Club Books, 1990), 44.
223. Keith Schneider, "Wasting Away," *New York Times Magazine*, Aug. 30, 1992, p. 42.

224. Keith Schneider, "U.S. Delays Start of Plant to Store Nuclear Wastes," *New York Times*, Sept. 14, 1988.
225. *New Mexico v. Watkins*, 783 F. Supp. 628, 633 (D.D.C. 1991).
226. *New Mexico v. Watkins*, 969 F.2d 1122 (D.C. Cir. 1992).
227. Waste Isolation Pilot Plant Land Withdrawal Act, Pub. L. No. 102–579, 106 Stat. 4777 (1992).
228. In 1990, EPA issued a No-Migration Determination for the site, 55 Fed. Reg. 47,700 (1990), pursuant to EPA regulations promulgated in 1985 under the Nuclear Waste Policy Act and struck down in part by a federal appeals court in 1987. See generally DOE, *5-Year Plan*, I-118.
229. "Waste Isolation Plant Update," *EM Progress*, (spring 1993): 4.
230. See Edwards, "Yucca Mountain," p. 273.
231. *Nevada v. Herrington*, 777 F.2d 529 (9th Cir. 1985); *Nevada v. Herrington*, 827 F.2d 1394 (9th Cir. 1987).
232. *Nevada v. Watkins*, 914 F.2d 1545 (1990), cert. denied, 499 U.S. 906 (1991).
233. *Nevada v. Burford*, 918 F.2d 854 (9th Cir. 1990), cert. denied sub nom. *Nevada v. Jamison*, 500 U.S. 932 (1991).
234. *County of Esmeralda, Nevada v. United States Dept. of Energy*, 925 F.2d 1216 (9th Cir. 1991).
235. *Nevada v. Watkins*, 939 F.2d 710 (9th Cir. 1991).
236. *Nevada v. Watkins*, 943 F.2d 1080 (9th Cir. 1991).
237. William J. Broad, "Experts Clash on Risk at Nuclear Waste Site," *New York Times*, Dec. 3, 1991.
238. Sandra Blakeslee, "Earthquake Raises Concern About Nuclear Waste Dump," *New York Times*, Jul. 4, 1992.
239. Nev. Rev. Stat. §445.224 (1991).
240. 40 C.F.R. §§144.6(d), 144.13.
241. *Natural Resources Defense Council, Inc. v. United States Environmental Protection Agency*, 824 F.2d 1258 (1st Cir. 1987). See case study, ch. 3, "Nuclear Weapons and Drinking Water."
242. Energy Policy Act of 1992, Pub. L. No. 102–486, §801(a), 106 Stat. 2921 (1992). See Matthew L. Wald, "Rules Rewritten on Nuclear Waste," *New York Times*, Oct. 11, 1992.
243. Pub. L. No. 102-486, §801(c), 106 Stat. 2921 (1992).
244. Ibid., §801(a)(2), 106 Stat. 2921.
245. A thoughtful analysis of the risks and their implications will be found in Kristin Shrader-Frechette, "Risk Estimation and Expert Judgment: The Case of Yucca Mountain," 3 *Risk: Issues in Health and Safety* (1992): 283.
246. *Public Service Co. of Colorado v. Andrus*, 1991 WL 87528 (D. Idaho).
247. *Idaho Department of Health and Welfare v. United States*, 959 F.2d 149 (9th Cir. 1992).
248. *Public Service Co. of Colorado v. Andrus*, 825 F. Supp. 1483 (D. Idaho 1993), as modified, 1993 WL 388312.
249. 825 F. Supp. at 1499, 1509.
250. 59 Fed. Reg. 32,688 (1994).
251. 42 U.S.C. §§2021d(b)(1)(A) and (b)(2).

252. See generally Office of Technology Assessment, *Dismantling the Bomb and Managing the Nuclear Materials* (OTA-0-572) (1993).
253. Thomas W. Lippman & R. Jeffrey Smith, "Weapons-Grade Uranium Exceeds Earlier Estimate: Energy Department Declassifies Figures," *Washington Post*, June 28, 1994.
254. Matthew L. Wald, "Nation Considers Means to Dispose of Its Plutonium," *New York Times*, Nov. 15, 1993. See generally Michael Renner, "Cleaning Up After the Arms Race," in *State of the World 1994*, ed. Lester R. Brown (New York: W. W. Norton & Co., 1994): 140–145; William J. Broad, "Deadly Nuclear Waste Piles Up with No Clear Solution at Hand," *New York Times*, Mar. 14, 1995.
255. Matthew L. Wald, "Stored Plutonium May Be a Danger, Government Says," *New York Times*, Dec. 7, 1994. Department of Energy, *Plutonium Vulnerability Management Plan* (DOE/EM-0199) (draft 1995), p. 2.
256. National Academy of Sciences, *Management and Disposition of Excess Weapons Plutonium* (Washington, D.C.: National Academy Press, 1994), 1.
257. Keith Schneider, "Nuclear Disarmament Raises Fear on Storage of 'Triggers,'" *New York Times*, Feb. 26, 1992.
258. Matthew L. Wald, "Surplus Plutonium Called Big Threat," *New York Times*, Jan. 25, 1994.
259. 59 Fed. Reg. 31,985-02 (1994).
260. William J. Broad, "Theory on Threat of Blast at Nuclear Waste Site Gains Support," *New York Times*, Mar. 25, 1995.
261. William J. Broad, "Experts Say U.S. Fails to Account for its Plutonium," *New York Times*, May 20, 1994.

5. Military Base Closures and Realignments (pp. 125–135)

1. Defense Base Closure and Realignment Commission, *1991 Report to the President*, pp. v, vi.
2. Some proposals for non-federal uses of closed base properties are described in Roger K. Lewis, "When 'Army Surplus' Is Real Estate," *Washington Post*, Mar. 5, 1994.
3. See generally Seth Shulman, *The Threat at Home: Confronting the Toxic Legacy of the U.S. Military* (Boston: Beacon Press, 1992); Lenny Siegel, Gary Cohen and Ben Goldman, *The U.S. Military's Toxic Legacy: America's Worst Environmental Enemy* (Boston: National Toxic Campaign Fund, 1991).
4. This case study is based on the court's opinion in *County of Seneca v. Cheney*, 806 F. Supp. 387, 389-390 (W.D. N.Y. 1992), preliminary injunction vacated, 12 F.3d 8 (2d Cir. 1994).
5. "Address to the Nation on United States Nuclear Weapons Reductions," 2 *Public Papers* (Sept. 27, 1991): 1220.
6. Department of Defense, *Defense Environmental Restoration Program: Annual Report to Congress for Fiscal Year 1992* (Apr. 1993), B-86.
7. 806 F. Supp. at 389–390.
8. 12 F.3d 8 (2d Cir. 1994).

9. *Breckinridge v. Rumsfeld*, 537 F.2d 864, 867 (6th Cir. 1976) (closing base and transferring eighteen military, 2,630 civilian employees does not affect the "human environment"). Accord, *National Association of Government Employees v. Rumsfeld*, 418 F. Supp. 1302 (E.D. Pa. 1976) (loss of 3,500 civilian jobs at facility in Philadelphia); *Concerned Citizens for the 442nd T.A.W. v. Bodycombe*, 538 F. Supp. 184 (W.D. Mo. 1982) (deactivation of Air Force Reserve unit).
10. 40 C.F.R. §1508.14.
11. *Jackson County, Mo. v. Jones*, 571 F.2d 1004 (8th Cir. 1978). See also *Shiffler v. Schlesinger*, 548 F.2d 96 (3d Cir. 1977) (transfer of 2,485 employees and $2 million payroll).
12. 40 C.F.R. §1508.14.
13. See, e.g., *Image of Greater San Antonio v. Brown*, 570 F.2d 517 (5th Cir. 1978). In *Conservation Law Foundation of New England v. General Service Admin.*, 707 F.2d 626, 629 (1st Cir. 1983), the government turned this principle on its head when it insisted that the negative socioeconomic and psychological effects of a delay or no action would far exceed the adverse impacts from non-military reuse of a closed naval facility.
14. No EIS required: *City and County of San Francisco v. United States*, 615 F.2d 498 (9th Cir. 1980); *National Association of Government Employees v. Rumsfeld*, 413 F. Supp. 1224 (D.D.C. 1976); *Maryland-National Capital Park and Planning Comm. v. Martin*, 447 F. Supp. 350 (D.D.C. 1978). EIS adequate: *Valley Citizens for a Safe Environment v. Aldridge*, 886 F.2d 458 (1st Cir. 1989); *Westside Property Owners v. Schlesinger*, 597 F.2d 1214 (9th Cir. 1979). EIS inadequate: *Davison v. Department of Defense*, 560 F. Supp. 1019 (S.D. Ohio 1982); *Prince George's County, Maryland v. Holloway*, 404 F. Supp. 1181 (D.D.C. 1975). See also *Bargen v. Department of Defense*, 623 F. Supp. 290 (D. Nev. 1985); *Smith v. Schlesinger*, 371 F. Supp. 559 (C.D. Cal. 1974).
15. See S. Rep. No. 125, 95th Cong., 1st Sess. 5 (1977), reprinted in 1977 *U.S. Code Congressional and Administrative News*, 542.
16. 10 U.S.C. §2687 (1988) (as amended).
17. The express reference to NEPA was deleted in 1985, Pub. L. No. 99-145, §1202(a), 99 Stat. 716, although in the absence of an explicit waiver, the environmental review prescribed by NEPA presumably must still be conducted.
18. News Release, Office of Assistant Secretary of Defense (Public Affairs), Jan. 29, 1990.
19. Defense Authorization Amendments and Base Closure and Realignment Act, Pub. L. No. 100-526, 102 Stat. 2627 (1988). The political history and practical considerations are summarized in *Base Closures and Realignments, Report of the Defense Secretary's Commission* (Dec. 1988).
20. Defense Base Closure and Realignment Act of 1990, Pub. L. No. 101-510, 104 Stat. 1808 (1990), as amended by Pub. L. No. 102–190, 105 Stat. 1345 (1991); Pub. L. No. 102–484, 106 Stat. 2502 (1992).
21. Defense Base Closure and Realignment Commission, *Report to the President 1991*.
22. Defense Base Closure and Realignment Commission, *1993 Report to the President*.

23. Pub. L. No. 100-526, §204(c)(3), 102 Stat. 2623 (1988); Pub. L. No. 101-510, §2905(c)(3), 104 Stat. 1815 (1990).
24. Other practical considerations are set out in Army Environmental Policy Institute, *Implementing Base Realignment and Closure Decisions in Compliance with the National Environmental Policy Act* (1991), 23.
25. See Raymond Takashi Swenson, Derence V. Fivehouse and Wayne Wisniewski, "Resolving the Environmental Complications of Base Closure," *Federal Facilities Environmental Journal* (autumn 1992): 279.
26. Clean Air Act §309, 42 U.S.C. §7609 (1988); 40 C.F.R. §1502.2(d).
27. Pub. L. No. 100–526, §204(b)(2)(D), 102 Stat. 2627 (1988).
28. *Conservation Law Foundation v. Department of the Air Force*, No. 1:92-CV-156-M (D.N.H. 1992).
29. 42 U.S.C. §7506(c)(1) (1988).
30. 42 U.S.C. §9620(h)(3) (1988).
31. This requirement was amended in 1992 to permit the transfer of contaminated properties. See below.
32. Pub. L. No. 102-426, §4, 106 Stat. 2177 (1992), amending §120(h)(3), 42 U.S.C. §9620(h)(3).
33. See generally Daniel C. Stepnick, Comment, "A Military Mess: CERCLA Liability and the Base Closure and Realignment Act," 59 *Journal of Air Law & Commerce* (1993): 449.
34. §120(h), 42 U.S.C. §9620(h).
35. §§120(h)(3) and (4)(D)(i), 42 U.S.C. §§9620(h)(3) and (4)(D)(i) (as amended by Pub. L. No. 102-426, §3, 106 Stat. 2175–2176 (1992)).
36. Pub. L. No. 102-484, §330, 106 Stat. 2371 (1992).
37. 10 U.S.C. §2687(b). However, no notice is required if the President certifies to Congress that the closure or realignment "must be implemented for reasons of national security or a military emergency." Id. §2687(c).
38. Pub. L. No. 100-526, §206(a), 102 Stat. 2627 (1988).
39. Pub. L. No. 101-510, §2907, 104 Stat. 1816 (1990).
40. Ibid., §2903(b), 104 Stat. 1810–1811.
41. 56 Fed. Reg. 6374–02 (1991).
42. Pub. L. No. 101-510, §2903(c)(4), 104 Stat. 1811 (1990); Pub. L. No. 102-910, §2821(i), 105 Stat. 1546 (1991).
43. 5 U.S.C. §704 (1988).
44. *Cohen v. Rice*, 992 F.2d 376, 380 (1st Cir. 1993).
45. *Specter v. Garrett*, 971 F.2d 936 (3d Cir. 1992), vacated and remanded, 113 S. Ct. 455 (1992), affirmed on remand, 995 F.2d 404 (3rd Cir. 1993), reversed sub nom. *Dalton v. Specter*, 114 S. Ct. 1719 (1994).
46. *Dalton v. Specter*, 114 S. Ct. 1719 (1994).
47. Pub. L. No. 101-189, §353, 103 Stat. 1423 (1989).
48. Pub. L. No. 101-510, §2923(c), 104 Stat. 1821 (1990). The first report was published a year later: Department of Defense, *Report of the Defense Environmental Response Task Force* (1991).
49. Pub. L. No. 101–511, §8120, 104 Stat. 1906–1907 (1990).

50. R. Lozar, et al., *Environmental Early Warning Systems (EEWS): User's Manual* (Army Corps of Engineers Tech. Rpt. N-86/06, July 1986).
51. Department of Defense, *Revitalizing Base Closure Communities Announced by President Clinton* (July 2, 1993).
52. See Memorandum from Deputy Secretary of Defense re *Fast Track Cleanup at Closing Installations* (Sept. 9, 1993).
53. These actions were to be carried out under the 1988 Base Closure and Realignment Act. Air Force Notice of Intent, 54 Fed. Reg. 6256, 33,265 (1989).
54. See Timothy Egan, "Pentagon, Facing Opposition, Suspends Land-Buying Plans," *New York Times*, Sept. 18, 1990.
55. Kenneth P. Doyle, "Cleaning Up Federal Facilities: Controversy Over an Environmental Peace Dividend," 23 *Environment Reporter* (BNA) (1993): 2659, 2664.
56. Pub. L. No. 100-526, §§204(a)(3), 207, 102 Stat. 2627 (1988); Pub. L. No. 101-510, §§2905(a), 2906, 104 Stat. 1808 (1990).

6. Environmental Protection During Wartime (pp. 136–152)

1. See Susan D. Lanier-Graham, *The Ecology of War* (New York: Walker & Co., 1993); J. P. Robinson, *The Effects of Weapons on Ecosystems* (Oxford: Pergamon Press, 1979); Stockholm International Peace Research Institute (SIPRI), *Warfare in a Fragile World: Military Impact on the Human Environment* (London: Taylor & Francis, 1980).
2. The threat of a "nuclear winter" is described in case study, ch. 2, "Buttoning Up for a Nuclear Winter."
3. An inventory and description of conventional weapons used by coalition forces in the Persian Gulf War may be found in William M. Arkin, Damian Durrant and Marianne Cherni, *On Impact: Modern Warfare and the Environment—A Case Study of the Gulf War*, App. A (Washington, D.C.: Greenpeace Intl., 1991) [hereinafter *On Impact*].
4. Paul F. Walker and Eric Stambler, ". . . And the Dirty Little Weapons," *Bulletin of the Atomic Scientists* (May 1991): 21.
5. Ibid.
6. Ibid., p. 22.
7. See Arthur H. Westing, ed., *Explosive Remnants of War: Mitigating the Environmental Effects* (London: Taylor & Francis, 1985).
8. Bureau of Political-Military Affairs, United States Department of State, *Hidden Killers: The Global Problem of Uncleared Landmines* (1993): 2. See also Donovan Webster, "One Leg, One Life at a Time," *New York Times Magazine*, Jan. 23, 1994, p. 26.
9. Jerry Gray, "Sihanouk, at U.N., Urges Ban on Land Mines," *New York Times*, Sept. 27, 1991.
10. See generally Arthur H. Westing, *Ecological Consequences of the Second Indochina War* (Stockholm: Almqvist & Wiksell, 1976); Malcolm W. Browne, "Battlefields of Khe Sanh: Still One Casualty a Day," *New York Times*, May 13, 1994.

11. Compare Eric Hoskins, "Making the Desert Glow," *New York Times*, Jan. 21, 1993, with Russell Seitz, "No Uranium Peril in Iraqi Desert," *New York Times*, Feb. 9, 1993 (letter).
12. See Arthur H. Westing, ed., *Environmental Hazards of War: Releasing Dangerous Forces in an Industrialized World* (London: Sage Publications, 1990): 38–47; B. Bowonder, Jeanne X. Kasperson and Roger E. Tognoni, "Avoiding Future Bhopals," *Environment* (September 1985): 6; Frank von Hippel and Thomas B. Cochran, "Chernobyl: the Emerging Story: Estimating Long-Term Health Effects," *Bulletin of the Atomic Scientists* (Aug./Sept. 1986): 18.
13. Merritt P. Drucker, "The Military Commander's Responsibility for the Environment," 11 *Environmental Ethics* (1989): 135, 149. See also Michael D. Diederich, Jr., "'Law of War' and Ecology—A Proposal for a Workable Approach to Protecting the Environment Through the Law of War," 136 *Military Law Review* (1992): 137, 156–160.
14. These impacts are extensively documented and analyzed in United Nations Environment Programme, *Report on the UN Inter-Agency Plan of Action for the ROPME Region, Phase I: Initial Surveys and Preliminary Assessment* (1991); *Kuwait: Report to the Secretary-General on the Scope and Nature of Damage Inflicted on the Kuwaiti Infrastructure During the Iraqi Occupation*, U.N. Sec. Coun. doc. no. S/22535 (1991); Arkin et al., *On Impact*; Jeremy Leggett, "The Environmental Impact of War: a Scientific Analysis and Greenpeace's Reaction," in *Environmental Protection and the Law of War,* ed. Glen Plant (London: Belhaven Press, 1992), p. 68.
15. R.D. Small, *Environmental Impact of Damage to Kuwaiti Oil Facilities* (Pacific-Sierra Research Corp., Jan. 11, 1991). See "Up in Flames," *Scientific American*, (May 1991): 17.
16. See, e.g., Matthew L. Wald, "Mines and Bombs Are Still a Threat in Kuwait," *New York Times*, May 12, 1991; *Kuwait: Report to the Secretary-General*, pp. 33–35.
17. Department of Defense, *Conduct of the Persian Gulf War: Final Report to Congress* (1992), xxii, 131–133, O-9 to -16; Arkin et al., *On Impact*, pp. 7, 83–85.
18. Steven Keeva, "Lawyers in the War Room," *ABA Journal* (Dec. 1991): 52, 58; Carlyle Murphy, "Kuwaitis Resisted, Survived," *Washington Post*, Mar. 4, 1991.
19. DOD, *Conduct of the Persian Gulf War*, O-10; Arkin et al., *On Impact*, p. 84.
20. See Walker and Stambler, ". . . And the Dirty Little Weapons," p. 21. Estimates of the accuracy of the precision guided ordnance vary widely. Arkin et al., *On Impact*, p. 80.
21. "Exotic Weapons Out, Sununu Says," *New York Times*, Jan. 28, 1991.
22. Walker and Stambler, ". . . And the Dirty Little Weapons," p. 22.
23. Paul Lewis, "After the War; U.N. Survey Calls Iraq's War Damage Near-Apocalyptic," *New York Times*, Mar. 22, 1991. Compare Paul Lewis, "Effects of War Begin to Fade in Iraq," *New York Times*, May 12, 1991.
24. Arkin et al., *On Impact*, pp. 46–47.
25. Compare Michael R. Gordon, "Pentagon Study Cites Problems With Gulf Effort," *New York Times*, Feb. 23, 1992; Jessica Mathews, "Acts of War and the Environ-

ment," *Washington Post*, Apr. 8, 1991; with DOD, *Conduct of the Persian Gulf War*, O-10 to -11; Ariane L. De Saussure, "The Role of the Law of Armed Conflict During the Persian Gulf War: An Overview," 37 *Air Force Law Review* (1994): 41.

26. Arkin et al., *On Impact*, pp. 10–13; John Horgan, "U.S. Gags Discussion of War's Environmental Effects," *Scientific American*, (May 1991): 24.
27. Leggett, *Environmental Impact of War*, pp. 75–76.
28. Arkin et al., *On Impact*, pp. 94–98. Coalition pilots reportedly used precision-guided weapons to collapse the buildings housing the reactors, in an effort to avoid damaging the fuel rods and dispersing radioactive materials. Larry L. Heintzelman, IV and Edmund S. Bloom, "A Planning Primer: Effective Legal Input into the War Planning and Combat Execution Process," 37 *Air Force Law Review* 5 (1994): 21.
29. See case study, ch. 2, "Defending Earth's Ozone Layer."
30. See Margaret T. Okorodudu-Fubara, "Oil in the Persian Gulf War: Legal Appraisal of an Environmental Warfare," 23 *St. Mary's Law Journal* (1991): 123; Nicholas A. Robinson, "International Law and the Destruction of Nature in the Gulf War," 21 *Environmental Policy and Law* (1991): 216.
31. See W. Hays Parks, "Air War and the Law of War," 32 *Air Force Law Review* (1990): 1, 13.
32. Deut. 20:19–20 (New Revised Standard Version).
33. Drucker, *Military Commander's Responsibility*, p. 144, describing Hugo Grotius, *On the Law of War and Peace*, trans. Francis W. Kelsey (New York: Oceana, 1964), pp. 746–756.
34. Grotius, *On the Law of War and Peace*, p. 747.
35. The law of war is surveyed in Morris Greenspan, *Modern Law of Land Warfare* (Berkeley: University of California Press, 1959); Myres S. McDougal and Florentino P. Feliciano, *Law and Minimum World Public Order: The Legal Regulation of International Coercion* (New Haven: Yale University Press, 1961); Adam Roberts and Richard Guelff, eds., *Documents on the Law of War* (Oxford: Oxford University Press, 1982); Dietrich Schindler and Jiri Toman, eds., *The Laws of Armed Conflicts: A Collection of Conventions, Resolutions, and Other Documents*, 3rd ed. (Norwell, Mass.: Kluwer Academic Publishers, 1988); United States Arms Control and Disarmament Agency, *Arms Control and Disarmament Agreements: Texts and Histories of Negotiations* (Washington, D.C.: Government Printing Office, 1982). See also Jozef Goldblat, "The Mitigation of Environmental Disruption by War: Legal Approaches," in Westing, *Environmental Hazards of War*, pp. 48–60; Arkin et al., *On Impact*, pp. 114–149.
36. See generally Bernard K. Schafer, "The Relationship Between the International Laws of Armed Conflict and Environmental Protection: The Need to Reevaluate What Types of Conduct Are Permissible During Hostilities," 19 *California Western International Law Journal* (1989): 287; Arthur H. Westing, "Environmental Warfare," 15 *Environmental Law* (1985): 645; Glen Plant, ed., *Environmental Protection and the Law of War* (London: Belhaven Press, 1992).
37. Convention on the Prohibition of Military or Any Other Hostile Use of Environmental Modification Techniques, May 18, 1977, 31 *U.S. Treaties* 333, 16 *Interna-*

tional Legal Materials 88. See Jerry Muntz, Developments, "Environmental Modification—United States Signs Convention on the Prohibition of Military or Any Other Hostile Use of Environmental Modification Techniques," 19 *Harvard International Law Journal* (1978): 384.

38. ENMOD Arts. I.1. and II.
39. See Roberts and Guelff, *Documents*, pp. 377–378. See generally Goldblat, *Mitigation*, pp. 52–53. Earlier constraints on the use of weather modification as a weapon of war are described in Ray Jay Davis, "Weather Warfare: Law and Policy," 14 *Arizona Law Review* (1972): 659.
40. Protocol Additional to the Geneva Conventions of 12 August 1949, and Relating to the Protection of Victims of International Armed Conflict, Dec. 12, 1977, 16 *International Legal Materials* 1391 [hereinafter Protocol I].
41. Guy B. Roberts, "The New Rules for Waging War: The Case Against Ratification of Additional Protocol I," 26 *Virginia Journal of International Law* (1985): 109, 145, 148, 165. For a contrasting view from the head of the U.S. delegation to the conference that adopted Protocol I, see George H. Aldrich, "Progressive Development of the Laws of War: A Reply to Criticisms of the 1977 Geneva Protocol I," 26 *Virginia Journal of International Law* (1986): 693.
42. John Embry Parkerson, Jr., "United States Compliance With Humanitarian Law Respecting Civilians During Operation Just Cause," 133 *Military Law Review* (1991): 31, 52.
43. Department of the Army Field Manual FM 27-10, *The Law of Land Warfare* (1956 and change 1976) [hereinafter FM 27-10].
44. Id. at paragraph 40c (change 1976).
45. Art. 52(2).
46. Treaty on the Prohibition of the Emplacement of Nuclear Weapons and Other Weapons of Mass Destruction on the Seabed and the Ocean Floor and in the Subsoil Thereof (Seabed Arms Control Treaty), Feb. 11, 1971, 23 *U.S. Treaties* 701, 10 *International Legal Materials* 146. See Pemmaraju Sree Nivasha Rao, "The Seabed Arms Control Treaty: A Study in the Contemporary Law of the Military Uses of the Seas," 4 *Journal of Maritime Law and Commerce* (1972): 67.
47. Treaty on the Principles Governing the Activities of States in the Exploration and Use of Outer Space, Including the Moon and Other Celestial Bodies, Jan. 27, 1967, 18 *U.S. Treaties* 2410, 610 *U.N. Treaty Series* 205.
48. Treaty for the Prohibition of Nuclear Weapons in Latin America (Treaty of Tlatelolco), Feb. 14, 1967, 22 *U.S. Treaties* 762, 6 *International Materials* 521.
49. Antarctic Treaty, Dec. 1, 1959, 12 *U.S. Treaties* 794, 402 *U.N. Treaty Series* 71.
50. See Richard Falk, "The Environmental Law of War: An Introduction," in Plant, *Environmental Protection and the Law of War*, p. 78; Schafer, *Relationship Between*.
51. Hague Convention IV Respecting the Laws and Customs of War on Land, Oct. 18, 1907, art. 22, 36 Stat. 2277, T.S. 539.
52. So proclaimed in St. Petersburg Declaration Renouncing the Use, In Time of War, of Explosive Projectiles Under 400 Grammes Weight, Dec. 11, 1868, 138 Parry's Treaty Series 297.

53. FM 27-10, paragraphs 3 and 3a. See Jonathan P. Tomes, "Legal Implications of Targeting for the Deep Attack," 64 *Military Review* (1984): 70. See also Walter D. Reed, "Teaching the Law of War in the Military," 16 *Air Force Law Review* (1974): 70.
54. *United States v. List*, 11 Trials of War Criminals Before the Nuremburg Military Tribunals Under Control Council No. 10 (1950), at 1253, 1272.
55. Protocol I, art. 57(2)(b); FM 27-10, paragraph 41 (change 1976). See Office of the Chief of Naval Operations, United States Department of the Navy, *The Commander's Handbook on the Law of Naval Operations*, NWP 9 (Rev. A), FMFM 1-10 (1989) [hereinafter NWP 9], Supp. 1989, at 8–2 n. 9.
56. Protocol I, Art. 54(2). See, e.g., NWP 9, Supp. 1989, at 8-4 n. 15.
57. FM 27-10, paragraph 504.
58. Washington Treaty on the Protection of Artistic and Scientific Institutions and Historic Monuments (Roerich Pact), Apr. 15, 1935, 49 Stat. 3267, Treaty Series 899; 1954 Hague Convention for the Protection of Cultural Property in the Event of Armed Conflict, May 14, 1954, 249 *U.N. Treaty Series* 240 (signed but not ratified by the United States).
59. Protocol I, Art. 56(1). See Kevin B. Jordan, "Petroleum Transport System: No Longer a Legitimate Target," 43 *Naval War Coll. Review* (spring 1990): 47.
60. Protocol I, Art. 56(2). See Department of the Air Force, "International Law—The Conduct of Armed Conflict and Air Operations," *AFP* 110-31 (1976), 5-9 to 5-11.
61. Protocol for the Prohibition of the Use in War of Asphyxiating, Poisonous or Other Gases, and of Bacteriological Methods of Warfare, June 17, 1925, 26 *U.S. Treaties* 571, 94 *League of Nations Treaty Series* 65.
62. See John Norton Moore, "Ratification of the Geneva Protocol on Gas and Bacteriological Warfare: A Legal and Political Analysis," 58 *Virginia Law Review* (1972): 419.
63. Executive Order No. 11,850, 40 Fed. Reg. 16,187 (1975). See also FM 27-10, paragraph 38c (change 1976).
64. Patrick E. Tyler, "Pentagon Said to Authorize U.S. Use of Nonlethal Gas," *New York Times*, Jan. 26, 1991.
65. Convention on the Prohibition of the Development, Production and Stockpiling of Bacteriological (Biological) and Toxin Weapons and on Their Destruction, Apr. 10, 1972, 26 *U.S. Treaties* 583, 1015 *U.N. Treaty Series* 163. See Note, "Establishing Violations of International Law: 'Yellow Rain' and the Treaties Regulating Chemical and Biological Warfare," 35 *Stanford Law Review* (1983): 259.
66. See *Foundation on Economic Trends v. Weinberger*, 610 F. Supp. 829 (D.D.C. 1985).
67. Convention on the Prohibition of the Development, Stockpiling and Use of Chemical Weapons and On Their Destruction, Jan. 13, 1993, 32 *International Legal Materials* 800, reproduced with extensive explanation in *Chemical Weapons Convention*, S. Treaty Doc. No. 21, 103d Cong., 1st Sess. (1993). See case study, ch. 3, "Regulating the Destruction of Chemical Weapons."
68. Convention on Prohibitions or Restrictions on the Use of Certain Conventional Weapons Which May be Deemed to be Excessively Injurious or to Have Indiscriminate Effects, Apr. 10, 1981, 20 *International Legal Materials* 1287.

69. Compare Elliott L. Meyerowitz, "The Laws of War and Nuclear Weapons" in Arthur S. Miller and Martin Feinrider eds., *Nuclear Weapons and Law* (Westport, Conn.: Greenwood Press, 1984); and Ved P. Nanda, "Nuclear Weapons and the Right to Peace Under International Law," 9 *Brooklyn Journal of International Law* (1983): 283, with Fred Bright, Jr., "Nuclear Weapons as a Lawful Means of Warfare," 30 *Military Law Review* 1 (1965); W. T. Mallison, Jr., "The Laws of War and the Juridical Control of Weapons of Mass Destruction in General and Limited Wars," 36 *George Washington Law Review* (1967): 308, 328–339; and John Norton Moore, "Nuclear Weapons and the Law: Enhancing Strategic Stability," 9 *Brooklyn Journal of International Law* (1983): 263. See generally Peter Raven-Hansen, ed., *First Use of Nuclear Weapons: Under the Constitution, Who Decides?* (Westport, Conn.: Greenwood Press, 1987).
70. See Department of the Army FM 101-31-1, *Staff Officers' Field Manual: Nuclear Weapons Employment Doctrine and Procedure* (1977). See also NWP 9, 10-1.
71. The request for an advisory opinion is described in Leonard M. Marks and Howard H. Weller, "Is the Use of Nuclear Weapons Illegal?" *New York Law Journal* (July 11, 1994), 1.
72. FM 27-10, paragraph 38b (change 1976).
73. Falk, *Environmental Law of War*, p. 79.
74. See Plant, *Environmental Protection and the Law of War*.
75. Proceedings of those trials are reported in Trials of War Criminals Before the Nuremberg Military Tribunals (1948).
76. See Roger Cohen, "Tribunal Charges Genocide by Serb," *New York Times*, Feb. 14, 1995.
77. Article 18 of the Uniform Code of Military Justice provides for trial of any person accused of violating the law of war. 10 U.S.C. §818 (1988).
78. See R. Peter Masterson, "The Persian Gulf War Crimes Trials," *Army Lawyer* (June 1991): 7; DOD, *Conduct of the Persian Gulf War*, O-22 to 26. See also John M. Broder and Robin Wright, "U.S. Building War Crimes Case Against Hussein," *Los Angeles Times*, Aug. 31, 1990; David Hoffman, "U.S.: No Plans to Try Saddam in Absentia," *Washington Post*, April 24, 1991.
79. S.C. Res. 687, paragraph 16, reprinted in 30 *International Legal Materials* 846 (1991).
80. U.N. Charter arts. 36(3), 92. The World Court is governed by the Statute of the International Court of Justice, included as an annex to the United Nations Charter, June 26, 1945, 59 Stat. 1031, T.S. No. 993.
81. Department of Defense, *Law of War Program* (DOD Dir. 5100.77) (July 10, 1979), at paragraph E(1)(a).
82. 10 U.S.C. §818 (1988). See Scott L. Silliman and Robinson O. Everett, "Forums for Punishing Offenses Against the Law of Nations," 29 *Wake Forest Law Review* (1994): 509.
83. See, e.g., Department of the Army Field Manual FM 27–2, *Your Conduct in Combat Under the Law of War* (1984); Dennis W. Shepherd, "A Bias-Free LOAC Approach Aimed at Instilling Battle Health in our Airmen," 37 *Air Force Law Review* (1994): 25; Department of the Army Field Manual FM 27-10, *The Law of Land Warfare* (1956, with change 1976); Department of the Air Force AFP

110-31, *International Law—The Conduct of Armed Conflict and Air Operations* (1976); Office of the Chief of Naval Operations, Department of the Navy NWP 9 (Rev. A), FMFM 1-10, *The Commander's Handbook on the Law of Naval Operations* (1989, with annotated supplement 1989).

84. The compliance review is ordered by DOD Instruction 5500.15. NWP 9, Supp. 1989, at 9-1 n. 1.
85. See Guy R. Phillips, "Rules of Engagement: A Primer," *Army Lawyer* (June 1993): 4; J. Ashley Roach, "Rules of Engagement," 36 *Naval War Coll. Review* 46 (Jan/Feb. 1983); W. Hays Parks, "Righting the ROE," *Proceedings* (May 1989): 83; George Bunn, "International Law and the Use of Force in Peacetime: Do U.S. Ships Have to Take the First Hit?" 39 *Naval War Coll. Review* (May-June 1986): 69. The planning process for the Persian Gulf War is described in Heintzelman and Bloom, "A Planning Primer"; John G. Humphries and Dorothy K. Cannon, "JA Wartime Planning: A Primer," 37 *Air Force Law Review* (1994): 69.
86. "All plans, rules of engagement, policies and directives shall be consistent with the DOD Law of War Program, domestic and international law . . . [and] be reviewed by the joint command legal advisor at each stage of preparation." Joint Chiefs of Staff Mem. (MJCS 59–83, June 1, 1983, and MJCS 0124–88, Aug. 4, 1988). The Department of Defense Law of War Program, DOD Dir. 5100.77 (July 10, 1979), calls on all services to ensure that their military operations comply with the law. See also FORSCOM message, Subject: Review of Operations Plans (Oct. 30, 1984).
87. Keeva, "Lawyers in the War Room," p. 52.
88. DOD, *Conduct of the Persian Gulf War*, p. xxiii.
89. Ch. 2, "Applying NEPA to Military Activities Abroad," and ch. 3, "Applying Environmental Regulations Abroad."
90. 32 C.F.R. pt. 197, Encl. 2, §C.3.a.(3).
91. Department of Defense, *Overseas Environmental Baseline Guidance Document* (1992): 1-1, 1-3.
92. Ocean Dumping Act, 33 U.S.C. §§1401–1445; Clean Water Act §312, 33 U.S.C. §1322.
93. "Sample Rules of Engagement," in 2 *Law for the Joint Warfighter* 41 (Carlisle Barracks, Pa.: U.S. Army War College, 1989): 46.
94. 7 U.S.C. §§136–136y (1988 & Supp. V 1993). See ch. 3, "Other Regulatory Statutes."
95. Science Applications International Corporation, *Mission Area Pollution Prevention Guide* (Dec. 10, 1993): 48–50 (unpublished manuscript submitted to the Army Environmental Policy Institute).
96. Letter from Assistant Secretary of Defense Colin McMillan to CEQ Chairman Michael Deland (Aug. 24, 1990).
97. 40 C.F.R. §1506.11. Compare DOD emergency regulations at 32 C.F.R. pt. 214, Encl. 1, §E.5. (environmental effects in the United States); 32 C.F.R. pt. 197, Encl. 1, §C.3. (environmental effects abroad); 32 C.F.R. Encl. 2, §C.3.b.(1) (same). According to the last regulation, an exemption may be granted by the head of a DOD component without prior approval from anyone else.

98. Letter from Mr. Deland to Mr. McMillan (Aug. 28, 1990). See Schneider, "Pentagon Wins Waiver of Environmental Rule," *New York Times*, Jan. 30, 1991.
99. Telephone interview with Carol Borgstrom, Department of Energy (Mar. 22, 1991); Memo for the Record from Timothy D. Pflaum, NEPA Compliance Officer/Defense Programs, Department of Energy (Jan. 16, 1991).
100. *Valley Citizens for a Safe Environment v. Aldridge*, 886 F.2d 458 (1st Cir. 1989).
101. Letter from Air Force Deputy Assistant Secretary Gary D. Vest to Cristobal Bonifaz, Esq. (Nov. 21, 1990).
102. Letter from Dinah Bear to Mr. Vest (Mar. 19, 1991).
103. *Valley Citizens for a Safe Environment v. Vest*, 1991 WL 330963, 22 *Environmental Law Reporter* (Environmental Law Institute) 20335 (D. Mass.), affirmed on other grounds sub nom. *Valley Citizens for a Safe Environment v. Aldridge*, 969 F.2d 1315 (1st Cir. 1992).
104. *Robertson v. Methow Valley Citizens Council*, 490 U.S. 332, 355 (1989); *Andrus v. Sierra Club*, 442 U.S. 347, 358 (1979).
105. Compare Robert Orsi, Comment, "Emergency Exceptions from NEPA: Who Should Decide?" 14 *Boston College Environmental Affairs Law Review* (1987): 481, 499–502 (CEQ lacks the authority to waive adherence to NEPA's plain statutory requirements), with Raymond Takashi Swenson, "Desert Storm, Desert Flood: A Guide to Emergency and Other Exemptions from NEPA and Other Environmental Laws," *Federal Facilities Environmental Journal* (spring 1991): 3 (emergency waiver important as "escape valve").
106. Letter from CEQ General Counsel Dinah Bear to the author (Apr. 9, 1991). Only two of those instances resulted in litigation. *Crosby v. Young*, 512 F. Supp. 1363 (E.D. Mich. 1981) (federal loan guarantee for destruction of an ethnic neighborhood in Detroit to make way for a new Cadillac assembly plant); *National Audubon Soc. v. Hester*, 801 F.2d 405 (D.C. Cir. 1986) (capture of all remaining California condors in the wild to prevent their extinction).
107. The leading case is *Flint Ridge Development Co. v. Scenic Rivers Association of Oklahoma*, 426 U.S. 776, 788 (1976).
108. Pub. L. No. 102–1, 105 Stat. 3 (1991).
109. Defendant's Mem. in Support of Motion for Summary Judgement, *Valley Citizens for a Safe Environment v. Rice*, Civ. No. 91–30077-F (D. Mass. filed Mar. 25, 1991).
110. 40 C.F.R. §§1501.8, 1506.10(d). See *Forelaws on Board v. Johnson*, 743 F.2d 677, 684 (9th Cir. 1985).
111. The classic case is *Youngstown Sheet & Tube Co. v. Sawyer*, 343 U.S. 579 (1952), and particularly Justice Jackson's concurring opinion in that case. Where Congress has acted within its own constitutional authority, as it plainly has in enacting the environmental laws, the President is bound by its mandate. Id., p. 634.

7. Environmental Protection in the Courts (pp. 153–170)

1. See Michael J. Van Zandt, "Defense of Environmental Issues in the Administrative Forum," 31 *Air Force Law Review* (1989): 183.

2. *Concerned About Trident v. Schlesinger*, 400 F. Supp. 454, 482 (D.D.C. 1975) (citations omitted), affirmed in part and reversed in part sub nom. *Concerned About Trident v. Rumsfeld*, 555 F.2d 817 (D.C. Cir. 1977). The passage echoes earlier statements in *McQueary v. Laird*, 449 F.2d 608 (10th Cir. 1971), and *Nielson v. Seaborg*, 348 F. Supp. 1369, 1372 (C.D. Utah 1972).
3. *United States v. United States District Court*, 407 U.S. 297, 320 (1972).
4. *Wisconsin v. Weinberger*, 745 F.2d 412, 427 (7th Cir. 1984).
5. Transcript at 26 (Sept. 19, 1990), *Don't Ruin Our Park v. Stone*, 1992 WL 220000 (M.D. Pa. 1992).
6. Michael Donnelly and James G. Van Ness, "The Warrior and Druid—The DOD and Environmental Law," 33 *Federal Bar News and Journal* (1986): 37, 42.
7. *Chevron, U.S.A., Inc. v. Natural Resources Defense Council, Inc.*, 467 U.S. 837, 843–844 (1984).
8. *Federal Election Commission v. Democratic Senatorial Campaign Comm.*, 454 U.S. 27, 32 (1981).
9. See Leslye A. Herrmann, Comment, "Injunctions for NEPA Violations: Balancing the Equities," 59 *University of Chicago Law Review* (1992): 1263.
10. *Wisconsin v. Weinberger*, 578 F. Supp. 1327, 582 F. Supp. 1489 (W.D. Wis. 1984).
11. *Wisconsin v. Weinberger*, 745 F.2d 412 (7th Cir. 1984).
12. Ibid., p. 425.
13. Ibid., p. 426. The ELF case is noted in Don Bruckner, Comment, "National Security and the National Environmental Policy Act," 25 *Natural Resources Journal* (1985): 467; Kenneth L. Rosenbaum, Comment, "Wisconsin v. Weinberger: The Chancellor's Foot and NEPA's Good Right Arm," 14 *Environmental Law Reporter* (Environmental Law Institute) (1984): 10,402.
14. *Legal Education Assistance Foundation v. Hodel*, 586 F. Supp. 1163, 1167, 1169 (E.D. Tenn. 1984).
15. *Weinberger v. Romero-Barcelo*, 456 U.S. 305, 310 (1982). See also *Aluli v. Brown*, 437 F. Supp. 602, 611 (D. Hawaii 1977) (citing "potential loss of military preparedness").
16. *Foundation on Economic Trends v. Weinberger*, 610 F. Supp. 829, 843–844 (D.D.C. 1985). See also *Friends of the Earth v. Hall*, 693 F. Supp. 904, 949–950 (W.D. Wash. 1988).
17. See Michael D. Axline, "Constitutional Implications of Injunctive Relief Against Federal Agencies in Environmental Cases," 12 *Harvard Environmental Law Review* (1988): 1; Zygmunt Plater, "Statutory Violations and Equitable Discretion," 70 *California Law Review* (1982): 524.
18. *Friends of the Earth v. Hall*, 693 F. Supp. 904, 949 (W.D. Wash. 1988). See also *Friends of the Earth v. United States Navy*, 841 F.2d 927 (9th Cir. 1988).
19. *Weinberger v. Romero-Barcelo*, 456 U.S. 305, 314 (1982).
20. *Public Interest Research Group of New Jersey, Inc. v. Rice*, 774 F. Supp. 317 (D.N.J. 1991). See also *Protect Key West, Inc. v. Cheney*, 795 F. Supp. 1552 (S.D. Fla. 1992); *Progressive Animal Welfare Society v. Department of the Navy*, 725 F. Supp. 475 (W.D. Wash. 1989).
21. *Natural Resources Defense Council, Inc. v. Nuclear Regulatory Commission*, 606 F.2d 1261 (D.C. Cir. 1979).

22. The quoted language, from the Safe Drinking Water Act, 42 U.S.C. §300j-6(a), is typical of the major environmental statutes.
23. Environmental Protection Agency, *Federal Facilities Compliance Strategy* (EPA 130/4–89–003) (1988) [hereinafter *Yellow Book*], at VI-3. See generally Nelson D. Cary, Note, "A Primer on Federal Facility Compliance with Environmental Laws: Where Do We Go From Here?" 50 *Washington & Lee Law Review* (1993): 801; Michael Herz, "United States v. United States: When Can the Federal Government Sue Itself?" 32 *William & Mary Law Rev.* (1991): 893; Stan Millan, "Federal Facilities and Environmental Compliance: Toward a Solution," 36 *Loyola Law Review* (1990): 319, 340–387; Michael W. Steinberg, "Can EPA Sue Other Agencies?" 17 *Ecology Law Quarterly* (1990): 317.
24. Letter from Robert A. McConnell, Assistant Attorney General to Representative John D. Dingell, Oct. 11, 1983; and *Hearings Before the Subcommittee on Oversight and Investigations of the House Committee on Energy and Commerce*, 100th Congress, 1st Session (1987) (testimony of F. Henry Habicht II, Assistant Attorney General). Both documents are reproduced in *Yellow Book*, App. H.
25. Executive Order 12,088, 43 Fed. Reg. 47,707 (1978). See also Executive Order No. 12,146, 44 Fed. Reg. 42,657 (1979) (calling for submission of legal disputes between agencies to the Attorney General for resolution).
26. *Yellow Book*, VI-1 to VI-23.
27. Pub. L. No. 102-386, §102(b), 106 Stat. 1506 (1992).
28. U.S. Const., art. I, §9, cl. 7; Anti-Deficiency Act, 31 U.S.C. §1341 (1988).
29. See W. Bradford Middlekauff, "Twisting the President's Arm: The Impoundment Control Act as a Tool for Enforcing the Principle of Appropriation Expenditure," 100 *Yale Law Journal* (1990): 209; Timothy R. Harner, "Presidential Power to Impound Appropriations for Defense and Foreign Relations," 5 *Harvard Journal of Law and Public Policy* (1982): 131.
30. *Yellow Book*, VI-3.
31. Ibid. (referring to CERCLA §120, 42 U.S.C. §9620).
32. Matthew L. Wald, "Energy Department to Pay Fine for Waste Cleanup," *New York Times*, May 14, 1991; Kenneth P. Doyle, "Cleaning Up Federal Facilities: Controversy Over an Environmental Peace Dividend," 23 *Environment Reporter* (BNA) (1993): 2659, 2665. Funds to pay the penalty were appropriated in Pub. L. No. 102-377, 106 Stat. 1315, 1336 (1993).
33. Many examples are cited in Herz, "United States," pp. 895–896.
34. Memo from Steven Herman, EPA Assistant Administrator for Enforcement, *EPA Enforcement Policy for Private Contractor Operators at Government-Owned/Contractor-Operated (GOCO) Facilities*, Jan. 7, 1994; *Yellow Book*, VI-14. The policy is criticized in Stanley Millan, "Environmental Dilemma of Defense Contractors—Cheshire Grin or Smile of Immunity?" 2 *Tulane Environmental Law Journal* (1989): 15.
35. EPA and authorized states can use their full enforcement powers against defense contractors at federal facilities. *Yellow Book*, VI-1.
36. See ch. 3, "Defense Programs for Compliance."
37. *Yellow Book*, VII-2.
38. On the subject of sovereign immunity and the ability of states to regulate federal

agency activities, see generally Barry Breen, "Supremacy and Sovereign Immunity Waivers in Federal Environmental Law," 15 *Environmental Law Reporter* (Environmental Law Institute) (1985): 10,326; Cary, *A Primer*; Richard E. Lotz, "Federal Facility Provisions of Federal Environmental Statutes: Waiver of Sovereign Immunity for 'Requirements' and Fines and Penalties," 31 *Air Force Law Review* (1989): 7; Millan, *Environmental Dilemma*, pp. 331–340; J. B. Wolverton, Note, "Sovereign Immunity and National Priorities: Enforcing Federal Facilities' Compliance with Environmental Statutes," 15 Harvard *Environmental Law Review* (1991): 565.

39. *Ohio v. United States Department of Energy*, 503 U.S. 607 (1992).
40. Pub. L. No. 102–386, §102(a)(3), 106 Stat. 1505.
41. See *United States v. Air Pollution Control Board of Tennessee Department of Health and the Environment*, No. 3:88–1030 (M.D. Tenn. March 2, 1990); *Ohio ex rel. Celebreeze v. United States Department of the Air Force*, 1987 WL 110,399, 17 *Environmental Law Reporter* (Environmental Law Institute) 21,210, 21,212 (S.D. Ohio 1987); *Alabama ex rel. Graddick v. Veterans Administration*, 648 F. Supp. 1208 (M.D. Ala. 1986). Cf. *California ex rel. State Air Resources Board v. Department of the Navy*, 431 F. Supp. 1271 (N.D. Cal. 1977).
42. *Maine v. Department of Navy*, 973 F.2d 1007 (1st Cir. 1992).
43. *Florida Dept. of Envtl. Reg. v. Silvex Corp.*, 606 F. Supp. 159, 162–163 (M.D. Fla. 1985).
44. *United States v. Pennsylvania Environmental Hearing Board*, 584 F.2d 1273 (3d Cir. 1978). Cf. *Goodyear Atomic Corp. v. Millar*, 486 U.S. 174 (1988).
45. Department of Defense, *Defense and the Environment: A Commitment Made* (1991).
46. §505(a), 33 U.S.C. §1365(a).
47. Such limitations may be found in Clean Water Act §505(b), 33 U.S.C. §1365(b).
48. 5 U.S.C. §§701–706 (1988).
49. *Lujan v. Defenders of Wildlife*, 504 U.S. 555, 560–561 (1992); *Valley Forge Christian College v. Americans United for Separation of Church and State*, 454 U.S. 464, 472 (1982).
50. *Sierra Club v. Morton*, 405 U.S. 727, 734 (1972).
51. *Warth v. Seldin*, 422 U.S. 490, 511 (1975).
52. *Natural Resources Defense Council, Inc. v. Watkins*, 954 F.2d 974 (4th Cir. 1992). Cf. *Animal Lovers Volunteer Association v. Weinberger*, 765 F.2d 937 (9th Cir. 1985) (no organizational standing to stop Navy from shooting feral goats on Navy property).
53. *Lujan v. National Wildlife Federation*, 497 U.S. 871 (1990).
54. See Karin Sheldon, "NWF v. Lujan: Justice Scalia Restricts Environmental Standing to Constrain the Courts," 20 *Environmental Law Reporter* (Environmental Law Institute) (1990): 10,557.
55. *Lujan v. Defenders of Wildlife*, 504 U.S. 555, 562–571 (1992).
56. Ibid., pp. 2143, 2145.
57. See Karin P. Sheldon, "Lujan v. Defenders of Wildlife: The Supreme Court's Slash and Burn Approach to Environmental Standing," 23 *Environmental Law Reporter* (Environmental Law Institute) (1993): 10,031; Gene R. Nichol, Jr.,

"Justice Scalia, Standing, and Public Law Litigation," 42 *Duke Law Journal* (1993): 1141.

58. *Heart of America Northwest v. Westinghouse Hanford Co.*, 820 F. Supp. 1265 (E.D. Wash. 1993).
59. *National Wildlife Federation v. Lujan*, 497 U.S. 871, 883.
60. *People ex rel. Hartigan v. Cheney*, 726 F. Supp. 219 (C.D. Ill. 1989).
61. Ibid., p. 225.
62. *Public Service Co. of Colorado v. Andrus*, 825 F. Supp. 1483, 1491–1494 (D. Idaho 1993).
63. Defendant's Mem. in Support of Motion for Judgment on the Pleadings, *Western Solidarity v. Reagan*, Nos. 84-L-280, 84-L-423 (D. Neb. 1984), at 25–26.
64. Brief for Appellants at 69, *Friends of the Earth v. Weinberger*, 725 F.2d 125 (D.C. Cir. 1984).
65. According to the Supreme Court, a political question might arise when there is

> a textually demonstrable constitutional commitment of the issue to a coordinate political department; or a lack of judicially discoverable and manageable standards for resolving it; or the impossibility of deciding without an initial policy determination of a kind clearly for nonjudicial discretion; or the impossibility of a court's undertaking independent resolution without expressing lack of the respect due coordinate branches of government; or an unusual need for unquestioning adherence to a political decision already made; or the potential of embarrassment from multifarious pronouncements by various departments on one question.

Baker v. Carr, 369 U.S. 186, 217 (1962).

66. *Concerned About Trident*, 400 F. Supp. at 482–483.
67. *Romer v. Carlucci*, 847 F.2d 445, 447, 461, 463 (8th Cir. 1988). See Douglas E. Baker, Note, "Anticipating Lamm v. Weinberger II (Romer v. Carlucci): The Political Question Doctrine, MX Missiles, and NEPA's Environmental Impact Statement," 21 *Creighton Law Review* (1988): 1245.
68. *No GWEN Alliance of Lane County, Inc. v. Aldridge*, 855 F.2d 1380, 1384 (9th Cir. 1988).
69. RCRA §3007(c), 42 U.S.C. §6927(c).
70. See, e.g., Clean Water Act §308, 33 U.S.C. §1318.
71. See generally Laurent R. Hourcle, "Military Secrecy and Environmental Compliance," 2 *N.Y.U. Environmental Law Journal* (1993): 316.
72. CERCLA §120(j)(2), 42 U.S.C. §9620(j)(2).
73. See, e.g., Clean Water Act §313(a), 33 U.S.C. §1323(a).
74. *Yellow Book*, V-6. The process is described in R. Bradford Stiles, "Environmental Law and the Central Intelligence Agency: Is There a Conflict Between Secrecy and Environmental Compliance?" 2 *N.Y.U. Environmental Law Journal* (1993): 347.
75. 42 U.S.C. §2168 (1988). See James Werner, "Secrecy and Its Effects on Environmental Problems in the Military: An Engineer's Perspective," 2 *N.Y.U. Environmental Law Journal* (1993): 351.
76. 5 U.S.C. §552 (1988). The operation of FOIA is described in Center for National

Security Studies, *Litigation Under the Federal Open Government Laws*, 18th ed., ed. Allan R. Adler (Washington, D.C.: American Civil Liberties Union Foundation 1993); Stephen Dycus, Arthur L. Berney, William C. Banks and Peter Raven-Hansen, *National Security Law* (Boston: Little, Brown, 1990), 543–586; James T. O'Reilly, *Federal Information Disclosure*, 2nd ed. (Colorado Springs, Colo.: Shepard's/McGraw-Hill, 1994); Justin T. Franklin and Robert F. Bouchard, eds., *Guidebook to the Freedom of Information and Privacy Acts*, 2nd ed. (Deerfield, Ill.: Clark, Boardman, Callaghan, 1994); United States Department of Justice, *Freedom of Information Act Guide and Privacy Overview*, ed. Pamela Maida (Washington, D.C.: Superintendant of Documents, 1993).

77. Exemption 1, 5 U.S.C. §552(b)(1).
78. Exemption 3, 5 U.S.C. §552(b)(3).
79. 42 U.S.C. §§2014(y), 2161–2162 (1988).
80. Exemption 5, 5 U.S.C. §552(b)(5).
81. 40 C.F.R. §1507.3(c).
82. *Weinberger v. Catholic Action of Hawaii/Peace Education Project*, 454 U.S. 139 (1981).
83. 5 U.S.C. §552(a)(4)(B) (1988) (de novo judicial review of FOIA exemption claims); *Ellsberg v. Mitchell*, 709 F.2d 51 (D.C. Cir. 1983) (in camera review of documents said to contain state secrets); *Halkin v. Helms*, 690 F.2d 977 (D.C. Cir. 1982) (same).
84. The leading case is *Phillippi v. CIA*, 546 F.2d 1009 (D.C. Cir. 1976) (requesting information about attempted recovery of a sunken Soviet submarine).
85. See generally Raoul Berger, *Executive Privilege: A Constitutional Myth* (Cambridge, Mass.: Harvard University Press, 1974).
86. *Committee for Nuclear Responsibility, Inc. v. Seaborg*, 463 F.2d 788, 793 (D.C. Cir. 1971).
87. Ibid., pp. 792–794.
88. Courts have taken this kind of initiative in several employment discrimination cases. See *Patterson v. Federal Bureau of Investigation*, 893 F.2d 595 (3d Cir. 1990); *In re United States*, 872 F.2d 472, cert. dismissed, 110 S. Ct. 398 (1989); and *Molerio v. Federal Bureau of Investigation*, 749 F.2d 815 (D.C. Cir. 1984), criticized in Frank Askin, "Secret Justice and the Adversary System," 18 *Hastings Constitutional Law Quarterly* (1991): 745. Cf. *Association for Reduction of Violence v. Hall*, 734 F.2d 63 (1st Cir. 1984) (ex parte determination on merits rejected).
89. *Hudson River Sloop Clearwater, Inc. v. Department of the Navy*, 1989 WL 50794 (E.D.N.Y.). See generally Stephen Dycus, "NEPA Secrets," 2 *N.Y.U. Environmental Law Journal* (1993): 300.
90. A good analogy is the Foreign Intelligence Surveillance Court, a federal court created by Congress to approve warrants for wiretaps and physical searches of foreign intelligence agents. See the Foreign Intelligence Surveillance Act, 50 U.S.C. §1801–1811 (1988), as amended by Pub. L. No. 103-359, §807, 108 Stat. 2442 (1994). Other models are suggested in Scott C. Whitney, "The Case for Creating a Special Environmental Court System," 14 *William & Mary Law Review* (1973): 473.
91. The Independent Counsel law, 28 U.S.C. §§591–599 (1988), provides a model.

92. *Bareford v. General Dynamic Corp.*, 973 F.2d 1138 (5th Cir. 1992); *Zuckerbraun v. General Dynamic Corp.*, 935 F.2d 544 (2d Cir. 1991).
93. *Weinberger v. Catholic Action of Hawaii/Peace Education Project*, 454 U.S. 139 (1981). See case study, chapter 2, "Nuclear Weapons in Paradise?"
94. Clean Water Act §505(d), 33 U.S.C. §1365(d). See also Clean Air Act §304(d), 42 U.S.C. §7604(d); RCRA §7002(e), 42 U.S.C. §6973(e); Safe Drinking Water Act, 42 U.S.C. §300j-8(d).
95. 28 U.S.C. §2412. See Susan Gluck Mezey and Susan M. Olson, "Fee Shifting and Public Policy: the Equal Access to Justice Act," 77 *Judicature* (1993): 13; Robert Hogfoss, Comment, "The Equal Access to Justice Act and Its Effect on Environmental Litigation," 15 *Environmental Law* (1985): 533.
96. §309(c), 33 U.S.C. §1319(c).
97. *United States v. Dee*, 912 F.2d 741 (4th Cir. 1990), *cert. denied*, 111 S. Ct. 1307 (1991).
98. *California v. Walters*, 751 F.2d 977 (9th Cir. 1985).
99. *United States v. Carr*, 880 F.2d 1550 (2nd Cir. 1989).
100. See James P. Calve, "Environmental Crimes: Upping the Ante for Noncompliance with Environmental Laws," 133 *Military Law Review* (1991): 279; Rami S. Hanash, "The Legal Grounds for Prosecuting Federal Employees for Environmental Law Violations," 1 *Federal Facilities Environmental Journal* (1990): 17; Stephen Herm, Note, "Criminal Enforcement of Environmental Laws on Federal Facilities," 59 *George Washington Law Review* (1991): 938; H. Allen Irish, "Enforcement of State Environmental Crimes on the Federal Enclave," 133 *Military Law Review* (1991): 249; Margaret K. Minister, "Federal Facilities and the Deterrence Failure of Environmental Laws: The Case for Criminal Prosecution of Federal Employees," 18 *Harvard Environmental Law Review* (1994): 137; Susan L. Smith, "Shields for the King's Men: Official Immunity and Other Obstacles to Effective Prosecution of Federal Officials for Environmental Crimes," 16 *Columbia Journal of Environmental Law* (1991): 1.
101. Pub. L. No. 102–386, §102(a)(4), 106 Stat. 1505, 1506 (1992).
102. *United States v. Curtis*, 988 F.2d 946 (9th Cir. 1993).
103. *Rockwell International Corp. v. United States*, 723 F. Supp. 176 (D.D.C. 1989).
104. Matthew L. Wald, "Judge Accepts Rockwell Plea On Arms Plant," *New York Times*, June 2, 1992.
105. Richard B. Stewart, "Confidentiality in Government Enforcement Proceedings," 2 *N.Y.U. Environmental Law Journal* (1993): 232, 243.
106. Matthew L. Wald, "Jury Battled Prosecutor on Nuke Plant," *New York Times*, Sept. 30, 1992; Matthew L. Wald, "New Disclosure Over Bomb Plant," *New York Times*, Nov. 11, 1992.

8. Liability for Environmental Damages (pp. 171–184)

1. CERCLA §§107(a)(4)(C), (f); 42 U.S.C. §§9607(a)(4)(C), (f).
2. *United States v. Mitchell*, 463 U.S. 206, 212 (1983).
3. U.S. Constitution art. I, §8.
4. An excellent history will be found in Donald N. Zillman, "Congress, Courts and

Government Tort Liability: Reflections on the Discretionary Function Exception to the Federal Tort Claims Act," 3 *Utah Law Review* (1989): 687. See also Barry R. Goldman, "Can the King Do No Wrong? A New Look at the Discretionary Function Exception to the Federal Tort Claims Act," 26 *Georgia Law Review* (1992): 837.

5. Zillman, "Congress, Courts and Government," p. 693 n. 39.
6. Act of Mar. 3, 1887, ch. 359, 24 Stat. 505, codified at 28 U.S.C. §1491 (1988).
7. Pub. L. No. 79-601, §§401–424, 60 Stat. 812, 842–847 (1946), codified at 28 U.S.C. §§1346(b), 2671–2680 (1988).
8. 28 U.S.C. §1346(b).
9. Military Claims Act, 10 U.S.C. §2733 (1988). See also 10 U.S.C. §§2734–2737 (1988).
10. 28 U.S.C. §§2680(b),(d),(j) and (k). See §2680 for additional exceptions.
11. 340 U.S. 135 (1950).
12. Ibid., pp. 136, 137, 146.
13. Ibid., pp. 135, 141, 143, 145.
14. See, e.g., *United States v. Johnson*, 481 U.S. 681 (1987) (recovery barred for widow of Coast Guard helicopter pilot killed on active duty); *United States v. Stanley*, 483 U.S. 669 (1987) (recovery barred for serviceman allegedly injured when involuntarily administered LSD as part of Army experiment).
15. *Jaffee v. United States*, 633 F.2d 1226, 1229 (11th Cir. 1980). However, the *Feres* doctrine was held not to bar recovery in two cases involving sailors who died from effects of radiation exposure, when the government failed to warn them afterward of the potential effects. *Molsbergen v. United States*, 757 F.2d 1016 (9th Cir. 1985); *Cole v. United States*, 755 F.2d 873 (11th Cir. 1985).
16. 28 U.S.C. §2680(a). See generally Barry Kellman, "Judicial Abdication of Military Tort Accountability: But Who Is to Guard the Guards Themselves?" 1989 *Duke Law Journal* 1597; Harold J. Krent, "Preserving Discretion Without Sacrificing Deterrence," 38 *UCLA Law Review* (1991): 871; David E. Seidelson, "From Feres v. United States to Boyle v. United Technologies Corp.: An Examination of Supreme Court Jurisprudence and a Couple of Suggestions," 32 *Duquesne Law Review* (1994): 219; Zillman, "Congress, Courts, and Government," pp. 703–715.
17. 28 U.S.C. §2680(a).
18. *United States v. Varig Airlines*, 467 U.S. 797, 808 (1984).
19. *Dalehite v. United States*, 346 U.S. 15 (1953).
20. Ibid., pp. 36, 42.
21. *United States v. Varig Airlines*, 467 U.S. at 813–814.
22. *Berkovitz v. United States*, 486 U.S. 531, 536–537 (1988). See Richard G. Rumrell, "Use of Circumstantial Evidence to Prove an Environmental Case Against the Federal Government Under the Federal Tort Claims Act," 7 *Cooley Law Review* (1990): 389. See also *United States v. Gaubert*, 499 U.S. 315 (1991).
23. The litigation is described in the first of nine Court of Appeals opinions reported seriatim and bearing the same name, beginning with *In re Agent Orange Product Liability Litigation*, 818 F.2d 145 (2d Cir. 1987).
24. Id. at 152. See Arthur H. Westing, *Ecological Consequences of the Second Indochina War* (Stockholm: Almqvist & Wiksell, 1976).

25. 818 F.2d at 150.
26. The settlement and payment program are described in the appellate court's opinions at 818 F.2d 145 and 818 F.2d 179 (2d Cir. 1987). See also Harvey P. Berman, "The Agent Orange Veteran Payment Program," 53 *Law & Contemporary Problems* (1990): 49; Warren E. Leary, "Herbicide Tied to 2 More Veterans' Ills," *New York Times*, July 28, 1993. A company required under the Defense Production Act to manufacture Agent Orange has been found not entitled to indemnification by the government for its contribution to the settlement fund. *Hercules, Inc. v. United States*, 24 F.3d 188 (Fed. Cir. 1994), cert. granted, 63 USLW3720 (1995).
27. 818 F.2d 187, 210.
28. 818 F.2d 198, 199, 210, 214, 215.
29. 818 F.2d 194, 201. The course of the litigation is traced in Peter Schuck, *Agent Orange on Trial* (Cambridge, Mass.: Harvard University Press, 1987).
30. Pub. L. No. 585, §6(a), 60 Stat. 755, 763 (1946), currently codified at 42 U.S.C. §2121(a).
31. See Howard Ball, *Justice Downwind* (New York: Oxford University Press, 1985); Jay B. Sorenson, "Venting, Shaking, and Zapping: The Environmental Effects of Nuclear Weapons Tests," in *Hidden Dangers: Environmental Consequences of Preparing for War*, eds. Anne H. Ehrlich and John W. Birks (San Francisco: Sierra Club Books, 1990), 96.
32. Pub. L. No. 585, §3(a), 60 Stat. 755, 758–759 (1946), currently reflected in 42 U.S.C. §§2012(d)-(e), 2013(d), 2051(d).
33. See William A. Fletcher, "Atomic Bomb Testing and the Warner Amendment: A Violation of the Separation of Powers," 65 *Washington Law Review* (1990): 285. The plight of military personnel involved in the testing is described in Harold L. Rosenberg, *Atomic Soldiers: American Victims of Nuclear Experiments* (Boston: Beacon Press, 1980); Michael Uhl and Tod Ensign, *GI Guinea Pigs, How the Pentagon Exposed Our Troops to Dangers More Deadly than War: Agent Orange and Atomic Radiation* (New York: Wideview Books, 1980).
34. *Allen v. United States*, 816 F.2d 1417, 1419, 1421, 1423 (10th Cir. 1987).
35. Ibid., p. 1424.
36. *In re Consolidated United States Atmospheric Testing Litigation*, 820 F.2d 982 (9th Cir. 1987).
37. Ibid., p. 995.
38. Ibid., p. 997 n. 17.
39. Pub. L. No. 98-525, §1631, 98 Stat. 2646 (1984), repealed and replaced by Pub. L. No. 101-510, §3140, 104 Stat. 1837 (1990), codified at 42 U.S.C. §2212 (Supp. V 1993). See also *Hammond v. United States*, 786 F.2d 8 (1st Cir. 1986).
40. The litigation is analyzed in Howard Ball, "The Problems and Prospects of Fashioning a Remedy for Radiation Injury Plaintiffs in Federal District Court: Examining Allen v. United States," *Utah Law Review* (1985): 267; Gisele C. DuFort, "All the King's Forces or The Discretionary Function Doctrine in the Nuclear Age: Allen v. United States," 15 *Ecology Law Quarterly* (1988): 477; Fletcher, "Atomic Bomb Testing"; A. Costandina Titus and Michael W. Bowers, "Konizeski and the Warner Amendment: Back to Ground Zero for Atomic Litigants," 1988 *Brigham Young University Law Review* 387.

41. Pub. L. No. 101-426, 104 Stat. 920, note following 42 U.S.C §2210 (Supp. V 1993).
42. Suits by injured uranium miners were defeated when the government invoked the discretionary function exception in *Barnson v. United States*, 816 F.2d 549 (10th Cir.), cert. denied, 484 U.S. 896 (1987); and *Begay v. United States*, 768 F.2d 1059 (9th Cir. 1985).
43. *Nevin v. United States*, 696 F.2d 1229, 1231 (9th Cir. 1983). See Leonard A. Cole, "Cloud Cover," *Common Cause*, Jan./Feb. 1988: 16.
44. *Kirchmann v. United States*, 8 F.3d 1273, 1274 (8th Cir. 1993).
45. Ibid., p. 1277.
46. *Daigle v. Shell Oil Company*, 972 F.2d 1527, 1538, 1540–1541 (10th Cir. 1992).
47. *Starrett v. United States*, 847 F.2d 539 (9th Cir. 1988).
48. Exec. Order No. 11258, §§4(a), 4(d), 30 Fed. Reg. 14,483 (1965).
49. *Santa Fe Pacific Realty v. United States*, 780 F. Supp. 687 (E.D. Cal. 1991).
50. *Dickerson, Inc. v. United States*, 875 F.2d 1577, 1581 (11th Cir. 1989).
51. *Roberts v. United States*, 887 F.2d 899 (9th Cir. 1989).
52. *Prescott v. United States*, 973 F.2d 696 (9th Cir. 1992). Cf. *Duff v. United States*, 999 F.2d 1280 (8th Cir. 1993) (fumes from painting military housing unit); *Jones v. United States*, 698 F. Supp. 826 (D. Haw. 1988) (injuries from termite control pesticide).
53. *United States v. Causby*, 328 U.S. 256 (1946). For a more recent case on similar facts, see *Branning v. United States*, 654 F.2d 88 (Ct. Cl. 1981), 784 F.2d 361 (Fed. Cir. 1986).
54. *Clark v. United States*, 8 Cl. Ct. 649 (1985).
55. *Clark v. United States*, 660 F. Supp. 1164 (W.D. Wash. 1987), affirmed, 856 F.2d 1433 (9th Cir. 1988).
56. *Clark v. United States*, 19 Cl. Ct. 220 (1990).
57. *Yearsley v. W.A. Ross Const. Co.*, 309 U.S. 18, 22 (1940).
58. *McKay v. Rockwell Intl. Corp.*, 704 F.2d 444, 448 (9th Cir. 1983). The government's immunity from liability for indemnification of the supplier was established in *Stencel Aero Engineering Corp. v. United States*, 431 U.S. 666 (1977). The government was insulated from the injured serviceman's claim by the *Feres* doctrine, described earlier in this chapter.
59. 704 F.2d at 451.
60. *In re Agent Orange Product Liability Litigation*, 818 F.2d at 150.
61. *Shaw v. Grumman Aerospace Corp.*, 778 F.2d 736, 741 (11th Cir. 1985).
62. *Boyle v. United Technologies Corp.*, 487 U.S. 500, 502 (1988). The litigation and its implications are analyzed in Barry Kellman, "Judicial Abdication of Military Tort Accountability: But Who is to Guard the Guards Themselves?" 1989 Duke *Law Journal* 1597; Emie Stewart, Comment, "The Government Made Me Do It!: Boyle v. United Technologies Extended the Government Contractor Defense Too Far?" 57 *Journal of Air Law & Commerce* (1992): 981.
63. 487 U.S. at 504, 507.
64. Ibid., p. 509.
65. At least one court has held that the government contractor defense applies only to products manufactured for the government and not to services provided to the

government. *Amtreco, Inc. v. O. H. Materials, Inc.*, 802 F. Supp. 443 (M.D. Ga. 1992).

66. Keith Schneider, "Uranium Dust at Ohio Plant Is Rated as High," *New York Times*, Dec. 19, 1991.
67. See Kenneth B. Noble, "U.S., For Decades, Let Uranium Leak at Weapon Plant," *New York Times*, Oct. 15, 1988; Aaron Freiwald, "Fallout Over Fernald," *The American Lawyer* (July/August, 1990): 77.
68. *Crawford v. National Lead Co.*, 784 F. Supp. 439, 441 (S.D. Ohio 1989). See R. Joel Ankney, Note, "'But I Was Only Following Orders': The Government Contractor Defense in Environmental Tort Litigation," 32 *William and Mary Law Review* (1991): 399.
69. Department of Energy, *Closing the Circle on the Splitting of the Atom* (1995), p. 72.
70. 33 U.S.C §407 (1988).
71. 10 C.F.R. §20.106 and App. B, Table II (1988).
72. Here the court cited the Supreme Court's earlier opinion in *Berkovitz v. United States*, 486 U.S. 531 (1988), described earlier in this chapter.
73. *In re Fernald Litigation*, 1989 WL 267039 (S.D. Ohio).
74. *McKay v. United States*, 703 F.2d 464 (10th Cir. 1983).
75. See *Cook v. Rockwell International Corp.*, 147 F.R.D. 237 (1993).

9. National Defense vs. Environmental Protection: We Can Have It Both Ways (pp. 185–194)

1. Letter to A. Hodges, April 4, 1864, *Collected Works* VIII, ed. R. Basler, 1953–1955: 281.
2. 42 U.S.C. §§13,101–13,109 (Supp. V 1993).
3. *Federal Compliance with Right-to-Know Laws and Pollution Prevention Requirements*, Exec. Order No. 12,856, 58 Fed. Reg. 41,981 (1993). See ch. 3, "Pollution Prevention."
4. *Train v. Colorado Public Interest Research Group*, 426 U.S. 1 (1976). See ch. 3, "Clean Water Act."
5. See ch. 3, "Resource Conservation and Recovery Act."
6. See ch. 3, "Safe Drinking Water Act."
7. See ch. 3, "Clean Air Act."
8. See ch. 4, "Comprehensive Environmental Response."
9. See case study, ch. 4, "Struggle for Control at Basin F."
10. See ch. 7, "Enforcement by the States."
11. This language appears in the federal facilities provision of several statutes, including Clean Air Act §118(a), 42 U.S.C. §7418(a).
12. See ch. 7, "Judicial Deference in National Security Disputes."
13. See case study, ch. 3, "Regulating the Disposal of Chemical Weapons."
14. See ch. 2, "Applying NEPA to Military Activities Abroad," and ch. 3, "Applying Environmental Regulations Abroad."
15. See ch. 3.

16. See John H. Cushman, Jr., "House Approves A New Standard for Regulations," *New York Times*, Mar. 1, 1995.
17. See ch. 3, "Defense Programs for Compliance."
18. See ch. 7, "Enforcement by EPA."
19. See ch. 7, "Criminal Sanctions for Statutory Violations."
20. Thomas W. Lippman, "Energy: Key Decisions Awaited for Bulk of Cuts," *Washington Post*, Dec. 20, 1994.
21. Jeff Erlich, "Budget Cutters Eye Base Cleanup Funds," *Defense News*, Jan. 16, 1995.
22. Treaty on the Non-Proliferation of Nuclear Weapons, July 1, 1968, 21 *U.S. Treaties* 483, 729 *U.N. Treaty Series* 161.
23. See generally Peter Raven-Hansen, ed., *First Use of Nuclear Weapons: Who Decides?* (New York: Greenwood Press, 1987).
24. Treaty Between the United States of America and the Union of Soviet Socialist Republics on the Limitation of Anti-Ballistic Missile Systems, May 26, 1972, 23 *U.S. Treaties* 3435, 11 *International Legal Materials* 784.
25. Case study, ch. 3, "Buttoning Up for a Nuclear Winter."
26. See case study, ch. 3, "Regulating the Disposal of Chemical Weapons."
27. 22 U.S.C. §2753(a) (1988).

Index

University Press of New England publishes books under its own imprint and is the publisher for Brandeis University Press, Dartmouth College, Middlebury College Press, University of New Hampshire, University of Rhode Island, Tufts University, University of Vermont, Wesleyan University Press, and Salzburg Seminar.

Library of Congress Cataloging-in-Publication Data

Dycus, Stephen.
National defense and the environment / Stephen Dycus.
p. cm.
Includes index.
ISBN 0–87451–675–7 (cloth : alk. paper). — ISBN 0–87451–735–4 (paper : alk. paper)
1. United States—Defenses—Environmental aspects. 2. Liability for environmental damages—United States. 3. World politics—1989— I. Title.
UA23.D95 1996
363.7—dc20 95–20878
∞